AI OFFICE
The Center of AI Excellence

Jitesh Goswami

ISBN-13: 978-0-9942477-2-8 (Paperback)
ISBN-13: 978-0-9942477-1-1 (eBook)

Published by Nainty Publishing
Sydney, Australia

First Edition: May 2025

Dedication

To my lovely wife, Manisha,
without whose support,
this effort wouldn't have been possible.

Table of Contents

Tables & Figures

List of Figures

List of Tables

List of Visual Elements Used Throughout

	Important		Best Practice
	Definition		Hint
	Tip		Note
	Remember		Insight
	Fact		Design Consideration
	Light-Bulb-Moment		RACI Clarity

Abbreviations

- **A/B Testing:** A method comparing two versions to determine which performs better.
- **ACID:** Atomicity, Consistency, Isolation, Durability (Database transaction properties).
- **ADF:** Azure Data Factory.
- **ADLS:** Azure Data Lake Storage.
- **AGI:** Artificial General Intelligence.
- **AIA:** AI Impact Assessment.
- **AIF360:** AI Fairness 360 (IBM Toolkit).
- **AIMS:** AI Management System (ISO/IEC 42001).
- **AIO:** AI Office, the central coordinating function for enterprise AI (often used interchangeably with CoE).
- **AI RMF:** AI Risk Management Framework (NIST).
- **AKS:** Azure Kubernetes Service.
- **ANN:** Artificial Neural Network.
- **AOV:** Average Order Value.
- **API:** Application Programming Interface.
- **APM:** Application Performance Management.
- **ART:** Adversarial Robustness Toolbox (IBM Library).
- **AUC:** Area Under the Curve (Classification metric).
- **AWS:** Amazon Web Services.
- **BCG:** Boston Consulting Group.
- **BERT:** Bidirectional Encoder Representations from Transformers (NLP Model).
- **BI:** Business Intelligence.
- **BU:** Business Unit.
- **CAIO:** Chief AI Officer.
- **CapEx:** Capital Expenditure.
- **CCPA:** California Consumer Privacy Act.
- **CDC:** Change Data Capture.
- **CDO:** Chief Data Officer.
- **CD:** Continuous Delivery or Continuous Deployment.
- **CES:** Customer Effort Score.
- **CI/CD/CT:** Continuous Integration / Continuous Delivery / Continuous Training.
- **CISO:** Chief Information Security Officer.
- **CLV:** Customer Lifetime Value.
- **CM:** Continuous Monitoring.
- **CNN:** Convolutional Neural Network.
- **CoE:** Center of Excellence (often used interchangeably with AI Office).
- **CoP:** Community of Practice.
- **COTS:** Commercial Off-The-Shelf.
- **CPRA:** California Privacy Rights Act.
- **CPU:** Central Processing Unit.

- **CRM:** Customer Relationship Management.
- **CRO:** Chief Risk Officer.
- **CSAT:** Customer Satisfaction Score.
- **CSO:** Chief Strategy Officer.
- **CT:** Continuous Training.
- **CTR:** Click-Through Rate.
- **CVPR:** Conference on Computer Vision and Pattern Recognition.
- **CX:** Customer Experience.
- **D&I:** Diversity & Inclusion.
- **DAG:** Directed Acyclic Graph (Used in workflow orchestration).
- **DAST:** Dynamic Application Security Testing.
- **DBMS:** Database Management System.
- **DevOps:** Development Operations.
- **DevSecOps:** Development Security Operations.
- **DQ:** Data Quality.
- **DPIA:** Data Protection Impact Assessment.
- **DSL:** Domain-Specific Language.
- **DVC:** Data Version Control.
- **EAMM:** Enterprise AI Maturity Matrix.
- **EAMI:** Enterprise AI Maturity Index.
- **EDA:** Exploratory Data Analysis.
- **EKS:** Elastic Kubernetes Service (AWS).
- **ELK:** Elasticsearch, Logstash, Kibana (Logging stack).
- **ELT:** Extract, Load, Transform.
- **E-MBA:** Executive Master of Business Administration.
- **ERP:** Enterprise Resource Planning.
- **ESG:** Environmental, Social, and Governance.
- **ETL:** Extract, Transform, Load.
- **EX:** Employee Experience.
- **F1-Score:** Harmonic mean of Precision and Recall (Classification metric).
- **FinOps:** Financial Operations (for Cloud).
- **FL:** Federated Learning.
- **FTE:** Full-Time Equivalent.
- **GCP:** Google Cloud Platform.
- **GDPR:** General Data Protection Regulation.
- **GenAI:** Generative AI.
- **GIGO:** Garbage In, Garbage Out.
- **GKS:** Google Kubernetes Engine.
- **GloVe:** Global Vectors for Word Representation.
- **GPT:** Generative Pre-trained Transformer (OpenAI Model Series).
- **GPU:** Graphics Processing Unit.
- **GRC:** Governance, Risk Management, and Compliance.
- **HBR:** Harvard Business Review.
- **HCI:** Human-Computer Interaction.
- **HITL:** Human-in-the-Loop.
- **HTTP:** Hypertext Transfer Protocol.

- **IaC:** Infrastructure as Code.
- **IaaS:** Infrastructure-as-a-Service.
- **ICLR:** International Conference on Learning Representations.
- **ICML:** International Conference on Machine Learning.
- **IDC:** International Data Corporation.
- **IDE:** Integrated Development Environment.
- **IDS:** Intrusion Detection System.
- **IoC:** Inversion of Control.
- **IoT:** Internet of Things.
- **IP:** Intellectual Property.
- **IPA:** Intelligent Process Automation.
- **IR:** Incident Response.
- **ISO:** International Organization for Standardization.
- **ITSM:** IT Service Management.
- **JMLR:** Journal of Machine Learning Research.
- **JSON:** JavaScript Object Notation.
- **K8s:** Kubernetes.
- **KPI:** Key Performance Indicator.
- **L&D:** Learning & Development.
- **LGPD:** Lei Geral de Proteção de Dados (Brazilian Data Protection Law).
- **LLMOps:** Large Language Model Operations.
- **LIME:** Local Interpretable Model-agnostic Explanations (XAI Technique).
- **MAE:** Mean Absolute Error.
- **MDM:** Master Data Management.
- **MLE:** Machine Learning Engineer.
- **MLOps:** Machine Learning Operations.
- **MLSecOps:** Machine Learning Security Operations.
- **MPP:** Massively Parallel Processing.
- **MTTR:** Mean Time to Recovery.
- **MVP:** Minimum Viable Product.
- **NLP:** Natural Language Processing.
- **NLTK:** Natural Language Toolkit.
- **NPS:** Net Promoter Score.
- **NeurIPS:** Neural Information Processing Systems (Conference).
- **NIST:** National Institute of Standards and Technology.
- **ONNX:** Open Neural Network Exchange.
- **OPA:** Open Policy Agent.
- **OpEx:** Operational Expenditure.
- **OSS:** Open Source Software.
- **OWASP:** Open Web Application Security Project.
- **PaaS:** Platform-as-a-Service.
- **PETs:** Privacy-Enhancing Technologies.
- **PII:** Personally Identifiable Information.
- **PMI:** Project Management Institute.
- **PMO:** Project Management Office or Program Management Office.
- **PoC:** Proof of Concept.

- **PR:** Public Relations.
- **PwC:** PricewaterhouseCoopers.
- **QA:** Quality Assurance or Question Answering.
- **QC:** Quality Control.
- **QML:** Quantum Machine Learning.
- **RACI:** Responsible, Accountable, Consulted, Informed (Matrix).
- **RBAC:** Role-Based Access Control.
- **RCA:** Root Cause Analysis.
- **RDBMS:** Relational Database Management System.
- **RDSMS:** Relational Data Stream Management System.
- **REST:** Representational State Transfer (API Style).
- **RL:** Reinforcement Learning.
- **RMSE:** Root Mean Squared Error.
- **RPA:** Robotic Process Automation.
- **SaaS:** Software-as-a-Service.
- **SAFe:** Scaled Agile Framework.
- **SAST:** Static Application Security Testing.
- **SBOM:** Software Bill of Materials.
- **SCA:** Software Composition Analysis.
- **SCC:** Standard Contractual Clauses (Data Transfer).
- **SDK:** Software Development Kit.
- **SecML / SecMLOps:** Secure Machine Learning / Secure Machine Learning Operations.
- **SHAP:** SHapley Additive exPlanations (XAI Technique).
- **SIEM:** Security Information and Event Management.
- **SLA:** Service Level Agreement.
- **SLI:** Service Level Indicator.
- **SLO:** Service Level Objective.
- **SMART:** Specific, Measurable, Achievable, Relevant, Time-bound (Objectives).
- **SME:** Subject Matter Expert or Small and Medium-sized Enterprise.
- **Snyk:** A commercial tool for finding and fixing vulnerabilities in code, dependencies, containers, and IaC. Mentioned for SCA.
- **SOC 2:** System and Organization Controls 2 (Compliance Standard).
- **SOW:** Statement of Work.
- **SQL:** Structured Query Language.
- **SRE:** Site Reliability Engineering.
- **SSDLC:** Secure Software Development Lifecycle.
- **SVM:** Support Vector Machine.
- **TA:** Talent Acquisition.
- **TCO:** Total Cost of Ownership.
- **TF Serving:** TensorFlow Serving.
- **TPU:** Tensor Processing Unit (Google Hardware).
- **UAT:** User Acceptance Testing.
- **URL:** Uniform Resource Locator.
- **UX:** User Experience.
- **VNET:** Virtual Network (Azure).
- **VPC:** Virtual Private Cloud.

- **W&B:** Weights & Biases (MLOps Platform).
- **WAF:** Web Application Firewall.
- **WIIFM:** What's In It For Me?
- **XAI:** Explainable AI.
- **XGBoost:** Extreme Gradient Boosting (ML Library).
- **XML:** Extensible Markup Language.
- **YAML:** YAML Ain't Markup Language.
- **YoY:** Year-over-Year.

Preface

The artificial intelligence revolution is no longer a distant forecast. It is reshaping industries and redefining competitive landscapes today. Enterprises globally are grappling with the immense potential and profound challenges of harnessing AI – moving beyond isolated experiments to strategic, scalable, and responsible implementation. Yet, the path from recognizing AI's promise to realizing tangible, sustainable business value is often fraught with difficulty, marked by stalled projects, unclear returns, governance pitfalls, and organizational friction.

This book was born from observing these challenges firsthand and recognizing the need for a structured, holistic approach. It is intended for the leaders, strategists, technologists, and change agents tasked with navigating this complex transformation – those responsible for building the capability within their organizations to leverage AI effectively and ethically. Whether you are establishing a dedicated AI Office or Center of Excellence (CoE), or seeking to mature your existing AI program, this guide provides a comprehensive blueprint.

At its core are two interconnected frameworks developed through extensive research and practical application: the AI Office Compass™, which provides an operational model across five critical domains (Strategy, Governance, Performance & Value, Resources, and Maturity Management), and the Enterprise AI Maturity Matrix (EAMM) paired with the quantitative Enterprise AI Maturity Index (EAMI), which offer a robust system for assessing capabilities and tracking progress across 14 key dimensions.

How to Use This Book

This work is designed not just to be read, but to be actively used. An initial read-through will provide a comprehensive overview of the frameworks, critical success factors, and the journey through different maturity levels, illustrated by the OmnioTech case study woven throughout the chapters.

However, its true value lies in its application. Consider this a guide to revisit frequently. Use the AI Office Compass™ to structure your team's focus and ensure balanced capability development. Employ the EAMM/ EAMI framework (detailed in Chapter 24 and Appendix B/C) as a practical diagnostic tool to assess your current state, identify critical gaps, and inform your strategic uplift initiatives (Chapter 25). Use the detailed discussions within each Compass domain (Parts 2-6) to refine your specific strategies and operational practices. The goal is to use these frameworks actively as you lead your organization's AI transformation, staying ahead of the curve and winning in the AI revolution.

A Note on Scope and Density

This book represents a significant condensation of a much larger body of knowledge on establishing and operating effective AI Offices. The aim is to provide a thorough, actionable blueprint covering the essential strategic, governance, technical, data, talent, financial, and cultural dimensions. While striving for clarity, the necessary detail across these interconnected areas within over 300 pages may make certain sections feel dense. It reflects the inherent complexity of building a truly mature enterprise AI capability.

Looking Ahead

The field of enterprise AI is vast and rapidly evolving. While this book provides the strategic framework for the AI Office, the practical implementation details vary significantly across industries and use cases. Readers interested in deeper dives into specific operational aspects may find value in a forthcoming series of practical implementation guides and playbook resources designed to complement this foundational work.

Embarking on the enterprise AI journey requires vision, discipline, and persistence. It is my hope that this book provides you with the clarity, structure, and confidence needed to navigate this exciting and transformative path successfully.

Jitesh Goswami
Sydney, Australia
May 2025

PART 1 - INTRODUCTION

CHAPTER 1

The Strategic Imperative for an AI Office: Orchestrating Enterprise Transformation

"The greatest danger in times of turbulence is not the turbulence itself, but to act with yesterday's logic."
— Peter Drucker

1.1 Introduction: The AI Revolution and the Enterprise Challenge

Artificial intelligence (AI) is no longer a futuristic abstraction but a powerful, present-day force reshaping industries, economies, and the very fabric of organizational operations. Across the globe, enterprises are leveraging AI's capabilities to achieve significant outcomes: optimizing intricate global supply chains with predictive analytics (e.g., Amazon reportedly saving billions annually), delivering hyper-personalized customer experiences that demonstrably boost engagement and loyalty (e.g., Stitch Fix driving retention), accelerating complex scientific research and development cycles (seen in pharmaceutical discovery), and automating previously manual, resource-intensive tasks across functions like finance and HR (often yielding efficiency gains reported in the 30-40% range). This technology offers transformative opportunities for innovation, efficiency, and securing a distinct competitive advantage in a dynamic global marketplace, with credible projections estimating AI could contribute upwards of $15.7 trillion to the global economy by 2030 (McKinsey, 2023). From nimble startups utilizing accessible open-source frameworks (like TensorFlow or PyTorch) to multinational corporations deploying sophisticated enterprise-grade AI platforms (such as AWS SageMaker, Google Vertex AI, or Microsoft Azure ML), the race to harness AI's potential is well underway.

However, this transformative potential is frequently accompanied by significant execution challenges. The path from recognizing AI's promise to realizing tangible, sustainable business value is often fraught with difficulty. Many organizations find themselves caught in a cycle of fragmented experimentation, where initial enthusiasm gives way to stalled projects, unclear returns, and mounting frustration.

A common failure pattern emerges: numerous AI projects, despite demonstrating impressive technical metrics in isolation (e.g., high model accuracy), ultimately fail to deliver meaningful business impact. This often stems from a fundamental lack of connection to core strategic objectives (a key aspect of the EAMM Strategic Alignment dimension), inadequate governance (EAMM Governance), poor data foundations (EAMM Data Management/Quality), or an inability to scale beyond limited pilot phases (often due to low EAMM MLOps maturity). Executives are left questioning the significant investments made, and the true potential of AI remains unrealized.

Initial excitement often leads organizations down a path of scattered, uncoordinated AI activities. Individual departments or teams, driven by local needs or technological curiosity, initiate isolated projects. Marketing might experiment with a chatbot, Operations with predictive maintenance, and Finance with fraud detection – all using different tools, datasets, and standards. While such exploration can provide valuable initial learning, this ad-hoc approach inherently limits enterprise-wide impact and introduces significant risks. It typically results in wasted resources due to duplicated efforts (industry reports suggest significant cost increases due to duplicated tools), inconsistent quality and standards leading to unreliable or untrustworthy AI systems, unmanaged ethical and security vulnerabilities (such as biased algorithms or data privacy breaches, potentially incurring massive fines like the €1.7 billion GDPR penalties seen in 2022), and substantial missed strategic opportunities where a coordinated AI approach could have driven far greater value through synergy or foundational capability building.

This chapter makes the compelling case for a fundamental shift: moving decisively away from fragmented, ad-hoc experimentation towards a deliberate, strategically aligned, and centrally orchestrated approach to enterprise AI. We will dissect the inherent limitations and significant risks associated with uncoordinated AI initiatives, using illustrative examples and data points. We will then establish the strong business case for creating a dedicated organizational capability – the AI Office (AIO) (also often referred to as an AI Center of Excellence or CoE) – specifically chartered to guide AI strategy, govern implementation responsibly, manage resources efficiently, maximize the value derived from AI investments, and systematically drive the organization's overall AI maturity. Understanding why such strategic orchestration is not just beneficial but essential is the critical foundation before delving into how to design and operate an effective AI Office (AIO).

'AI Office (AIO)' will be used several times in this initial chapter to firmly establish 'AIO' as a recognized abbreviation for the 'AI Office'. For consistency and ease of reading in subsequent chapters, the full term 'AI Office' will generally be used.

To provide a practical structure for this journey, this chapter also introduces two interconnected frameworks central to this book's methodology (detailed fully in later chapters):

1. **The AI Office (AIO) Compass™ Framework:** A management model organizing the AI Office's re-

sponsibilities across five critical, interconnected domains: Strategic Alignment (North), Governance (East), Performance & Value Delivery (West), Resource Management (South), and orchestrating it all, Maturity Management (Center). (Detailed in Chapter 3).

2. **The Enterprise AI Maturity Matrix (EAMM) and Enterprise AI Maturity Index (EAMI):** An assessment framework evaluating AI capabilities across key dimensions (like Strategy, Governance, Data, MLOps, Value, etc.) and defining five distinct maturity levels (from Level 1: Initial to Level 5: Transformative). The EAMM provides the qualitative descriptions, while the EAMI serves as the corresponding quantitative index, providing an objective score (e.g., 0-100%) based on measurable Key Performance Indicators (KPIs) to track maturity progress over time. (Detailed in Chapter 24; comprehensive KPIs detailed in Appendix B). While the detailed mechanics of the EAMM/EAMI assessment framework are covered in Chapter 24, and comprehensive KPI examples are provided in Appendix B, the core concepts and capability dimensions will be referenced throughout the book to provide context and illustrate the journey towards higher AI maturity.

OmnioTech Case Study Context: Throughout this book, we will follow the journey of OmnioTech, a fictional mid-sized technology firm specializing in enterprise software, founded in 2020 with 2,000 employees, producing smart appliances, wearables, and personal electronics. OmnioTech serves as a practical case study, illustrating how the concepts, frameworks, challenges, and solutions discussed apply within a realistic organizational context. Any resemblance to real entities is coincidental.

By the end of this chapter, readers will grasp the compelling reasons why a strategically managed AI Office (AIO) is indispensable in the modern enterprise and understand how structured frameworks like the AI Office (AIO) Compass™ and EAMM/EAMI provide the necessary foundation for achieving successful, scalable, and responsible AI adoption.

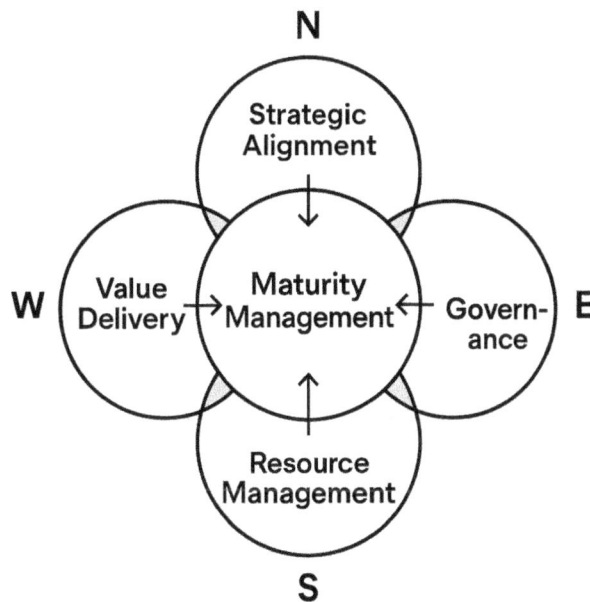

Figure 1.1 - AI Office (AIO) Compass™ Conceptual Overview Diagram

1.2 The Modern AI Landscape: Opportunities and Hurdles

Successfully navigating the path to AI value requires understanding the current technological landscape, which presents both immense opportunities and significant hurdles:

- **Generative AI Evolution:** The rapid rise of powerful Large Language Models (LLMs like GPT-4, Claude 3) and multimodal systems (handling text, image, audio, video) continues, driving widespread experimentation in content creation, summarization, coding assistance, conversational interfaces, and increasingly, more complex reasoning and agentic capabilities (where AI can plan and execute multi-step tasks). Managing cost, ensuring factual accuracy, and governing responsible use remain key challenges (Domain East).
- **Predictive AI Maturity:** Traditional machine learning (ML) techniques (e.g., using algorithms like XGBoost, Random Forests, deep learning for specific tasks) deliver substantial operational value in forecasting, anomaly detection (e.g., fraud, predictive maintenance), and customer analytics (segmentation, churn prediction). These often form the bedrock of initial value delivery (Domain West).
- **Cloud AI Platforms:** Major providers (AWS SageMaker, Google Vertex AI, Azure ML) offer integrated platforms lowering technical barriers with managed infrastructure, pre-built models, and MLOps services. Strategic selection and cost management (FinOps, Domain South) are critical.
- **Data as Foundation (and Bottleneck):** AI remains fundamentally data-dependent. Accessing high-quality, relevant, well-governed data at scale is often the primary challenge (EAMM Data Management, EAMM Data Quality). Overcoming data silos and ensuring data readiness (Chapters 18, 21-23, Domain South) are prerequisites for trustworthy AI. Poor data foundations undermine sophisticated algorithms (It is widely observed that data silos affect a significant number of enterprises - Gartner suggested 60% in 2022).
- **MLOps Imperative:** Reliable, scalable deployment necessitates disciplined Machine Learning Operations (MLOps) practices (Chapter 20, Domain South), extending DevOps to manage the unique ML lifecycle (data/model versioning, drift monitoring, automated pipelines). This is key to achieving higher EAMI MLOps maturity.
- **Ethical & Regulatory Scrutiny:** Concerns around bias, fairness, transparency, privacy (GDPR, CCPA), and security drive emerging regulations (e.g., EU AI Act) and demand robust, proactive governance frameworks (Part 3, Domain East).
- **Talent Scarcity:** Demand for specialized AI skills (ML Engineering, MLOps, AI Ethics) often outstrips supply (analysts frequently report significant gaps), requiring strategic talent management (Chapter 16, Domain South).

1.3 The High Cost of Uncoordinated AI: Common Pitfalls

Attempting to navigate this complex landscape through uncoordinated, ad-hoc efforts exposes organizations to significant risks and inefficiencies. The table below summarizes common pitfalls frequently observed in organizations lacking a strategic AI approach, often characteristic of low EAMI maturity (Level 1-2):

Table 1.1: Common Pitfalls of Uncoordinated AI

Pitfall	Description	Common Consequence
Strategic Misalignment	Projects lack clear, demonstrable links to core business strategy or key performance indicators (KPIs). Low EAMM Strategic Alignment maturity.	Investment wasted on "science projects" with limited business value; difficulty proving ROI; low contribution to EAMI Value Realization score.
Duplication & Wasted Resources	Different teams unknowingly build similar models, data pipelines, or procure redundant tools/datasets.	Significant cost overruns (industry reports suggest 30%+), inefficient use of scarce talent and compute resources, slowed progress.
Inconsistent Standards & Quality	Absence of central guidelines for data preparation, model development, validation, testing, or deployment (reflecting low EAMM MLOps/Data Quality maturity).	Variable model reliability, accumulation of technical debt, difficult integration, results cannot be trusted.
Increased Governance & Ethical Risks	Lack of unified oversight for privacy (GDPR/CCPA adherence), bias detection/mitigation, fairness, security protocols, and model transparency (low EAMM Governance/Ethics/Security maturity).	High risk of large regulatory fines, lawsuits, significant reputational damage, erosion of customer trust; low EAMI Governance score.
Scalability Challenges ("Pilot Purgatory")	Promising prototypes fail to transition to robust, scalable production systems (analysts report high failure rates, sometimes >70%) due to inadequate engineering, data access issues, platform incompatibility, or lack of MLOps maturity.	Potential value demonstrated in pilot phase remains unrealized at scale; innovation cycle stalls.
Lack of Organizational Learning	Valuable insights, code libraries, model components, best practices, and lessons learned remain isolated within project teams or departments (low EAMM Collaboration/Culture maturity).	Slowed enterprise-wide AI capability development (low EAMI Talent/Culture score), repeated mistakes, constant reinvention.
Difficulty Demonstrating Value	Inability to aggregate the impact of disparate initiatives into a clear, quantifiable narrative of business value delivered (low EAMM Value Realization maturity).	Difficulty justifying sustained AI investment beyond initial hype cycles; AI perceived as a cost center, not a value driver.
Competitive Blind Spots	An inward-looking, tactical focus on isolated point solutions prevents recognition of broader market trends or strategic AI moves by competitors.	Risk of being disrupted or losing market share to organizations leveraging AI more cohesively and strategically.

Conduct a simple diagnostic within your organization. Ask leaders of different AI initiatives: 1. What specific corporate strategic objective does this project directly support? 2. How is its business value being measured? 3. Is there a standard process for ethical and security review before deployment? 4. Are you leveraging common platforms or data assets used by other teams? Consistent difficulty in answering these questions strongly suggests the presence of these ad-hoc adoption pitfalls and likely low EAMI maturity.

1.4 The Strategic Imperative: Orchestrating for Value and Responsibility

The prevalence and high cost of these pitfalls underscore a clear strategic imperative: organizations seeking to truly leverage AI for sustainable advantage must move beyond fragmented experimentation towards a deliberate, integrated, and centrally orchestrated approach. This requires treating AI not merely as a collection of technologies, but as a fundamental enterprise capability that needs systematic development and management across multiple dimensions, as structured by the AI Office (AIO) Compass™ Framework (Strategy - North, Governance - East, Resources - South, Performance - West, Maturity - Center).

This imperative demands that AI efforts are purposefully aligned with and driven by critical business outcomes. The focus must shift from simply exploring what AI can do, to strategically deploying AI to achieve what the business needs to do. This involves:

- **Prioritizing ruthlessly (North):** Focusing investments and talent on AI applications that offer the highest potential impact on core strategic goals (e.g., improving operational efficiency via intelligent automation, enhancing revenue through hyper-personalization, mitigating critical risks using predictive analytics).
- **Governing proactively (East):** Establishing clear ethical guidelines, robust security protocols, reliable compliance mechanisms, and transparent processes before deploying AI systems, especially those impacting customers or employees.
- **Building foundational capabilities (South):** Investing systematically in the necessary data infrastructure, technology platforms, MLOps practices, and talent development required to support scalable and reliable AI deployment.
- **Measuring what matters (West):** Implementing frameworks to track not just technical performance, but the tangible business value delivered by AI initiatives, contributing to positive EAMI Value Realization scores.

Achieving this strategic integration necessitates strong, visible leadership commitment and a proactive approach to managing the associated organizational changes (Center - Culture/Change Management). Overcoming cultural inertia, addressing skill gaps through training and development (Chapter 26), navigating departmental politics regarding data or resource ownership, and maintaining executive sponsorship through clear communication of value (Chapter 15) are critical success factors. Managing these human and organizational dynamics effectively often requires a dedicated function with the mandate, expertise, and influence to drive change across silos – the AI Office (AIO).

1.5 The AI Office (AIO) / Center of Excellence (CoE): The Orchestrator

The AI Office, or AI CoE, emerges as the essential organizational construct designed to implement this strategic, orchestrated approach.

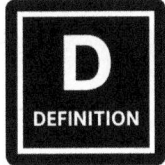

D **DEFINITION**

AI Office (AIO) / CoE - The central coordinating function chartered with developing enterprise AI strategy (North), establishing and enforcing governance frameworks (East), enabling efficient resource utilization (South), ensuring AI initiatives deliver measurable business value (West), and ultimately driving enterprise AI maturity(Center), often guided by frameworks like the AI Office (AIO) Compass™ and assessed by tools like EAMM/EAMI.

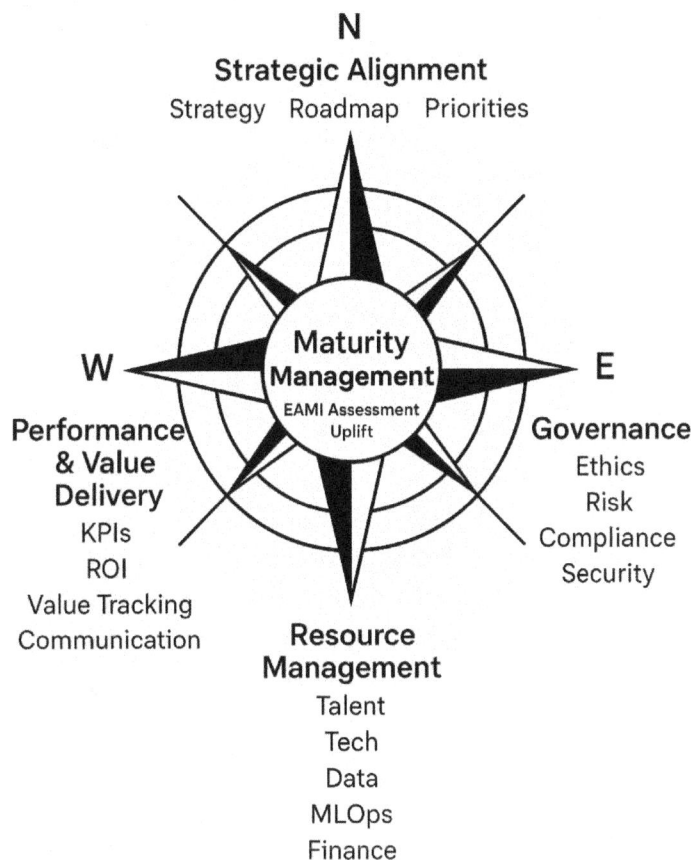

Figure 1.2: The AI Office Compass™ Framework Diagram

By centralizing key functions and coordinating efforts across the enterprise, the AI Office (AIO) directly addresses the pitfalls of ad-hoc adoption and delivers substantial benefits, driving improvements across EAMI dimensions:

- **Ensures Strategic Alignment (North):** Connects AI initiatives directly to business goals (EAMM Strategy).
- **Mitigates Risk (East):** Implements robust governance, ethics, security, and compliance frameworks (EAMM Governance, Ethics, Security, Compliance).

- **Optimizes Resources (South):** Reduces duplication, promotes reuse, manages talent/tech/data/finance strategically (EAMM Talent, Tech, Data, Financial).
- **Improves Quality & Reliability (South/East):** Establishes development standards, data quality processes, and mature MLOps practices (EAMM Data Quality, MLOps).
- **Builds Enterprise Capability (Center/South):** Facilitates knowledge sharing, training, and AI literacy programs (EAMM Culture, Talent/Skills).
- **Accelerates Scaling (South/West):** Provides infrastructure, expertise, and processes to move validated pilots into production reliably (EAMM MLOps, Tech).
- **Fosters Responsible Innovation (Center/East):** Enables exploration of new AI possibilities within defined ethical and strategic boundaries (EAMM Innovation, Ethics).
- **Demonstrates Value (West):** Implements systematic tracking and reporting of business impact and ROI (EAMM Value Realization, Performance Metrics).

Applicability Across Scales: While the specific size and structure may vary (as discussed in Chapter 2), the function of strategic orchestration provided by an AI Office (AIO) is crucial for organizations of all sizes aiming for serious AI adoption. Startups benefit from focus and discipline; mid-sized firms need structure to scale effectively; large enterprises require it to manage inherent complexity, ensure coherence across diverse units, and leverage potential economies of scale. The AI Office (AIO) function is essential, even if its specific organizational form adapts to the context.

1.6 OmnioTech Case Study: Recognizing the Need for Change

Setting the Scene:

OmnioTech, our fictional $2B tech firm, faced mounting challenges from its uncoordinated AI efforts. Marketing's personalization pilots yielded poor results due to inability to access unified customer data from Product's systems (a Data Management gap). Product's predictive maintenance prototypes for smart appliances remained stuck in the lab, unable to leverage operational data or deploy via existing IT infrastructure due to lack of MLOps maturity (an MLOps gap). Supply Chain's forecasting initiatives failed due to reliance on siloed, poor-quality data feeds (Data Quality/Management gaps). Cumulative spend across these efforts exceeded $2M with no demonstrable ROI (a Value Realization failure), while competitors launched integrated AI features. An initial EAMI assessment confirmed a low Level 1.5 maturity.

The Challenge & Diagnosis:

During a quarterly business review, CEO Julia Weber highlighted the disconnect. The symptoms were undeniable (referencing Table 1.1 pitfalls):

- *Duplication:* Multiple teams attempting customer behavior modeling.
- *Pilot Purgatory:* Valuable predictive maintenance feature stalled indefinitely.
- *Unclear ROI:* Significant spend ($2M+) without measurable business impact.
- *Emerging Risks:* CISO flagged security risks; Legal warned of potential GDPR violations (Governance gaps).
- *Strategic Drift:* No unified plan guiding AI investment priorities (Strategy gap).

The Decision & Imperative:

Facing pressure from the board and observing competitors gaining ground, the executive team acknowledged the urgent need for change. Julia Weber declared, "Our ad-hoc approach isn't working. We need a coordinated, strategic, and governed way to leverage AI, or we risk falling further behind." The projected cost of inaction (continued waste, missed opportunities, potential compliance fines) was estimated at over $1M annually. The decision was made: establish a central AI Office (AIO) Hub. The COO and CTO were formally tasked with designing this function (Chapter 2), giving it a clear mandate to establish governance (East), align initiatives with strategy (North), enable foundational capabilities (South), and drive OmnioTech towards a target EAMI Level 3 (Established) within two years.

1.7 Conclusion: Embracing Orchestration for AI Success

The journey to becoming an AI-driven enterprise demands more than just technological prowess; it requires strategic vision, disciplined execution, and robust governance. The common pitfalls associated with fragmented, ad-hoc AI initiatives – wasted resources, increased risks, stalled progress, and unrealized value – highlight the critical need for a centralized, orchestrating function: the AI Office (AIO) or Center of Excellence. Guided by comprehensive frameworks like the AI Office (AIO) Compass™ and utilizing objective assessment tools like EAMM/EAMI, the AI Office (AIO) provides the structure necessary to align AI efforts with business strategy (North), manage resources effectively (South), ensure responsible deployment (East), demonstrate tangible impact (West), and drive continuous maturity improvement (Center).

> ✅ Establishing an AI Office (AIO) signifies a crucial shift from treating AI as a series of isolated technical projects to managing it as a core, strategic enterprise capability. This shift in perspective and approach is fundamental to unlocking AI's transformative potential responsibly and sustainably.

OmnioTech's realization and commitment to change illustrate the typical catalyst for forming an AI Office (AIO). This book provides the detailed blueprint for designing, implementing, and operating this vital function effectively, starting with the architectural considerations for the AI Office (AIO) itself in Chapter 2.

Key Takeaways:

- AI offers immense potential but ad-hoc approaches frequently fail due to common pitfalls like misalignment, duplication, governance gaps, and scaling issues (often indicating low EAMI maturity).
- A dedicated AI Office (AIO)/CoE is essential for strategic orchestration, risk mitigation, resource optimization, and demonstrating value.
- The AI Office (AIO) Compass™ framework guides operations across Strategy (North), Governance (East), Performance & Value Delivery (West), Resource Management (South), and Maturity Management (Center).
- The EAMM/EAMI framework provides objective maturity assessment and progress tracking (qualitative

matrix and quantitative index).

- Success requires an integrated approach, strong leadership, and managing AI as a core enterprise capability.

Food for Thought / Application Exercise:

- Reflect on your organization's current AI initiatives. Do they exhibit any of the pitfalls listed in Table 1.1? Which ones are most prevalent?
- Consider your organization's primary strategic objectives for the next 1-2 years. Where might strategically aligned AI provide the most significant leverage or address key challenges?
- If your organization has disparate AI efforts, what are the 1-2 most critical first steps needed to move towards better coordination or establishing a central guiding function like an AI Office (AIO)?

~ CHAPTER 2 ~

Designing the AI Office: Mandate, Models, and Structure

"Structure follows strategy. Organizational design is not merely an administrative exercise; it's the blueprint for how strategy gets executed."
— Adapted from Alfred D. Chandler Jr.

2.1 Introduction: Architecting the AI Engine

Chapter 1 established the compelling strategic imperative for organizations to move beyond ad-hoc AI experimentation towards an orchestrated approach, facilitated by a dedicated AI Office or Center of Excellence (CoE). Having accepted this necessity, the critical next step is designing this entity effectively. Simply creating a new organizational unit is insufficient; the AI Office's structure, mandate, operating model, and leadership must be carefully architected to grant it the authority, resources, and strategic alignment required for success. An ill-conceived design can lead to friction, inefficiency, and ultimately, failure to achieve the transformative potential of AI, potentially stalling progress toward higher EAMI maturity levels (as defined in Chapter 24).

This chapter delves into the fundamental design considerations for establishing an effective AI Office. We will explore how to precisely define its core mandate and objectives, examine different operating models (Centralized, Decentralized, Hybrid/Federated) and their suitability in various contexts (often correlating with EAMI maturity), discuss key structural decisions regarding reporting lines and internal organization, outline essential initial roles and leadership qualities, and detail the crucial interfaces with other corporate functions.

Designing the AI Office isn't just about drawing an org chart. It's about consciously building the organizational capability to execute AI strategy effectively and responsibly across the enterprise, aligning structure with strategic intent. Getting this initial architecture right, informed by the principles of the AI Office Compass™ Framework (introduced conceptually in Chapter 1, detailed in Chapter 3) and aiming for measurable maturity uplift tracked by EAMI KPIs (e.g., progressing from EAMI Level 2 towards Level 3), is crucial for setting the AI Office on a path to long-term impact and value creation.

2.2 Defining the Mandate: Purpose, Scope, and Objectives

Before determining structure or staffing, the most critical first step is defining the AI Office's mandate: its core purpose, scope of authority, and primary objectives. A clear, formally documented, and leadership-endorsed mandate provides direction, aligns stakeholder expectations, and empowers the office to act effectively, preventing confusion and turf wars that can hinder progress towards higher EAMI levels. Key questions to address include:

- **Primary Purpose:** Is the AI Office primarily focused on driving cutting-edge innovation (aligning with EAMM Innovation Ecosystem)? Establishing robust governance and ethical guardrails (aligning with EAMM Governance, Ethics criteria)? Delivering specific AI solutions (e.g., predictive models)? Building enterprise-wide AI capabilities and literacy (EAMM Talent/Culture)? Or a strategic combination? Clarity on the primary objective for the initial phase (often targeting higher EAMI Level capabilities) is vital.

- **Scope of Authority:** Does the mandate cover all AI-related activities, including basic Robotic Process Automation (RPA) or standard Business Intelligence (BI), or focus on ML and advanced AI? Does the AI Office own the enterprise data strategy/platforms (critical for EAMM Data Management) or primarily influence them via the CDO? Does it have authority to set mandatory standards (e.g., Responsible AI), approve/veto projects, and control AI budgets, or is its role mainly advisory?

- **Scope Clarity Examples:** Likely In Scope often includes enterprise AI strategy support (Compass North), the core AI governance framework (ethics, risk, compliance per EAMM - Compass East), AI platform strategy/enablement (Compass South), oversight/execution of high-risk projects, and promoting AI literacy (Compass Center). Potentially Out of Scope might include routine BI reporting, simple RPA, direct ownership of all enterprise data (usually CDO), or executing every minor departmental experiment.

- **Key Objectives (Initial 1-3 Years):** What specific, measurable, achievable, relevant, and time-bound (SMART) outcomes are expected, potentially linked to improving specific EAMI dimension scores? Examples: Implement V1.0 Governance Framework across X units (Target: EAMI Governance L3); Launch & validate value for 3 pilot projects delivering $Y (Target: EAMI Value Realization improvement); Establish core cloud ML platform (e.g., Azure ML) with baseline MLOps (Target: EAMI Technology & MLOps L3); Train Z managers in AI literacy (Target: EAMI Culture/Talent improvement). (Refer to Appendix B for more KPI examples).

Define the AI Office mandate collaboratively with executive sponsors and key stakeholders. Start with a focused, achievable mandate for the initial phase (e.g., foundational governance, enabling high-priority pilots, core platform setup) targeting progress towards higher EAMI Level capabilities. Plan for the mandate to evolve as the office demonstrates value and organizational AI maturity grows (Part 6). Ensure success criteria are clearly defined and measurable from the outset.

2.3 Operating Models: Centralized, Decentralized, or Hybrid?

The AI Office's operating model defines how it functions and interacts with the rest of the organization. Choosing the right model depends on factors like organizational size, culture, current EAMI maturity level, strategic goals, and industry context. The table below compares the common models:

Table 2.1: Comparison of AI Office Operating Models

Feature	Centralized Model	Decentralized (Embedded) Model	Hybrid / Federated (Hub-and-Spoke) Model
Structure	Most AI expertise, resources, platform ownership, project authority reside within a single, central CoE team.	AI talent (if any) and project responsibility are fully embedded within individual BUs or functional departments. Minimal/no central function.	A central "Hub" (the AI Office/CoE) provides strategy, governance frameworks, core platforms/tools, shared services, deep expertise, and enablement programs. Embedded "Spokes" (dedicated AI specialists or teams within BUs/functions) focus on domain-specific solutions, leveraging Hub resources and adhering to standards.
Pros	Strong initial control over standards, governance, and security; efficient talent pooling; easier knowledge sharing; focused platform investment.	Very close alignment with specific BU needs and domain context; potentially faster local iteration; strong BU ownership.	Balances central coordination & standards with BU agility & ownership; enables scalable deployment; fosters collaboration; optimizes expertise use.
Cons	Can disconnect from BU needs; potential bottleneck; perceived "ivory tower".	High risk of fragmented strategy, inconsistent standards, duplicated efforts; difficult knowledge sharing; inefficient talent use.	Requires very clear definition of roles/ interfaces; complex initial setup; needs strong coordination mechanisms.

Best Suited For	Organizations starting their AI journey (EAMI Level 1-2); AI-core businesses; highly regulated industries (initially).	Highly diversified conglomerates with extreme BU autonomy (rare); organizations with very mature, pre-existing embedded AI skills (rare).	Most medium-to-large orgs scaling AI (EAMI 3-5); target state for evolution from Centralized. Common across diverse sectors.
EAMI Alignment	Supports establishing foundational capabilities for EAMI Levels 2-3 (e.g., Governance, Technology standards).	Often reflects low EAMI maturity (Level 1-2 across Strategy, Governance, Standards, and Talent dimensions.	Typically represents higher EAMI maturity (Levels 3-5), enabling both standardization (Hub) and scaled, domain-specific application (Spokes).

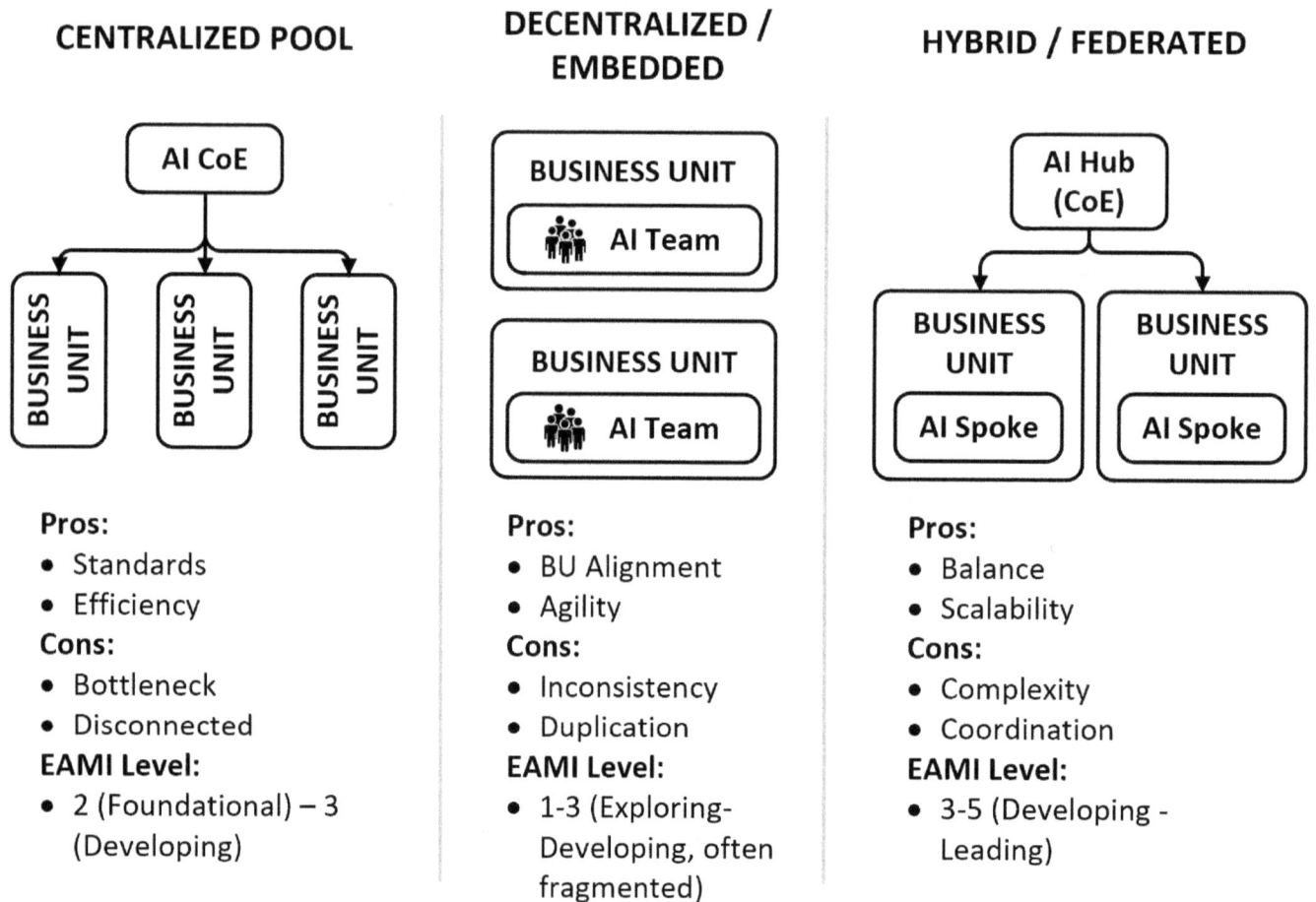

CENTRALIZED POOL

AI CoE

BUSINESS UNIT BUSINESS UNIT BUSINESS UNIT

Pros:
- Standards
- Efficiency

Cons:
- Bottleneck
- Disconnected

EAMI Level:
- 2 (Foundational) – 3 (Developing)

DECENTRALIZED / EMBEDDED

BUSINESS UNIT
AI Team

BUSINESS UNIT
AI Team

Pros:
- BU Alignment
- Agility

Cons:
- Inconsistency
- Duplication

EAMI Level:
- 1-3 (Exploring-Developing, often fragmented)

HYBRID / FEDERATED

AI Hub (CoE)

BUSINESS UNIT
AI Spoke

BUSINESS UNIT
AI Spoke

Pros:
- Balance
- Scalability

Cons:
- Complexity
- Coordination

EAMI Level:
- 3-5 (Developing - Leading)

Figure 2.1: AI Office Operating Models Diagram

> The "right" operating model isn't static. Plan for evolution. Organizations frequently start more centralized to build foundations (EAMI Levels 2-3) and then shift towards a hybrid model as maturity increases (EAMI Levels 4-5) and business units develop stronger AI capabilities. Document potential triggers for model evolution (e.g., specific EAMI score threshold, persistent central bottleneck).

2.4 Organizational Structure and Reporting Lines

The AI Office's placement within the broader organizational structure significantly impacts its influence, authority, and effectiveness in driving maturity across EAMM dimensions.

- **Reporting Line:** The choice of reporting line signals strategic priority and shapes focus:
  - » **Reporting to CTO/CIO:** Emphasizes the technological foundation (EAMM Technology/Data), integration with IT systems, and platform engineering aspects. May risk being perceived as purely technical if business alignment (EAMM Strategy/Value) isn't actively managed through strong partnerships.
  - » **Reporting to CDO:** Highlights the critical role of data strategy, governance, and quality (EAMM Data/Governance) as enablers for AI. May need strong partnerships to drive platform/operational aspects (EAMM Technology/MLOps).
  - » **Reporting to Chief Strategy Officer / Head of Transformation:** Positions AI clearly as a driver of business change and strategic initiatives (EAMM Strategy). Requires strong technical partnership to ensure feasibility.
  - » **Reporting directly to CEO/COO:** Signals the highest level of strategic importance and provides maximum authority for driving cross-functional change needed for higher EAMI maturity. Often seen in AI-first organizations or during major transformations.
  - » **Joint Reporting (e.g., to COO & CTO):** Aims to balance business/operational goals (EAMM Value/Strategy) with technical feasibility and platform strategy (EAMM Technology/Data). Requires exceptionally clear charters, defined decision processes (often via a steering committee involving both executives), and strong, unified sponsorship from both executives to mitigate potential conflicts arising from differing priorities (e.g., speed vs. platform stability).
- **Internal Structure:** Within the AI Office itself (especially the Hub in hybrid models), teams can be structured functionally (e.g., Governance team, Platform team, Data Science pool), project-based (teams formed for specific initiatives), or matrixed (individuals contribute expertise across multiple projects/functions). Hybrid models often utilize functional experts in the Hub supporting matrixed project teams that include embedded Spoke members. The structure should facilitate efficient execution across EAMM capability areas.
- **Relationship with Other Functions:** Establishing clear boundaries, collaboration protocols (e.g., documented in RACI charts), and potentially Service Level Agreements (SLAs) with interfacing functions like IT Infrastructure, Data Management, Cybersecurity, Legal, Compliance, HR, and Finance is vital to avoid gaps, friction, or duplicated efforts that hinder maturity progression.

2.5 Leadership and Initial Core Roles

Effective leadership and a well-defined core team are crucial for launching the AI Office successfully and driving progress across EAMI dimensions.

- **Leadership Qualities:** The Head of the AI Office requires a rare blend of skills: strategic vision (to align with EAMM Strategy), sufficient technical literacy (understanding AI capabilities/limits - EAMM Technology), strong business acumen (connecting AI to value - EAMM Value), exceptional communication and influence skills (to engage diverse stakeholders across EAMM dimensions), proven change leadership capabilities (EAMM Culture/Change), and an unwavering commitment to responsible and ethical AI principles (EAMM Governance).

- **Important:** AI Leadership Hiring Pitfalls Avoid hiring leaders who are solely technical experts lacking strategic vision, purely business-focused individuals without adequate technical understanding to guide platform/MLOps decisions, those who ignore the critical change management aspects needed for cultural adoption (EAMM Culture), or those lacking the influence to secure cross-functional buy-in and navigate organizational politics effectively. Look for a balanced profile with strong leadership and communication skills first and foremost.

- **Initial Core Team (Hub Focus):** To establish foundational capabilities (e.g. targeting EAMI Level 3), especially when starting centralized or initializing a hybrid model, the initial team often needs to cover these critical functions. The table below outlines common roles:

Table 2.2: Initial Core AI Office Roles

Role	Key Responsibilities	Critical EAMM Alignment
Head of AI Office	Overall leadership, strategy formulation, executive stakeholder management, budget oversight, cross-functional coordination.	Strategy, Leadership, Value, Financial
AI Governance Lead	Design & implement policies/standards (ethics, risk, compliance), manage review processes, liaise w/ Legal/Compliance.	Governance, Ethics, Compliance, Risk
Platform/MLOps Engineer Lead	Select, implement, manage core AI/ML platform & foundational MLOps tooling (CI/CD, registry, monitoring).	Technology, MLOps, Infrastructure
Lead Data Engineer (Liaison)	Ensure data architecture, pipelines, quality meet AI needs; collaborate closely with central Data/CDO function.	Data Management, Data Quality
AI Translator / Program Mgr	Bridge business & tech, manage AI portfolio/roadmap, track value, facilitate communication, manage projects/programs.	Project Management, Value, Collaboration

Deep data science/modeling skills might be initially sourced per project or borrowed via consultants, focusing the core team on enablement and governance foundations first, depending on the initial roadmap priorities and budget constraints.

2.6 Key Interfaces and Collaboration Modes

The AI Office cannot operate in isolation; its success in driving maturity across EAMM dimensions depends on strong partnerships and clearly defined interactions with other key organizational functions:

- **IT Infrastructure & Operations:** Collaboration on platform hosting, network requirements, compute provisioning (cloud cost management - FinOps), operational support, monitoring integration.
- **Cybersecurity (CISO):** Integration into security reviews (DevSecOps/MLSecOps), threat modeling for AI systems, defining security standards for AI platforms/data, incident response collaboration (EAMM Security).
- **Data Management & Governance (CDO):** Close alignment on data strategy, data access policies/processes, data quality standards/monitoring (EAMM Data Quality), metadata/catalog integration, master data management (EAMM Data Management).
- **Legal & Compliance:** Partnership on interpreting regulations (GDPR, AI Act), drafting/approving AI policies, reviewing vendor contracts, managing privacy requirements (DPIAs), IP protection (EAMM Compliance).
- **Human Resources (HR):** Collaboration on defining AI roles/career paths, sourcing/recruiting AI talent, developing/delivering AI literacy and technical training (EAMM Talent/Culture), supporting change management communications (EAMM Change Management).
- **Finance:** Partnership on defining funding models, project budgeting processes, TCO estimation (EAMM Financial), ROI validation methodologies (EAMM Value Realization), cost tracking/optimization (FinOps alignment).
- **Business Units (BUs):** Primary source for identifying strategic opportunities, providing essential domain expertise, sponsoring specific initiatives, championing user adoption (EAMM Culture), and ultimately realizing/reporting the business value achieved (EAMM Value Realization).

Formalize critical interactions early. Use mechanisms like:
- **Joint Committees:** Establish an AI Governance Council or Steering Committee with representation from key functions.
- **Defined Processes:** Document handoffs and review gates within workflows (e.g., using BPMN or integrating into tools like Jira).
- **RACI Charts:** Clearly define who is Responsible, Accountable, Consulted, and Informed for key AI governance and development activities.
- **Shared Templates:** Use standardized templates for project proposals, risk assessments, model documentation (Model Cards), etc.
- **Service Level Agreements (SLAs):** Define expectations for critical shared services (e.g., platform uptime, data pipeline delivery times).

2.7 Common Pitfalls in Initial AI Office Design

Avoid these frequent design errors when establishing the AI Office:

- **Vague Mandate:** Unclear purpose, scope, decision rights, or objectives lead to confusion, inaction, inability to prioritize, and turf wars.

- **Model/Structure Mismatch:** Choosing an operating model (e.g., fully decentralized too early for an organization at EAMI Level 2) or reporting line that clashes with the organization's culture, existing maturity level, or strategic needs.
- **Lack of Genuine Sponsorship:** Executive support is merely passive ("lip service") or lacks the tangible commitment (resources, political capital) to drive necessary cross-functional changes or resolve conflicts.
- **Unclear Roles & Interfaces:** Ambiguity between the AI Office's responsibilities and those of existing functions (IT, Data, BUs) causes turf wars, critical governance or operational gaps (e.g., who owns production model monitoring?), or duplicated efforts.
- **Under-resourcing:** Launching with insufficient budget or lacking critical core staff (e.g., no dedicated governance or platform expertise) prevents the office from establishing foundational capabilities (needed for EAMI Level 3) and demonstrating early value, leading to a loss of momentum.
- **Ignoring Change Management:** Focusing solely on technology platforms and governance policies without a concurrent, well-resourced plan for managing the significant cultural shifts, skill development (EAMM Talent/Culture), and process redesign required across the organization for successful AI adoption (EAMM Change Management).

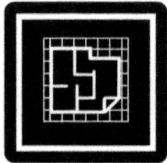

Careful, collaborative design involving key executive sponsors and leaders from interfacing functions from the outset, guided by an initial EAMI assessment, significantly mitigates these risks.

2.8 OmnioTech Case Study: Designing the AI Office Hub

Setting the Scene:

Following the executive decision to establish a central AI coordinating function (Chapter 1), OmnioTech's COO (Michael Vance) and newly appointed CTO (Stefan Ritter) are tasked with designing the initial AI Office Hub ($2M initial budget), aiming to lift OmnioTech from its EAMI Level 1.5 baseline.

The Challenge/Opportunity:

Design an AI Office providing strategic direction, governance, and enablement for initial pilots (Smart Appliance Predictive Failure, Hyper-Personalized Marketing Engine) without creating excessive bureaucracy, fitting their mid-sized context, and targeting EAMI Level 3 maturity within 18 months.

Action & Application:

The COO and CTO lead design workshops involving Priya Sharma (confirmed as Head of AI Office Hub), plus reps from IT, Data Architecture, Legal, HR, Product Management, and Marketing:

- **Mandate Defined:** Initial focus: Develop AI Strategy/Roadmap linked to product innovation & customer engagement goals (Target EAMI Strategy L3); Establish V1.0 Governance (esp. consumer data privacy/ethics - Target EAMI Governance L3); Select/Implement core Azure ML Platform (Target EAMI Technology L3); Enable 2 high-priority pilots (Target EAMI Project Management L3); Begin AI awareness building (Target EAMI Culture L2).
- **Operating Model Chosen:** Select Hybrid ("Hub-and-Spoke"), starting Hub-heavy ("Hybrid-Initializing"). Rationale: Need central control for governance & platform standards initially (reflecting current low EAMI Level 2 maturity). Plan to embed "Spokes" (AI Liaisons from Catalyst Program - Ch 16) into Product/Marketing teams in Year 2 as maturity grows towards EAMI Level 4. (Evolution Trigger Considered: EAMI score reaching 3.5 overall, increasing BU demand).
- **Structure & Reporting Determined:** Head of AI Office Hub (Priya Sharma) reports jointly to COO (Vance - business value focus) and CTO (Stefan - tech feasibility focus). Rationale & Mitigation: Balances EAMM Value/Strategy with Technology/Data dimensions. Potential conflicts addressed via clear charter for Priya, defined AI Review Team decision rights, and strong joint COO/CTO sponsorship demonstrated in kickoff meetings.
- **Initial Core Roles Identified (See Table 2.2):** Head of AI (Priya Sharma); external hires sought for Governance Lead and Platform/MLOps Lead; 75% allocation confirmed for Data Engineer Liaison from IT Data team; AI Translator function initially split, planned for dedicated hire in Year 2.
- **Key Interfaces & Governance Body Defined:** Mapped collaboration needs: Weekly syncs with IT Platform Eng; Bi-weekly with Legal/Privacy; Project-based engagement with Product/Marketing data stewards. Established the "AI Review Team" (Sponsors, Priya, Gov Lead, Platform Lead, Legal, HR, key BU reps) as the primary V1.0 governance body, meeting monthly + for key project gate reviews (targeting EAMI Governance L3 processes). Basic platform support SLA discussed with IT Ops, RACI drafted for pilot governance.

Outcome & Progress:

OmnioTech creates a pragmatic V1.0 design for its AI Office Hub. The mandate targets EAMI Level 3 capabilities. The hybrid-initializing model allows future scaling. Joint reporting, managed via clear roles and the AI Review Team, addresses potential conflicts. Core roles are identified, leveraging internal liaisons initially. Key interfaces are mapped. This design provides Priya Sharma the structure needed to launch the Hub and start executing strategy and governance, aiming to demonstrably improve specific EAMI dimension scores within the first year.

2.9 Conclusion: Laying the Architectural Foundation

Designing the AI Office is a critical strategic exercise that shapes the trajectory of an organization's entire AI journey. A well-defined mandate provides purpose and clarity. Selecting the appropriate operating model—often evolving from centralized towards hybrid as maturity increases—balances control with agility. Thoughtful structural design, including reporting lines and clear interfaces, enables effective execution and collaboration. Staffing the office with the right leadership and core expertise ensures the necessary capabilities are in place from the start. By carefully considering these architectural elements, informed by an understanding of current AI maturity (EAMI level) and strategic goals, organizations can avoid common pitfalls and create an AI Office empowered to navigate complexity, drive strategic alignment, ensure responsible innovation, and ultimately unlock significant business value. The OmnioTech case study illustrates how these design choices translate into a practical starting point for building this crucial capability.

With the foundational architecture established, the next step, detailed in Chapter 3, is to explore the operational blueprint provided by the AI Office Compass™ Framework, which guides the day-to-day activities and performance domains of the newly designed AI Office.

Key Takeaways:

- Designing the AI Office (mandate, model, structure) is a critical step beyond recognizing the need for one.
- The mandate must define purpose, scope, and measurable objectives (linked to EAMI targets).
- Operating models (Centralized, Hybrid, Decentralized) have trade-offs; Hybrid often suits scaling maturity.
- Structure, reporting lines, and clear interfaces with other functions are vital for effectiveness.
- Strong leadership and covering core roles (Gov, Platform, Data, Translator) are essential, avoiding common pitfalls.

Food for Thought / Application Exercise:

- Assess your organization's current AI operating model (even if informal). Is it Centralized, Decentralized, or Hybrid? Is it optimal for your current maturity and goals?
- Draft a potential high-level mandate (Purpose, Scope, 2-3 SMART Objectives) for an AI Office within your organization's context.
- Identify the top 3 potential pitfalls (from Section 2.7) that your organization might face when establishing or scaling an AI Office. What proactive steps could mitigate them?

~ CHAPTER 3 ~

The AI Office Compass Framework: A Blueprint for Excellence

"If you don't know where you are going, any road will get you there."
— Lewis Carroll

3.1 Introduction: Navigating the Complexities of Enterprise AI

Chapters 1 and 2 established the strategic necessity for an AI Office and explored the critical design considerations for its mandate, operating model, and structure. However, simply creating the office is insufficient; it needs a clear operational blueprint to guide its activities, ensure comprehensive coverage of essential functions, and drive progress towards higher AI maturity. Managing enterprise AI effectively requires orchestrating a complex interplay of strategy, governance, technology, data, talent, performance management, and continuous improvement – neglecting any one area can jeopardize the entire initiative.

To provide this essential navigational aid, this chapter introduces the AI Office Compass™ Framework. This proprietary model organizes the AI Office's responsibilities into five interconnected performance domains, visualized as points on a compass: Strategic Alignment (North), Governance (East), Performance & Value Delivery (West), Resource Management (South), and at the heart of it all, Maturity Management (Center). Each domain represents a critical cluster of capabilities and activities the AI Office must master and integrate to deliver sustained value and achieve transformative AI maturity, as measured by the EAMM/EAMI framework (introduced conceptually in Chapter 1, detailed in Chapter 24).

This chapter will detail each of the five Compass domains, explaining their scope, criticality, common challenges, and key focus areas for the AI Office. Understanding this framework provides a holistic perspective on the AI Office's role and equips leaders with a structured approach to managing its multifaceted responsibilities effectively.

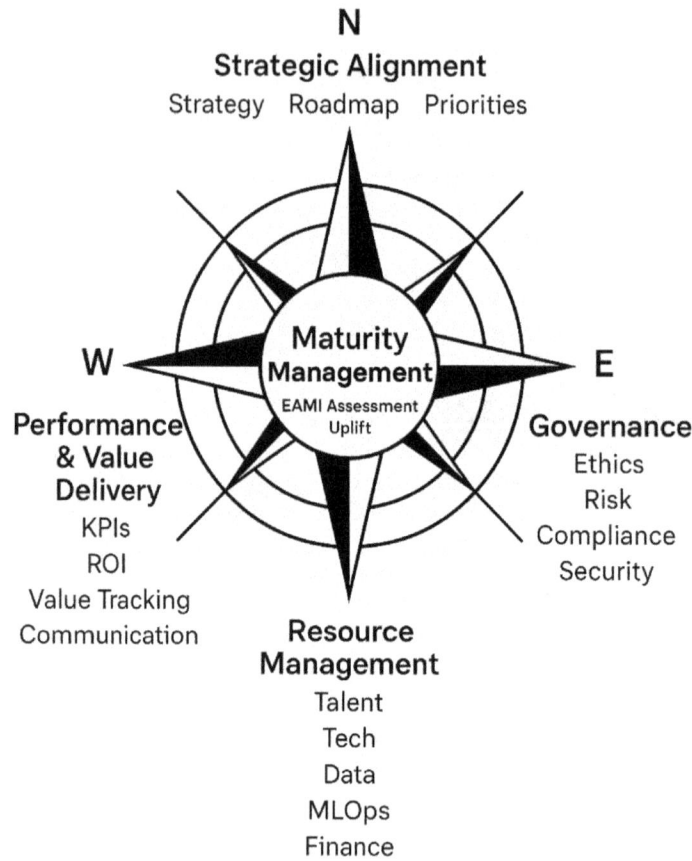

Figure 3.1: The AI Office Compass™ Framework Diagram

3.2 Overview of the Five Compass Domains

The AI Office Compass™ Framework provides a structured way to conceptualize and manage the diverse responsibilities required for successful enterprise AI. It ensures that critical areas are not overlooked and highlights the dependencies between them.

> **D**
> **DEFINITION**
>
> **AI Office Compass™ Framework -** A management model defining five critical, interconnected performance domains for an AI Office: Strategic Alignment (North), Governance (East), Performance & Value Delivery (West), Resource Management (South), and Maturity Management (Center). It serves as an operational blueprint guiding the AI Office's activities and ensuring holistic capability development, aligning directly with EAMM/EAMI criteria.

The five domains and their key focus areas, along with primary alignment to EAMM dimensions, are summarized below:

Table 3.1: AI Office Compass Domains and EAMM Alignment

Compass Domain	Direction	Primary Focus	Key Activities	EAMM Alignment (Primary)
Strategic Alignment	North	Ensuring AI initiatives align with and drive business strategy.	Strategy analysis, opportunity ID & prioritization, roadmap development, executive alignment.	Strategy, Project Management
Governance	East	Establishing ethical, legal, security, and operational guardrails for responsible AI.	Policy development, risk management, compliance adherence, ethical reviews, security protocols, bias mitigation.	Governance, Ethics, Compliance, Security, Risk Management
Performance & Value Delivery	West	Measuring AI performance, quantifying business value (ROI), and communicating impact.	KPI definition, value tracking frameworks, ROI/TCO calculation, stakeholder reporting, benefits realization.	Performance Metrics, Value Realization
Resource Management	South	Managing the essential inputs: talent, technology, data, MLOps, and finances.	Talent strategy (acquire, develop, retain), platform/tool management, data enablement/quality, MLOps implementation, budget/cost optimization.	Talent (Acquisition/Skills), Technology Selection, Infrastructure, Data Management, Data Quality, MLOps, Financial
Maturity Management	Center	Orchestrating continuous improvement, cultural adoption, and innovation across all AI capabilities.	EAMI assessment, uplift strategy execution, change management, AI literacy programs, CoP facilitation, innovation scouting & management.	Maturity Assessment, Uplift Strategy, Culture, Change Management, Innovation Ecosystem, Collaboration

EAMM Dimension and AI Office Compass™ Domain Alignment

EAMM Dimension	Primary Compass™ Domain
Strategic Alignment	North
Operational Efficiency	West
Customer Experience	West
Innovation & Product Development	Center
Financial Performance	West
Workforce Productivity	South
Data Management	South
Risk Management	East
Market Positioning	West
Compliance & Ethics	East
Stakeholder Engagement	Center
Sustainability	Center
Organizational Agility	Center
Change Management	Center

Figure 3.2: EAMM-Compass Mapping Table

The following sections explore each domain in more detail.

3.3 Strategic Alignment (North): Charting the Course

- **Scope:** Defines the 'Why' and 'What' of enterprise AI. Activities include deeply understanding corporate strategy and objectives, identifying specific AI use cases with high strategic impact (EAMM Strategy), effectively translating business needs into clear AI requirements, developing the overarching enterprise AI vision and a multi-year, prioritized roadmap (EAMM Project Management), and securing ongoing executive alignment and sponsorship for that roadmap (Chapters 4-6 detail this).

- **Criticality:** This is the foundational domain. It ensures that all AI investments, resources, and efforts are focused on solving the right business problems and contributing directly and measurably to organizational success, rather than pursuing disconnected technological novelties or departmental pet projects. Provides essential direction and focus for all other AI Office activities and resource allocation decisions.
- **Challenges:** Common hurdles include difficulty translating high-level business goals into specific, actionable AI opportunities; overcoming resistance from business units comfortable with existing processes; establishing and consistently applying clear prioritization frameworks leading to resource dilution; ensuring the AI roadmap remains dynamic and adapts to changing market conditions.
- **AI Office Focus:** Facilitating strategic planning workshops, employing structured opportunity assessment and prioritization frameworks (Chapter 5), developing and actively governing the enterprise AI roadmap (Chapter 6), ensuring clear communication of the AI vision and its link to business value.

> Effective Strategic Alignment isn't a one-time task. The AI Office must foster continuous dialogue between business and AI teams to ensure the roadmap evolves dynamically, remaining tightly coupled with shifting corporate priorities. It's about steering, not just setting a course.

3.4 Governance (East): Establishing the Guardrails

- **Scope:** Defines the 'How' of responsible AI deployment. Activities encompass developing, implementing, and enforcing policies, standards, and processes covering data privacy (e.g., GDPR compliance - EAMM Compliance), ethical AI (bias mitigation using tools like Fairlearn, fairness assessment, transparency via XAI techniques like SHAP - EAMM Ethics), regulatory compliance (e.g., EU AI Act readiness), AI security (threat modeling, secure development - EAMM Security), risk management frameworks (EAMM Risk Management), and establishing oversight bodies (Chapters 7-12 detail this).
- **Criticality:** Builds crucial trust; mitigates potentially catastrophic financial and reputational damage from ethical lapses or compliance failures; ensures AI systems are reliable and fair. Foundational for achieving higher EAMI Governance/Ethics/Compliance scores.
- **Challenges:** Keeping pace with rapidly evolving regulations and ethical norms; balancing governance rigor with innovation speed; effectively embedding controls into MLOps workflows (Chapter 20); ensuring consistent policy application across diverse use cases; managing transparency challenges of complex models.
- **AI Office Focus:** Developing a pragmatic, risk-tiered governance framework (Chapter 8), implementing robust review processes, selecting appropriate governance tooling, promoting ethical awareness training, collaborating closely with Legal, Compliance, Privacy, and Cybersecurity teams.

> Adopt a 'Governance-by-Design' approach. Integrate ethical, privacy, security, and compliance considerations early and continuously throughout the AI development lifecycle (within MLOps workflows), rather than treating governance solely as a late-stage review gate.

3.5 Performance & Value Delivery (West): Demonstrating the Impact

- **Scope:** Defines the 'So What?' of AI investments – rigorously measuring and clearly communicating the value delivered. Activities include establishing relevant Key Performance Indicators (KPIs) beyond technical metrics (e.g., impact on revenue, cost reduction, CSAT, EAMI score uplift - EAMM Performance Metrics), defining clear baseline performance, implementing structured frameworks for tracking value realization throughout the project lifecycle (EAMM Value Realization), calculating Return on Investment (ROI) and Total Cost of Ownership (TCO) using transparent methodologies, and effectively communicating performance results and quantified value stories to diverse stakeholders (Chapters 13-15 detail this). (See Appendix B for detailed KPI examples).
- **Criticality:** Justifies ongoing AI investment; builds essential credibility for the AI Office; provides data for prioritizing future initiatives; enables continuous improvement by highlighting what works. Directly drives improvement in the EAMI Value Realization dimension.
- **Challenges:** Accurately attributing business outcomes solely to AI initiatives; defining meaningful KPIs for complex projects; establishing reliable baselines; quantifying intangible benefits; communicating complex results clearly to non-technical audiences.
- **AI Office Focus:** Developing standardized value measurement frameworks (Chapter 13), implementing robust performance tracking mechanisms and dashboards (Chapter 14), ensuring rigorous ROI/TCO analysis, crafting compelling value narratives tailored to different audiences (Chapter 15).

3.6 Resource Management (South): Fueling the Engine

- **Scope:** Strategically manages the essential inputs required for successful AI execution. Activities encompass comprehensive talent strategy (Chapter 16 - EAMM Talent Acquisition, Skill Development), technology stack management (platforms, tools, infrastructure - Chapter 17 - EAMM Technology Selection, Infrastructure), data enablement (accessibility, quality, pipelines - Chapters 18, 21-23 - EAMM Data Management, Data Quality), MLOps implementation (Chapter 20 - EAMM MLOps), and financial stewardship (budgeting, cost optimization - Chapter 19 - EAMM Financial).
- **Criticality:** Ensures the AI Office and project teams possess the necessary people, platforms, data, processes, and funding to execute the AI strategy effectively and reliably. Inadequate resources are a primary cause of AI failure and impede EAMI progress across multiple dimensions.
- **Challenges:** Intense competition for AI talent; rapid technology evolution requiring constant assessment; overcoming data silos and ensuring quality at scale; implementing disciplined MLOps practices consistently; securing adequate and sustainable funding; managing cloud costs effectively (FinOps).

- **AI Office Focus:** Developing the AI talent strategy (Chapter 16), defining the technology architecture (Chapter 17), collaborating with the CDO on data enablement (Chapter 18), driving MLOps adoption (Chapter 20), establishing financial governance (Chapter 19).

3.7 Maturity Management (Center): Orchestrating Progress

- **Scope:** Serves as the central orchestrator, integrating insights and driving improvements across the other four domains to systematically advance the enterprise's overall AI capability. Activities include conducting regular AI maturity assessments using EAMM/EAMI (Chapter 24 - EAMM Maturity Assessment), analyzing results to identify gaps, defining targeted uplift strategies (Chapter 25 - EAMM Uplift Strategy), fostering an AI-aware culture (Chapter 26 - EAMM Culture), managing organizational change (EAMM Change Management), facilitating collaboration (EAMM Collaboration), and driving responsible innovation (Chapter 27 - EAMM Innovation Ecosystem).
- **Criticality:** Provides the engine for structured, continuous improvement, ensuring the AI program adapts, scales responsibly, and stays aligned with strategic goals. Prevents stagnation and ensures evolution towards higher EAMI levels and transformative impact.
- **Challenges:** Defining objective maturity metrics (addressed by EAMI); securing buy-in for cross-functional uplift; overcoming cultural resistance; balancing operational demands with strategic improvement; managing the innovation pipeline effectively; ensuring proactive change management.
- **AI Office Focus:** Owning and managing the EAMI assessment framework (Chapter 24), developing and overseeing maturity uplift plans (Chapter 25), leading culture and change initiatives (Chapter 26), fostering innovation (Chapter 27), continuously facilitating collaboration and knowledge sharing.

> **(!)** Maturity Management (Center) is the conductor of the AI orchestra. It uses the EAMI assessment as its score sheet to understand where performance across Strategy (North), Governance (East), Performance (West), and Resources (South) needs improvement, directing targeted interventions to create a harmonious and continuously improving enterprise AI capability.

3.8 The Interconnectedness of the Compass Domains

The true power and practical utility of the AI Office Compass™ Framework lie in recognizing and actively managing the deep interdependence of the five domains. Success in one area often depends on capabilities in others, and weaknesses in one can undermine the entire structure. This synergy is crucial for achieving holistic AI maturity:

- Effective **Strategic Alignment (North)** requires robust Resource Management (South) (talent, tech, data) for execution and clear Performance Management (West) to demonstrate strategic value achieved.
- Strong **Governance (East)** relies on capable Resources (South) (e.g., governance tools, trained reviewers) and must be integrated into the Strategic Alignment (North) process to ensure responsible innovation from the outset.
- Meaningful **Performance Management (West)** needs reliable data access and stable platforms from Re-

source Management (South) and needs clear strategic objectives defined in the North against which to measure success.

- Efficient **Resource Management (South)** must be strategically guided by North priorities (e.g., investing in platforms critical for the roadmap) and operate within the ethical and security guardrails established by Governance (East).
- **Maturity Management (Center)** is inherently integrative; it requires comprehensive inputs from assessments covering all other domains (North, East, West, South) using tools like EAMI, and, in turn, it drives targeted improvements across all other domains based on identified gaps and the strategic priorities set in the North.

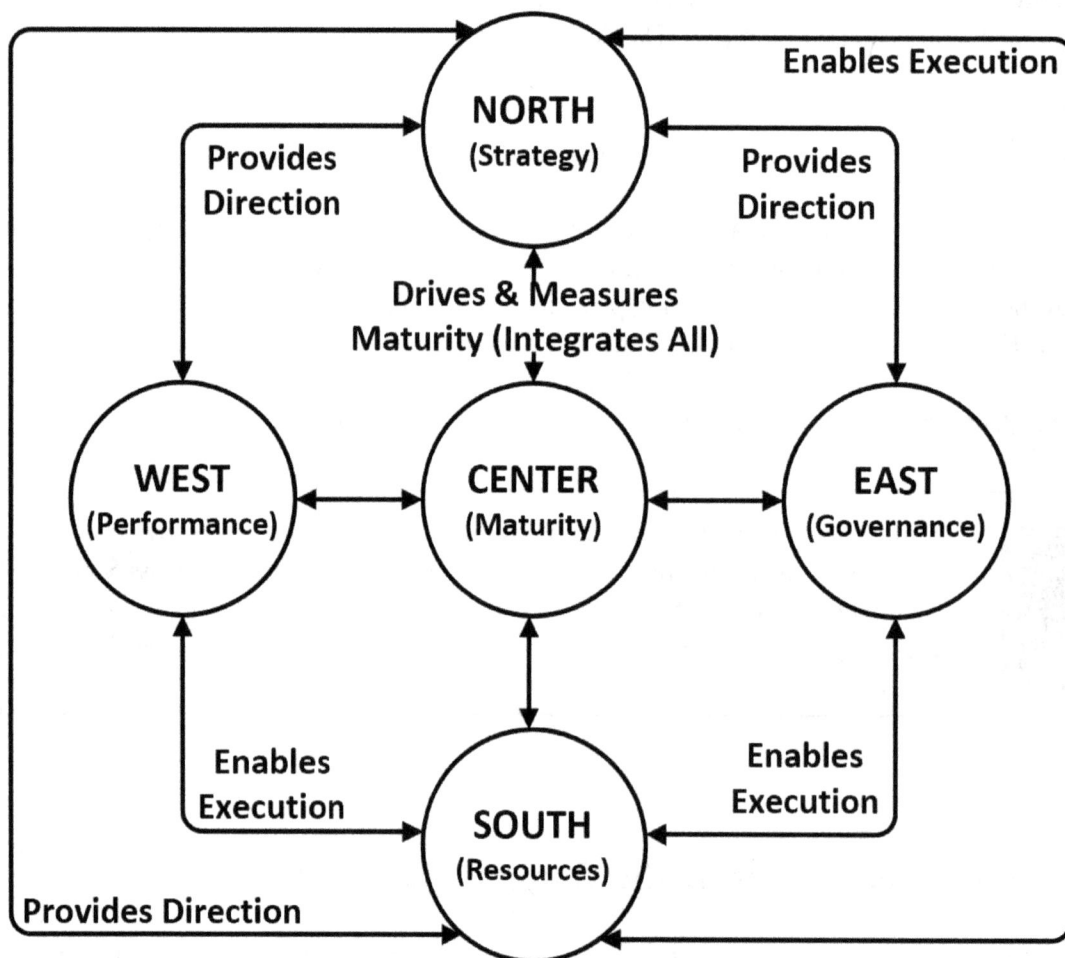

Figure 3.3: Interconnected Domains Graphic

Neglecting any single domain creates significant risk and hinders overall progress towards higher EAMI levels. The AI Office must cultivate a holistic perspective, fostering synergy and managing trade-offs across all five Compass points to achieve sustainable enterprise AI success.

3.9 OmnioTech Case Study: Applying the Compass Framework

Setting the Scene:

Following the AI Office Hub design (Chapter 2), Priya Sharma (Head of AI Office Hub) and her nascent team begin operationalizing their mandate, using the AI Office Compass™ Framework to structure their initial activities and ensure a balanced approach aligned with their goal of reaching EAMI Level 3.

Action & Application Across Domains

- **North (Strategic Alignment):** Priya facilitates workshops using a Value vs. Effort matrix to confirm the prioritization of the Smart Appliance Predictive Failure and Hyper-Personalized Marketing Engine pilots, ensuring direct linkage to 2025 corporate objectives. The initial 2-year AI roadmap is formally documented and approved.
- **East (Governance):** The newly hired Governance Lead partners with Legal/Privacy to draft V1.0 policies using a risk-tiered approach, focusing initially on consumer data usage (GDPR), ethical guidelines for personalization, and a mandatory risk assessment template. The AI Review Team is formally convened.
- **West (Performance & Value Delivery):** Priya collaborates with pilot sponsors to define specific, measurable success KPIs linked to business outcomes (e.g., "% reduction in unscheduled maintenance calls", "% uplift in personalized offer conversion rate" - referencing examples similar to those in Appendix B). Baseline metrics are established before pilot launch. A simple value tracking template is created.
- **South (Resource Management):** The Platform Lead finalizes the initial Azure ML workspace setup with core MLOps tooling integrated. The Data Liaison works with IT Data Engineering to establish secure data pipelines using Azure Data Factory. HR actively sources candidates for open roles. An initial budget allocation is approved.
- **Center (Maturity Management):** Priya uses the initial EAMI assessment results (confirming Level 1.5) to validate immediate priorities identified in the mandate (Strategy, Governance, Tech foundations). She schedules the first AI awareness webinar and plans a follow-up EAMI assessment in 6 months to track progress towards Level 3 targets.

Outcome & Progress

Applying the Compass Framework provides immediate structure and focus for OmnioTech's AI Office Hub. It ensures foundational activities are initiated across all critical domains simultaneously, preventing the siloed efforts and critical omissions characteristic of their earlier ad-hoc approach. The framework helps Priya communicate the Hub's integrated mandate and approach to stakeholders and explicitly align initial activities with the strategic goal of achieving EAMI Level 3 maturity.

3.10 Conclusion: The Guiding Compass for AI Success

The AI Office Compass™ Framework provides an essential operational blueprint for navigating the multifaceted challenge of building a mature, effective, and responsible enterprise AI capability. By structuring activities across the five interconnected domains of Strategic Alignment (North), Governance (East), Performance & Value Delivery (West), Resource Management (South), and Maturity Management (Center), it ensures a holistic, integrated, and strategically grounded approach. This framework helps the AI Office prioritize actions, allocate resources effectively, manage risks responsibly, demonstrate value clearly, and drive continuous improvement towards higher levels of AI maturity and impact, as measured by the Enterprise AI Maturity (i.e. EAMM/EAMI) framework. It provides the necessary structure to manage complexity and avoid critical omissions on the path to AI-driven transformation.

With this foundational operational model established, the subsequent parts of this book will delve deeper into the specific strategies, processes, and best practices required to achieve excellence within each individual Compass domain, beginning with a detailed exploration of Strategic Alignment (North) in Part 2 (Chapters 4-6).

Key Takeaways:

- **The AI Office Compass™ Framework structures AI Office responsibilities into five interconnected domains:** North (Strategy), East (Governance), West (Performance/Value), South (Resources), and Center (Maturity Management).
- Each domain addresses critical capabilities needed for enterprise AI success and aligns with specific EAMM dimensions.
- Success requires a holistic approach, managing activities and addressing challenges across all five domains synergistically.
- Neglecting any domain creates significant risks and hinders overall AI maturity progress (EAMI score).
- The Compass provides a practical blueprint for the AI Office to orchestrate AI initiatives effectively.

Food for Thought / Application Exercise:

- Consider your organization's current AI efforts (or lack thereof). Which of the five Compass domains receives the most attention? Which domains might be currently neglected?
- Think about a specific AI initiative in your organization (or a potential one). How do activities related to that initiative map across the five Compass domains? Are there clear owners or processes for each domain aspect?
- How could visualizing AI responsibilities using the Compass framework help improve communication and coordination between different teams (e.g., Business, IT, Data, Legal) involved in AI within your organization?

PART 2 - AI STRATEGIC ALIGNMENT

~ CHAPTER 4 ~

Integrating AI with Business Strategy: Setting the Direction

"Strategy without tactics is the slowest route to victory. Tactics without strategy is the noise before defeat."
— Sun Tzu

4.1 Introduction: The North Star of AI - Business Strategy

Welcome to Part 2, where we delve into the first and arguably most critical domain of the AI Office Compass™ Framework: Strategic Alignment (North). While the potential of AI technology is undeniably exciting, its true enterprise value is only unlocked when it is purposefully directed towards achieving core business objectives. Too often, organizations, captivated by the allure of cutting-edge technology like Generative AI, chase the latest trends or allow disparate, technically interesting experiments to bloom without a clear connection to overall corporate strategy. This frequently results in wasted resources, disillusioned stakeholders, and minimal measurable impact – the very pitfalls Chapter 1 warned against, often reflecting low EAMI Strategic Alignment maturity.

This chapter focuses on the foundational principle within the Strategic Alignment (North) domain: ensuring that the enterprise AI vision and all subsequent initiatives are deeply integrated with, and fundamentally driven by, the overarching business strategy. We will explore why AI strategy must rigorously follow, not attempt to lead, the established business strategy. We will discuss practical methods for analyzing corporate goals to identify where AI can provide the most significant leverage and competitive advantage. We will also define the crucial role of the AI Office as a strategic integrator and translator, bridging the often-significant gap between business needs and technical possibilities. Finally, we will highlight the tangible risks and opportunity costs associated with strategic misalignment. Establishing this strong strategic foundation is paramount before moving to opportunity prioritization (Chapter 5) and detailed roadmap development (Chapter 6), and it directly contributes to improving the EAMM Strategic Alignment maturity dimension.

Learning Objectives:

- Understand why enterprise AI strategy must be derived from and fully aligned with overall business strategy ("business-pull" vs. "technology-push").
- Learn practical methods for analyzing corporate goals and value chains to identify high-impact AI opportunities.
- Recognize the AI Office's vital role as a strategic integrator and translator between business leadership and technical teams.
- Identify the common pitfalls and substantial risks associated with pursuing misaligned AI strategies.
- Appreciate how strong strategic alignment forms the essential foundation for effective opportunity prioritization and roadmap development.

4.2 Why AI Strategy Must Follow Business Strategy

It's a common temptation, especially within technically oriented teams or dedicated innovation labs excited by new possibilities (e.g., the latest advancements in LLMs or computer vision), to develop AI capabilities in relative isolation and then search for business problems these new tools might solve. This "technology-push" approach, driven by technological capability rather than strategic need, frequently leads to elegantly engineered solutions desperately seeking relevant problems. The result is often projects that, while technically impressive showcases, fail to address genuine, prioritized business challenges or deliver measurable, strategic value (Domain West). They become expensive "science projects" or "solutions in search of a problem," indicative of low EAMI Strategic Alignment.

> The most consistently successful enterprise AI programs adopt a fundamentally different orientation: a "business-pull" approach. They begin by deeply understanding the organization's core strategic objectives as defined by executive leadership — whether those objectives relate to accelerating market share growth in specific segments, achieving defined operational efficiency improvements, meeting critical customer retention targets, enabling entry into new geographic markets or product categories, or mitigating specific high-priority enterprise risks. Only after these strategic priorities are crystal clear do they ask the critical question: "How can AI uniquely help us achieve these specific objectives more effectively, more efficiently, or perhaps in entirely new ways, compared to traditional methods?".

This "strategy-first, AI-second" alignment ensures several critical benefits:

- **Focus on Impact:** Scarce resources (talent, budget, compute power - Domain South) are directed towards initiatives that matter most to the organization's success.
- **Executive Buy-in & Sponsorship:** Leadership is significantly more likely to actively sponsor, fund, and champion AI projects that are clearly and explicitly linked to the strategic priorities they themselves have set and are accountable for delivering.
- **Clear Success Metrics:** Well-defined business goals provide the essential context for establishing meaningful AI project KPIs and accurately measuring the delivered business value (Domain West, EAMM Per-

formance Metrics, EAMM Value Realization). Success is defined in business terms, not just technical ones. (Refer to Appendix B for detailed KPI examples).

- **Rational Prioritization:** A clear strategic lens allows the AI Office and governance bodies (Domain East) to make defensible, transparent decisions about which opportunities among many deserve investment and inclusion on the roadmap (Chapter 5).
- **Sustainability & Integration:** When AI solutions directly address core business needs, they are far more likely to be adopted, integrated into operational workflows, maintained over time, and become part of the organization's enduring capabilities (Domain Center - Culture/Change), rather than being treated as transient technological experiments.

The table below contrasts these two fundamental approaches:

Table 4.1: Technology-Push vs. Business-Pull Approach

Feature	Technology-Push Approac	Business-Pull Approach
Starting Point	AI Capability / New Technology	Business Strategy / Strategic Objectives / Key Challenges
Primary Question	"What problems can this cool AI solve?"	"How can AI best help us achieve this critical business goal?"
Focus	Technical feasibility, innovation for its own sake	Business impact, solving prioritized problems, achieving goals
Risk Profile	High risk of misalignment, low adoption, unclear ROI	Lower risk of misalignment, higher likelihood of adoption & value
Outcome	Often "solutions looking for problems", stalled pilots	AI initiatives directly supporting strategic execution
EAMI Alignment	Weak EAMI Strategic Alignment	Strong EAMI Strategic Alignment

Technology–Push vs. Business–Pull

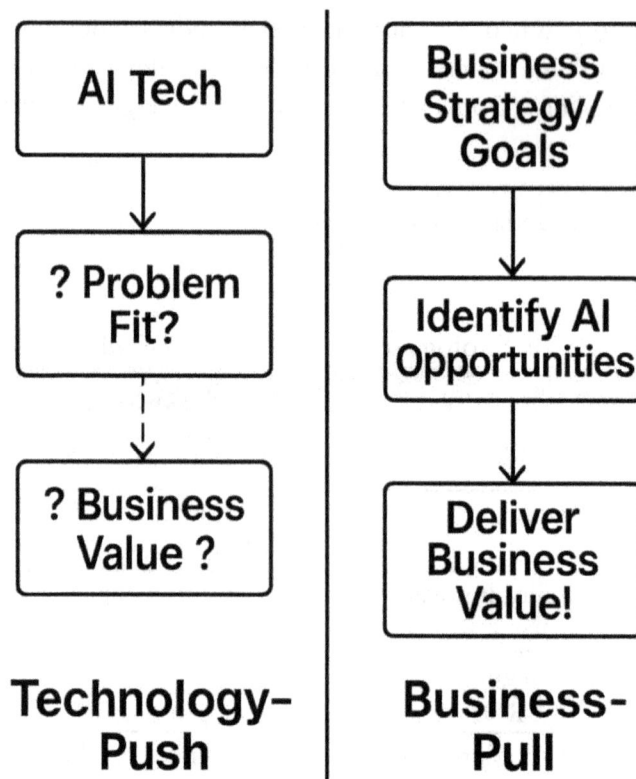

```
Technology-Push          Business-Pull

   AI Tech              Business
      |                 Strategy/
      v                  Goals
  ? Problem               |
    Fit?                  v
      |                Identify AI
      v               Opportunities
  ? Business              |
   Value ?                v
                       Deliver
                      Business
                       Value!
```

Technology-Push | **Business-Pull**

Figure 4.1: Technology-Push vs. Business-Pull Flow Diagram

4.3 Analyzing Business Goals for AI Opportunities

Integrating AI effectively requires a systematic process for translating high-level corporate strategy into specific, actionable AI use cases where AI can provide unique leverage. The AI Office plays a crucial role in facilitating this translation (Compass North activity), working collaboratively across business and technology functions. Methods for Strategic Analysis include:

- **Deconstruct Strategic Documents:** Methodically review the company's formal strategic plan, annual reports, investor communications, board meeting minutes, and key functional strategy documents (e.g., marketing plan, digital transformation roadmap, supply chain strategy). Identify and list explicit goals, strategic priorities, key performance indicators (KPIs), and identified risks or challenges. Map these to potential areas where AI might intervene to support EAMM Strategic Alignment.

- **Conduct Targeted Executive & Leadership Interviews:** Engage in structured conversations with senior executives (C-suite) and functional leaders (VPs, Directors) across the business (Sales, Marketing, Operations, R&D, Finance, HR, Product). Go beyond generalities; ask specific questions designed to uncover strategic pain points and high-leverage opportunities:
  - » "What are your top 2-3 strategic objectives for the next 18-24 months, and what are the biggest obstacles

to achieving them?"

- » "Where do you see the greatest opportunities for significant improvement in efficiency, cost reduction, or revenue generation within your area?" (Link to potential KPIs in Appendix B).
- » "What critical decisions are currently made based on intuition or incomplete data where better insights could significantly improve outcomes?"
- » "Are there manual, repetitive, or data-intensive processes that consume significant resources or are prone to errors?"
- » "What emerging competitive threats or market shifts are causing the most concern, and could AI play a role in addressing them?"

- **Analyze Value Chains & Core Business Processes:** Map out the organization's primary value chain (e.g., R&D → Supply Chain → Manufacturing → Marketing & Sales → Service) and key supporting processes. Analyze each stage to identify steps where AI could potentially introduce significant improvements through:
  - » **Automation:** Automating manual tasks (e.g., invoice processing, report generation, data entry).
  - » **Prediction:** Forecasting key variables (e.g., customer demand, equipment failure, market trends, credit risk - link to EAMM Performance Metrics).
  - » **Optimization:** Finding the best solutions to complex problems (e.g., optimizing logistics routes, dynamic pricing, marketing spend allocation).
  - » **Insight Generation:** Uncovering hidden patterns or providing enhanced decision support (e.g., identifying drivers of customer churn, providing real-time sales recommendations).
  - » **Personalization:** Tailoring experiences or offerings to individual users (e.g., personalized product recommendations, adaptive learning paths - link to EAMM Value Realization). Tools like process mining (using platforms such as Celonis or UiPath Process Mining) can be extremely valuable in objectively identifying process bottlenecks and inefficiencies ripe for AI intervention.
- **Analyze Competitive & Industry Landscape:** Systematically study how direct competitors, industry leaders, and even adjacent industries are leveraging AI. While direct imitation is rarely the best strategy, understanding competitor actions can reveal potential threats (e.g., a competitor using AI to drastically lower costs), identify emerging best practices, validate potential use cases, and inform the organization's unique strategic positioning regarding AI (informing Compass North).
- **Leverage the EAMM Framework:** Explicitly use the EAMM Strategic Alignment dimension criteria (Does the organization have a defined AI strategy? Is it clearly aligned with business goals? Is there visible C-level sponsorship and communication? Are AI KPIs linked to business KPIs?) as a diagnostic checklist during stakeholder interviews and document reviews. Identifying gaps against these maturity criteria points directly to areas requiring immediate focus within the Strategic Alignment domain.

Focus on identifying opportunities where AI offers a distinctive advantage. Don't just look for problems AI can potentially solve; prioritize problems where AI can deliver a solution that is significantly faster, more accurate, more efficient, more scalable, or fundamentally different from what can be achieved through traditional methods or simpler analytics. Always ask: "Is AI truly the best tool for this specific strategic job?". Quantify the potential impact clearly whenever possible (e.g., "Reducing customer service resolution time by 30% through an AI chatbot directly supports our strategic goal of improving CSAT by 10 points," referencing KPIs from Appendix B).

4.4 The AI Office as Strategic Integrator and Translator

Given its unique position, the AI Office is ideally suited to act as the crucial bridge between high-level business strategy and the practical realities of AI implementation. It must perform two vital functions:

- **Integrator:** The AI Office synthesizes the diverse AI needs, opportunities, and existing initiatives identified across different business functions and technical teams into a single, coherent, enterprise-wide AI vision and roadmap (Compass North activity). It ensures that individual projects are not pursued in isolation but are aligned with the overall strategy, leverage common platforms and data assets where appropriate (Domain South), adhere to enterprise governance standards (Domain East), and collectively contribute to the organization's strategic objectives. This integration role is vital for preventing the silos and fragmentation discussed in Chapter 1.

- **Translator:** Perhaps one of the most critical, yet often underestimated, roles of the AI Office is bridging the significant communication and understanding gap that frequently exists between business stakeholders (who understand strategic needs but may lack deep AI knowledge) and technical AI/data science teams (who understand AI capabilities but may lack deep business context). This translation works in both directions:
  - » **Business to Tech Translation:** The AI Office helps business leaders articulate their strategic problems and goals in a way that technical teams can understand. It translates these needs into clear, well-defined AI use cases, specific requirements, measurable success criteria (business KPIs), and context regarding data availability and operational constraints. This ensures technical teams build the right solutions.
  - » **Tech to Business Translation:** The AI Office explains complex AI capabilities, potential benefits, inherent limitations, associated risks (ethical, security, operational - Domain East), and realistic timelines in clear, concise, non-technical business language to executives and functional leaders. This involves managing expectations effectively (avoiding over-hype), demystifying the technology, and facilitating informed decision-making regarding AI investments and deployments.

Cultivate or formally establish roles like "AI Translator," "AI Product Manager," or "Business Relationship Manager" within or closely aligned with the AI Office. These individuals, possessing a blend of strong business acumen, domain knowledge, and sufficient AI/data literacy (understanding basic concepts without needing to code), are invaluable for facilitating continuous dialogue, ensuring requirements are accurately captured, managing stakeholder expectations, and maintaining tight alignment between business strategy (Compass North) and AI execution throughout the project lifecycle. The "Spokes" in a hybrid model (Chapter 2) often fulfill this crucial translation function within their respective business units.

4.5 Pitfalls of Misaligned AI Strategies

Failure to rigorously align AI initiatives with overarching business strategy from the outset (low EAMI Strategic Alignment) can lead to significant negative consequences, wasting resources and undermining the entire AI program, directly impacting EAMI maturity progress:

- **Wasted Investment & Resources:** Significant budget, valuable compute resources, and scarce AI talent are poured into developing technically sophisticated models or platforms that ultimately deliver little or no demonstrable business value, resulting in a poor return on investment (low EAMI Value score).

- **"Shiny Object" Syndrome:** The organization gets caught up in chasing the latest AI fads or implementing complex technologies (e.g., deploying a computationally expensive deep learning model where a simpler, more interpretable regression model would suffice) without a clear strategic justification or business case.
- **Low Adoption & Integration Failure:** AI solutions developed without genuine business need, clear value proposition, or direct input from end-users often face resistance and fail to gain traction or become embedded into existing operational workflows, remaining isolated tools rather than integrated capabilities (low EAMI Culture/Change score).
- **Erosion of Credibility & Future Funding:** Repeated failures to deliver tangible strategic impact severely undermine the AI Office's credibility and make it increasingly difficult to secure executive sponsorship and funding for future, potentially more ambitious, AI initiatives.
- **Significant Opportunity Cost:** Focusing limited resources on low-impact or misaligned projects means neglecting high-value opportunities where AI could genuinely create significant competitive advantage or solve critical business problems.
- **Increased Unmanaged Risk:** Misaligned projects, potentially viewed as less strategic, might inadvertently bypass necessary governance reviews (Domain East) or security assessments, introducing unforeseen ethical, compliance, or security risks into the organization.

4.6 OmnioTech Case Study: Setting the Strategic Direction

Setting the Scene:

Following the design of the AI Office Hub (Chapter 2), Priya Sharma (Head of AI Office) recognizes the immediate need to ensure all AI efforts are laser-focused on OmnioTech's explicit corporate strategy for the next two years (Compass North activity). This strategy prioritizes: (1) Increasing market share in the highly competitive smart home segment through differentiated product innovation, and (2) Enhancing customer lifetime value (CLV) through significantly improved engagement and post-sale service.

Action & Application:

Priya and her nascent team undertake a focused strategic alignment process:

- **Analyze Strategy & Interview Leaders:** They meticulously review the detailed corporate strategic plan documents and conduct structured interviews with the VP of Product Development, the VP of Marketing & Sales, and the Head of Customer Support. Key measurable goals extracted: Launch 3 new AI-enabled smart appliance features differentiating on user experience; Reduce customer churn rate by a target of 10% within 18 months; Increase average cross-sell revenue per existing customer by 15% over 2 years.
- **Identify & Map AI Alignment:** The team facilitates brainstorming sessions linking potential AI use cases directly to these strategic goals:
  - » **Goal 1 (Product Innovation):** Predictive maintenance alerts (identified earlier, strong link), AI-optimized energy consumption management for appliances, personalized device interfaces adapting to user habits.
  - » **Goal 2 (CLV Enhancement):** Hyper-personalized marketing offers based on usage data (identified earlier, strong link), AI-powered customer support chatbot for faster issue resolution, proactive customer outreach triggered by AI identifying potential issues or upgrade opportunities.

- **Initial Strategic Prioritization Filter:** Evaluate these brainstormed opportunities primarily through the lens of direct strategic impact (EAMM Strategic Alignment) and urgency, before detailed feasibility assessment (Chapter 5). The previously identified pilots (Predictive Maintenance, Hyper-Personalization) clearly remain top priorities due to their strong alignment with both primary strategic goals and the existence of some (albeit flawed) initial exploratory work. The AI-powered support chatbot is flagged as a high-potential initiative directly supporting Goal 2, warranting immediate feasibility assessment alongside the primary pilots. Other ideas (like energy optimization) are logged but deemed lower priority for the initial roadmap based on current strategic focus.
- **Develop Initial AI Vision Statement:** Based on the strategy and prioritized areas, Priya drafts a concise, compelling vision statement for internal alignment: "OmnioTech will leverage AI to create uniquely intelligent products and deliver hyper-personalized experiences that demonstrably drive market leadership and enduring customer loyalty".
- **Secure Executive Endorsement:** Priya presents the strategic analysis findings, the proposed AI vision statement, and the rationale for prioritizing the initial focus areas (Predictive Maintenance, Hyper-Personalization, AI Support Chatbot feasibility) to the AI Review Team (including COO Michael Vance, CTO Stefan Ritter) and subsequently to CEO Julia Weber. She explicitly links each initiative back to the corporate strategic goals and associated KPIs (referencing Appendix B potential KPIs). Formal endorsement is secured, confirming C-level sponsorship and alignment (a key EAMM criterion).

Outcome & Progress:

This structured strategic alignment process ensures OmnioTech's initial AI efforts (Compass North) are directly and demonstrably tied to core business strategy, preventing resource diversion to less critical projects. It provides a clear justification for focusing initial resources and establishes a shared understanding and executive backing for the program's direction. This explicitly addresses the foundational EAMM Strategic Alignment criteria, positioning OmnioTech to begin improving its maturity in this crucial dimension (moving towards Level 3).

4.7 Conclusion: Strategy First, Technology Second

Successfully integrating artificial intelligence into the fabric of an enterprise fundamentally begins with a clear understanding of, and rigorous alignment with, the organization's core business strategy. By consciously adopting a "business-pull" approach, systematically analyzing strategic goals to identify high-leverage AI opportunities, and positioning the AI Office as a vital integrator and translator between business needs and technical capabilities, organizations can ensure their AI investments are focused, impactful, and sustainable. This strategic grounding is the essential first step in the Strategic Alignment (North) domain; it prevents costly missteps, builds crucial executive sponsorship, and provides the necessary foundation for effectively prioritizing opportunities (Chapter 5) and developing a coherent, actionable Enterprise AI Roadmap (Chapter 6). Only by putting strategy first can organizations truly harness the power of AI to achieve their most important business objectives and advance confidently on their AI maturity journey.

Key Takeaways:

- AI strategy must follow business strategy ("business-pull") to ensure relevance and impact.
- Analyzing business goals, value chains, and competitive landscapes helps identify high-value AI opportunities.
- The AI Office acts as a crucial integrator (synthesizing needs) and translator (bridging business and tech).

- Misalignment leads to wasted resources, low adoption, lost credibility, and missed opportunities.
- Strong strategic alignment is the foundation for effective AI prioritization and roadmap development.

Food for Thought / Application Exercise:

- Review your organization's current top 2-3 strategic objectives. Where specifically could AI potentially offer the most significant leverage in achieving them?
- Consider a recent technology initiative (AI or otherwise) in your organization. Was it primarily driven by a business need ("pull") or by the availability of the technology ("push")? What was the outcome?
- Who currently performs the "translator" role between business needs and technical AI possibilities in your organization? How effective is this translation process?

~ CHAPTER 5 ~

Identifying and Prioritizing Strategic AI Opportunities

"The main thing is to keep the main thing the main thing."
— Stephen Covey

5.1 Introduction: Focusing AI Efforts for Maximum Impact

Chapter 4 established the critical principle that enterprise AI initiatives must be driven by, and aligned with, overall business strategy (Compass North). However, even with clear strategic alignment, organizations often generate far more potential AI ideas than they can realistically pursue with limited resources (Domain South - talent, budget, time). Simply having a long list of strategically relevant ideas is not enough; the AI Office must implement a disciplined process to systematically identify, assess, and prioritize these opportunities to ensure focus remains on the initiatives offering the highest potential value and strategic contribution (EAMM Project Management).

This chapter, continuing within the Strategic Alignment (North) domain of the AI Office Compass™ Framework, details the process for building and managing a robust AI opportunity pipeline. We will explore methods for discovering potential use cases across the organization, introduce frameworks for assessing opportunities based on key criteria like business value (Domain West), technical feasibility (Domain South), and associated risks (Domain East), and discuss techniques for prioritizing the most promising initiatives for inclusion in the AI roadmap (Chapter 6). This structured approach ensures that the AI Office channels resources effectively, maximizing the return on AI investments and driving progress on relevant EAMI dimensions like Value Realization and Strategic Alignment.

Learning Objectives:
- Understand the concept of an AI opportunity funnel and methods for populating it.
- Learn frameworks for assessing AI opportunities based on business value, technical feasibility, and risk.
- Explore different prioritization techniques (e.g., Value vs. Effort matrix, Weighted Scoring) to select the most impactful initiatives.

- Recognize the importance of balancing the AI portfolio across different types of initiatives (e.g., quick wins vs. strategic bets).
- Appreciate the AI Office's role in facilitating the identification and prioritization process.

5.2 Building the AI Opportunity Funnel: Discovery and Ideation

The first step is creating a system to capture potential AI use cases from across the organization. This "opportunity funnel" should be continuously fed through various discovery mechanisms facilitated or coordinated by the AI Office (Compass North / Center activities):

- **Top-Down Strategic Alignment:** Directly deriving opportunities from the strategic analysis conducted in Chapter 4. Executive workshops focused on strategic goals are a primary source.
- **Bottom-Up Ideation:** Encouraging employees across different departments to submit ideas based on their domain expertise and understanding of operational pain points or customer needs. This can be done through:
  - » **Ideation Platforms:** Using digital tools (e.g., dedicated channels in Slack/Teams, platforms like Brightidea or Miro) for structured idea submission, discussion, and initial voting.
  - » **Workshops & Brainstorming Sessions:** Facilitating targeted sessions with specific business units (e.g., Supply Chain, Customer Service) or functional teams focused on identifying AI applications within their specific context and challenges.
  - » **Communities of Practice (CoPs):** Leveraging established AI or Data Science CoPs as active forums for sharing challenges encountered in project work and brainstorming potential AI-driven solutions (related to EAMM Collaboration).
- **Process Mining & Analysis:** Employing process discovery and mining tools (e.g., Celonis, UiPath Process Mining, SAP Signavio) to objectively analyze existing end-to-end business processes (like order-to-cash or procure-to-pay) and identify specific bottlenecks, variations, or inefficiencies where AI-driven automation, prediction, or optimization could yield significant measurable benefits.
- **Data Exploration:** Allocating dedicated time (e.g., "innovation sprints" or a percentage of data scientist time) for exploratory data analysis (EDA) on key enterprise datasets to uncover previously unknown patterns, correlations, or potential predictive opportunities that might not be immediately obvious from a purely business perspective.
- **External Scanning:** Systematically monitoring industry trends (e.g., via analyst reports from Gartner, Forrester), competitor AI deployments, academic research publications (e.g., from conferences like NeurIPS, ICML), emerging startup technologies, and vendor solution offerings to identify novel AI capabilities and potential applications relevant to the organization's strategic context (this links closely to the Innovation Hub function discussed in Chapter 27, related to EAMM Innovation Ecosystem).

Cast a wide net initially but filter efficiently. Encourage broad participation in ideation across levels and functions. However, establish a clear, lightweight intake process and initial screening mechanism (perhaps managed by the AI Office or designated "AI Champions") to quickly filter out ideas that are clearly out of scope, technically infeasible, or strategically misaligned, preventing the assessment process from becoming overwhelmed by low-potential submissions. A simple submission template requiring information on the problem, proposed solution, expected impact, potential data sources, and strategic alignment helps standardize initial vetting.

5.3 Assessing Opportunities: Value, Feasibility, and Risk

Once potential opportunities are captured in the funnel, they need rigorous and consistent assessment before they can be meaningfully prioritized. The AI Office typically leads or facilitates this assessment process, ensuring input is gathered from relevant stakeholders including the business sponsor, domain experts, data scientists, ML engineers, data engineers, governance/risk specialists (Domain East), and security experts. Key assessment dimensions usually fall into three categories, aligning with EAMM criteria, summarized in the table below:

Table 5.1: AI Opportunity Assessment Dimensions

Assessment Dimension	Key Aspects	Example Questions & Considerations	EAMM Alignment Focus
1. Business Value	Strategic alignment, quantifiable impact (financial/operational KPIs - refer to Appendix B), qualitative benefits, urgency/timing.	How strongly does this support a core strategic goal? What is the estimated annual revenue uplift or cost saving ($ range)? Does it significantly improve CSAT/NPS? Does it provide a key competitive differentiator? Is there a critical market window?	Strategy, Value Realization
2. Technical Feasibility	Data availability & quality, algorithmic complexity & maturity, platform/infra needs, integration complexity, required technical skills.	Is the necessary data accessible and of sufficient quality? Is the proposed AI technique well-understood and proven for this problem? Can it run on our existing platform (e.g., Azure ML)? How complex is integration with CRM/ERP? Do we have the ML engineering skills?	Data Mgmt, Data Quality, Technology, MLOps, Talent/Skills
3. Risk Profile	Ethical risks (bias, fairness, transparency), security risks (attacks, breaches), compliance/legal risks (privacy, IP), operational/execution risks.	Could this model produce discriminatory outcomes? What are the data privacy implications (GDPR)? What are the potential security vulnerabilities? How reliable does this system need to be? What is the risk of project failure due to complexity?	Governance, Ethics, Security, Compliance, Risk Mgmt

Develop and consistently use a standardized AI Opportunity Assessment Scorecard or template. This template should guide the assessment across the key dimensions (Value, Feasibility, Risk) and prompt for specific evidence or ratings (e.g., using a 1-5 scale for impact, feasibility, risk levels; or High/Medium/Low ratings). Requiring sign-off from key functional experts (e.g., Data Lead for data feasibility, Security Lead for security risk) ensures comprehensive evaluation and provides a consistent, documented basis for comparing diverse opportunities during prioritization discussions. For high-potential but technically uncertain opportunities, consider funding small, time-boxed Proofs of Concept (PoCs) or feasibility studies specifically to reduce uncertainty around technical viability before committing to a full-scale project on the roadmap.

5.4 Prioritizing Initiatives: Making Strategic Choices

With opportunities rigorously assessed, the AI Office must facilitate a transparent and data-driven prioritization process to select which initiatives will receive funding and resources and earn a place on the Enterprise AI Roadmap (Chapter 6). Simply picking the ideas with the highest potential value might not be optimal if they are extremely difficult or risky. Conversely, only pursuing easy, low-risk projects might fail to deliver significant strategic impact. Effective prioritization requires balancing multiple factors. Common techniques include:

- **Value vs. Effort Matrix:** A widely used, visually intuitive method plotting assessed Business Value (typically on the Y-axis, from Low to High) against estimated Implementation Effort or Complexity (often incorporating technical feasibility and resource needs, on the X-axis, from Low to High). This divides opportunities into four quadrants:
  - » **High Value / Low Effort (Quick Wins):** Often prioritized first to build momentum, demonstrate value quickly, and gain organizational confidence.
  - » **High Value / High Effort (Strategic Bets / Major Projects):** Require significant investment, careful planning, strong executive sponsorship, and dedicated resources. Pursue selectively based on strategic importance.
  - » **Low Value / Low Effort (Fill-ins / Incremental Improvements):** Can be pursued opportunistically if resources allow or delegated to business units for local implementation, but shouldn't consume prime AI Office resources.
  - » **Low Value / High Effort (Avoid / Deprioritize):** Generally avoided unless mandated by regulation or essential for enabling a future high-value initiative.
- **Weighted Scoring Model:** A more quantitative approach where different assessment criteria (e.g., Strategic Alignment score, Financial Value score, Feasibility score, Risk score – potentially inverted) are assigned weights reflecting current organizational priorities (e.g., Strategic Alignment might be weighted 40%, Value 30%, Feasibility 20%, Risk 10%). Each opportunity receives a score for each criterion, which is then multiplied by the weight. The sum of these weighted scores provides a total priority score, allowing opportunities to be ranked objectively. This method allows for more nuance than the simple 2x2 matrix.
- **Portfolio Balancing:** Beyond individual project scores, consider the overall mix of initiatives in the prioritized portfolio across several dimensions to ensure strategic coherence and manage overall risk:
  - » **Risk Profile:** Deliberately balance high-risk/high-reward exploratory or innovative projects with lower-risk, more certain operational efficiency improvements.
  - » **Time Horizon:** Include a mix of short-term initiatives delivering value within months (quick wins) and longer-term strategic projects building foundational capabilities or tackling major transformations over 1-3 years.
  - » **Business Areas / Strategic Goals:** Ensure the portfolio addresses priorities across different key business functions or strategic pillars, demonstrating broad impact (unless the strategy dictates a deep focus in one specific area initially).
  - » **AI Techniques / Maturity:** Balance projects leveraging mature, well-understood AI techniques with selective, controlled explorations of newer technologies (e.g., Generative AI, Reinforcement Learning) aligned with innovation goals (Chapter 27, EAMM Innovation Ecosystem).
- **Capacity & Dependency Constraints:** Overlay the prioritized list with realistic estimates of available resources (critical talent capacity, allocated budget per quarter/year - Domain South) and map key dependencies between initiatives. This practical check ensures the selected portfolio is feasible to execute within the planning horizon. An initiative might score highly but be deferred if a prerequisite project or required skillset is not yet available.

AI Opportunity Prioritization Value vs. Effort

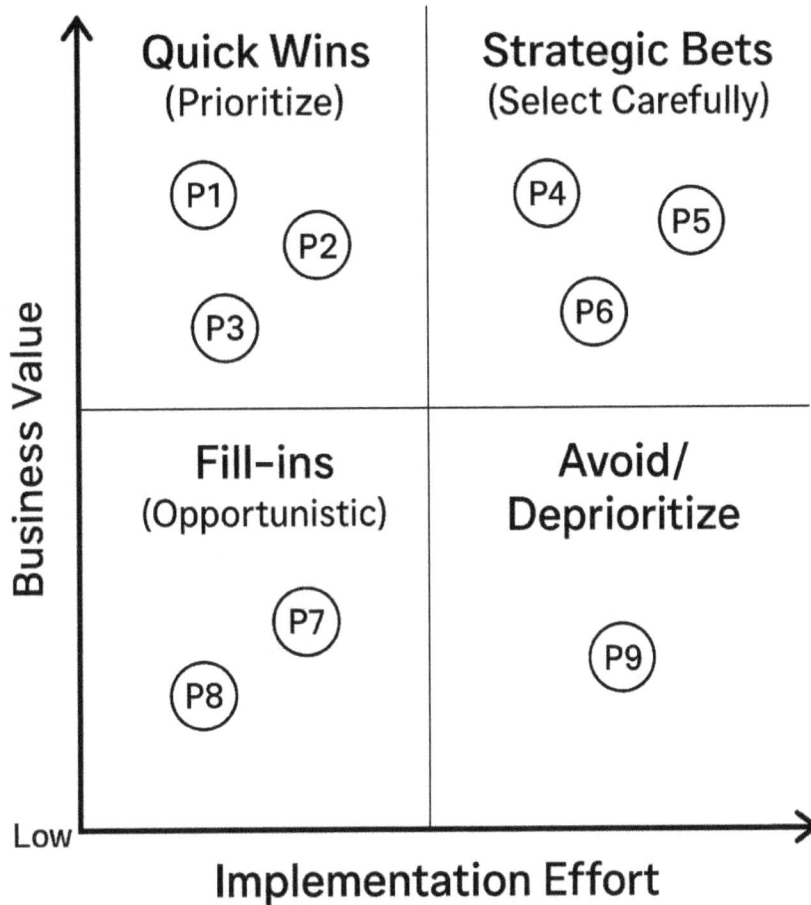

Figure 5.1: Value vs. Effort Matrix Example

The table below briefly compares common prioritization frameworks:

Table 5.2: Comparison of Prioritization Frameworks

Prioritization Framework	How it Works	Pros	Cons
Value vs. Effort Matrix	Plots opportunities on a 2x2 grid based on qualitative/ quantitative ratings of value and effort/complexity.	Simple, visual, good for initial screening and identifying quick wins.	Can oversimplify complex trade-offs; relies on accurate subjective ratings.

Weighted Scoring Model	Assigns weights to multiple criteria (value, feasibility, risk, alignment), calculates a total score for ranking.	More nuanced, data-driven, allows tailoring to strategic priorities via weights.	Requires careful definition of criteria and weights; can be more complex to implement.
ICE Score (Impact, Confidence, Ease)	Scores opportunities on three factors (1-10 scale), often multiplied or averaged for a priority score.	Relatively simple quantitative method, balances multiple factors.	Confidence/Ease scores can be subjective; might not fully capture strategic alignment.
RICE Score (Reach, Impact, Confidence, Effort)	Similar to ICE but adds 'Reach' (number of users/ customers affected). Often used in product management.	Incorporates scale of impact explicitly.	Confidence/Effort still subjective; primarily product-focused.

> **(!)** Prioritization is a dynamic and ongoing process, not a one-time event. The AI Office should facilitate regular reviews (e.g., quarterly or semi-annually) of the opportunity backlog and the active AI roadmap priorities. New ideas should be assessed, existing priorities potentially re-evaluated based on new market information or changing business conditions, and the roadmap adjusted accordingly through a defined governance process (Compass East/Center). Agility in prioritization is key..

5.5 The Role of the AI Office in Facilitation

The AI Office typically doesn't own all the AI ideas, but it plays a critical facilitation and coordination role (Compass North/Center) in the identification and prioritization process:

- **Establishing the Process:** Defining and communicating the standardized end-to-end process for opportunity submission, assessment criteria, scoring/weighting logic, and prioritization meetings.
- **Providing Frameworks & Tools:** Supplying standardized templates (submission forms, assessment scorecards), methodologies (like Value vs. Effort or Weighted Scoring guidelines), and potentially tools (like a shared backlog in Jira or an ideation platform) to support the process.
- **Coordinating Assessments:** Orchestrating the assessment activities, ensuring the right stakeholders (business sponsors, domain experts, technical leads, governance reviewers) provide timely input for each opportunity.
- **Maintaining the Opportunity Backlog:** Acting as the central custodian and system of record for all identified AI opportunities, their assessment status, scores, and prioritization decisions.
- **Facilitating Prioritization Meetings:** Preparing assessment summaries, presenting data objectively, guiding discussions within the relevant governance body (e.g., AI Review Team), ensuring criteria are applied consistently, and documenting the final prioritization decisions and rationale.
- **Ensuring Transparency:** Communicating the prioritization criteria, process, and outcomes clearly across

the organization to manage expectations and build understanding.

5.6 OmnioTech Case Study: Prioritizing the Initial Opportunities

Setting the Scene:

Following the strategic alignment exercise (Chapter 4), Priya Sharma (Head of AI Office Hub) and the newly formed AI Review Team at OmnioTech must formally prioritize the initial set of high-potential opportunities: Predictive Maintenance (PredMaint), Hyper-Personalization (HyperPers), and the AI Support Chatbot, alongside other ideas gathered (e.g., Supply Chain Logistics AI, Code Documentation AI).

Action & Application:

- **Assessment Scorecard Used:** Each opportunity is formally assessed using a standardized scorecard with weighted criteria agreed upon by the AI Review Team: Strategic Alignment (30%), Potential Value (30%), Technical Feasibility (20%), and Risk Level (Inverse Weight, 20%). Input gathered from Product, Marketing, Support, IT, Data, Legal, and Security representatives.
  - » **PredMaint:** Scores High on Alignment (Product Innovation goal), High on Value ($1.5M est. annual warranty savings), Medium on Feasibility (IoT data integration challenge identified), Low on Risk (well-understood technique). Weighted Score: 85.
  - » **HyperPers:** Scores High on Alignment (CLV goal), High on Value (15% cross-sell uplift potential), Medium-High on Feasibility (requires Customer Data Platform integration), Medium on Risk (GDPR/ethics require careful controls - Domain East). Weighted Score: 82.
  - » **AI Support Chatbot:** Scores Medium-High on Alignment (CLV/Support Efficiency), Medium on Value ($500k est. cost reduction), Medium on Feasibility (requires integration with knowledge base & CRM), Low-Medium on Risk. Weighted Score: 68.
  - » **Supply Chain Logistics AI:** Scores Medium on Alignment (Operational Efficiency - currently secondary strategy), High on Value (potential $2M+ savings), Medium-Low on Feasibility (requires complex global data integration, significant effort), Low on Risk. Weighted Score: 65.
  - » **Code Documentation AI (GenAI):** Scores Low on Alignment (internal developer efficiency), Low-Medium on Value, High on Feasibility, Low on Risk. Weighted Score: 45.
- **Prioritization Decision (AI Review Team):**
  - » **Tier 1 (Roadmap - Year 1):** PredMaint (Score 85), HyperPers (Score 82). Approved for immediate inclusion in the initial roadmap based on high scores and strong EAMM Strategic Alignment.
  - » **Tier 2 (Roadmap - Year 2 Consideration):** AI Support Chatbot (Score 68), Supply Chain Logistics AI (Score 65). Placed on the roadmap for potential initiation in Year 2, pending successful delivery of Tier 1 projects and foundational capabilities (platform, data - Domain South).
  - » **Tier 3 (Backlog):** Code Documentation AI (Score 45). Placed in the opportunity backlog; IT/Engineering encouraged to explore independently if desired, but not prioritized for AI Office resources at this time.
- **Portfolio Balance Check:** The initial portfolio focuses on the two primary strategic goals (Product Innovation, CLV) and represents 'Strategic Bets' requiring dedicated effort. Foundational work (Platform, Data Quality - Domain South) needed for these will also be prioritized on the roadmap (Chapter 6).

Outcome & Progress

The structured assessment and prioritization process, facilitated by the AI Office Hub and executed by the cross-functional AI Review Team, provides a clear, data-driven, and defensible basis for focusing OmnioTech's initial AI investments (EAMM Project Management best practice). It confirms the strategic importance of the PredMaint and HyperPers initiatives while managing expectations for other promising ideas by placing them explicitly on a future roadmap or backlog. This disciplined approach aligns strongly with achieving higher EAMI maturity.

5.7 Conclusion: From Ideas to Impactful Initiatives

Identifying and prioritizing AI opportunities (Compass North activity) is a crucial step in translating high-level strategy into an actionable plan. By establishing a robust opportunity funnel fed by diverse sources, implementing rigorous assessment frameworks (evaluating value, feasibility, and risk - Domains West, South, East), and utilizing structured prioritization techniques, the AI Office can effectively guide the organization to focus its limited resources on the initiatives most likely to deliver significant strategic impact and advance AI maturity (EAMM Project Management, Strategy, Value). This disciplined approach moves beyond ad-hoc experimentation towards building a coherent portfolio of AI initiatives purposefully chosen to drive business success. With clear priorities now established, the next logical step is to assemble these prioritized initiatives into a cohesive, multi-year Enterprise AI Roadmap, the subject of Chapter 6.

Key Takeaways:

- A structured process is needed to move from many potential AI ideas to a prioritized set of initiatives.
- Building an opportunity funnel involves both top-down strategic alignment and bottom-up ideation.
- Assess opportunities rigorously across Business Value, Technical Feasibility, and Risk Profile dimensions.
- Use prioritization techniques (Value/Effort, Weighted Scoring) and portfolio balancing to make strategic choices.
- The AI Office plays a key facilitation role in the identification, assessment, and prioritization process.

Food for Thought / Application Exercise:

- How are potential AI opportunities currently identified and prioritized in your organization? Is there a formal process? If not, what initial steps could be taken to establish one?
- Consider the Value vs. Effort matrix. Think of one potential AI project in your area and try to place it qualitatively in one of the four quadrants. Does this placement suggest it should be prioritized?
- What weights would you assign to Strategic Alignment, Potential Value, Technical Feasibility, and Risk Mitigation if creating a Weighted Scoring model for AI projects in your organization today? Why?

~ CHAPTER 6 ~

Developinng the Enterprise AI Roadmap: Charting the Course

"A vision without a roadmap is just a hallucination."
— Adapted from Thomas Edison

6.1 Introduction: From Priorities to an Actionable Plan

Chapters 4 and 5 focused on aligning AI efforts with business strategy and prioritizing the most impactful opportunities (Compass North). With a clear understanding of why certain AI initiatives are important and which ones offer the most promise, the next crucial step within the Strategic Alignment (North) domain is to translate these priorities into a coherent, actionable plan: the Enterprise AI Roadmap. This roadmap serves as the master strategic plan, outlining the sequence of prioritized initiatives, key milestones, critical dependencies, high-level resource requirements (budget, talent - Domain South), and clear ownership over a defined timeframe, typically spanning 1 to 3 years.

Developing a well-structured AI roadmap is essential for effective execution, coordination, and communication across the enterprise. It provides clarity and a shared vision for stakeholders, guides the strategic allocation of often-scarce resources (Domain South), enables systematic progress tracking and performance measurement (Domain West), and ensures that individual AI projects collectively contribute to achieving the overarching strategic AI vision and targeted EAMI maturity goals (Domain Center). This chapter details the essential components of an effective AI roadmap, explores key considerations for sequencing initiatives thoughtfully, discusses the critical importance of securing broad stakeholder buy-in, and outlines the AI Office's central role in creating, maintaining, and governing this vital strategic artifact (EAMM Project Management / Strategy).

Learning Objectives:

- Understand the purpose and key components of a comprehensive Enterprise AI Roadmap.
- Identify critical planning considerations for developing the roadmap, including realistic phasing, dependency management, and resource allocation alignment.
- Learn guiding principles for sequencing AI initiatives effectively to balance quick wins with long-term goals and manage risk.
- Recognize the importance of maintaining a balanced portfolio view within the roadmap.
- Appreciate the necessity of securing broad stakeholder buy-in and implementing clear communication strategies for the roadmap.
- Understand the AI Office's facilitating and governing role in roadmap development and ongoing management.

6.2 Purpose and Key Components of the AI Roadmap

The Enterprise AI Roadmap is far more than just a list of projects plotted on a timeline; it functions as a dynamic strategic communication tool that visualizes the intended journey of enterprise AI adoption, capability development, and value creation. Its primary purposes are to:

- **Communicate the Strategic Plan:** Provide a clear, concise, and shared view of planned AI initiatives, their intended outcomes, target timelines, and explicit linkage to business objectives for all relevant stakeholders, from the C-suite to delivery teams.
- **Align and Allocate Resources:** Serve as the primary input for guiding the allocation of critical resources – budget appropriations, assignment of scarce AI talent (data scientists, ML engineers), prioritization of technology platform investments (Domain South) – towards the most strategically important initiatives.
- **Manage Interdependencies:** Proactively identify, visualize, and manage critical dependencies between different AI projects (e.g., Model A output feeds Model B), data availability prerequisites (e.g., need unified customer data before personalization can scale - Domain South), required platform capabilities (e.g., MLOps pipeline needed for deployment - Domain South), and alignment with other major corporate initiatives (e.g., a core system migration).
- **Enable Progress Tracking & Governance:** Establish a clear baseline against which the AI Office and governance bodies can monitor execution progress, measure success against defined milestones and KPIs (Domain West, reference Appendix B), identify potential delays or roadblocks early, and make informed decisions regarding adjustments or interventions (Domain East/Center).
- **Build Momentum & Confidence:** Demonstrate a clear, coherent, and strategically grounded path forward for enterprise AI, building confidence among stakeholders, justifying ongoing investment, and sustaining momentum for the overall program.

A comprehensive and effective AI roadmap typically includes the following components, often visualized using roadmap software (like Aha!, ProductPlan, Jira Advanced Roadmaps) or structured documents:

Table 6.1: Key Components of an AI Roadmap

Roadmap Component	Description	Example
Strategic Themes/Goals	High-level objectives derived from corporate strategy that the roadmap aims to achieve.	Enhance Customer Experience through Personalization, Optimize Supply Chain Efficiency, Build Foundational AI Governance Capability.
Key Initiatives/ Projects	Specific, prioritized AI projects (from Chapter 5) supporting the strategic themes, each with a defined scope, objective, and business sponsor.	Hyper-Personalization Engine Pilot, Predictive Maintenance for Fleet, Develop Responsible AI Policy V1.0.
Phasing & Timelines	Visual timeline (e.g., Gantt chart) showing planned start/end dates or target quarters for initiatives over the roadmap horizon (1-3 years).	Initiative A (Pilot: Q1-Q2 Y1, Scale: Q3Y1-Q2Y2), Initiative B (Pilot: Q2-Q3 Y1).
Major Milestones	Significant checkpoints, deliverables, or decision gates for each key initiative.	Pilot Go/No-Go Decision, Platform V1 Ready for Use, First Business Unit Rollout Complete, EAMI Level 3 Achieved.
Dependencies	Explicit identification of critical prerequisites or interconnections between initiatives, data, platforms, or external factors.	Scale Phase depends on Platform V1, Data Quality Project must complete before Model Retraining, Requires CRM Upgrade.
Resource Estimates	High-level (order of magnitude) estimates of required resources for major initiatives or phases (linked to Domain South - Finance/ Talent).	Budget: $X-$Yk, Requires: 2 ML Engineers, 1 Data Scientist (FTEs), Needs Cloud Compute Tier Z.
Ownership	Clearly assigned owners responsible for the successful delivery of each strategic theme and key initiative.	Theme Owner: VP Marketing; Initiative Sponsor: Director of Ops; Project Lead: AI Office Program Manager.
Success Metrics/KPIs	Key metrics linked to business value that will measure the success of initiatives and the overall roadmap (Domain West - details in Ch 13, Appendix B).	Increase Cross-Sell Rate by X%, Reduce Maintenance Costs by Y%, Achieve EAMI Value Score > Z.

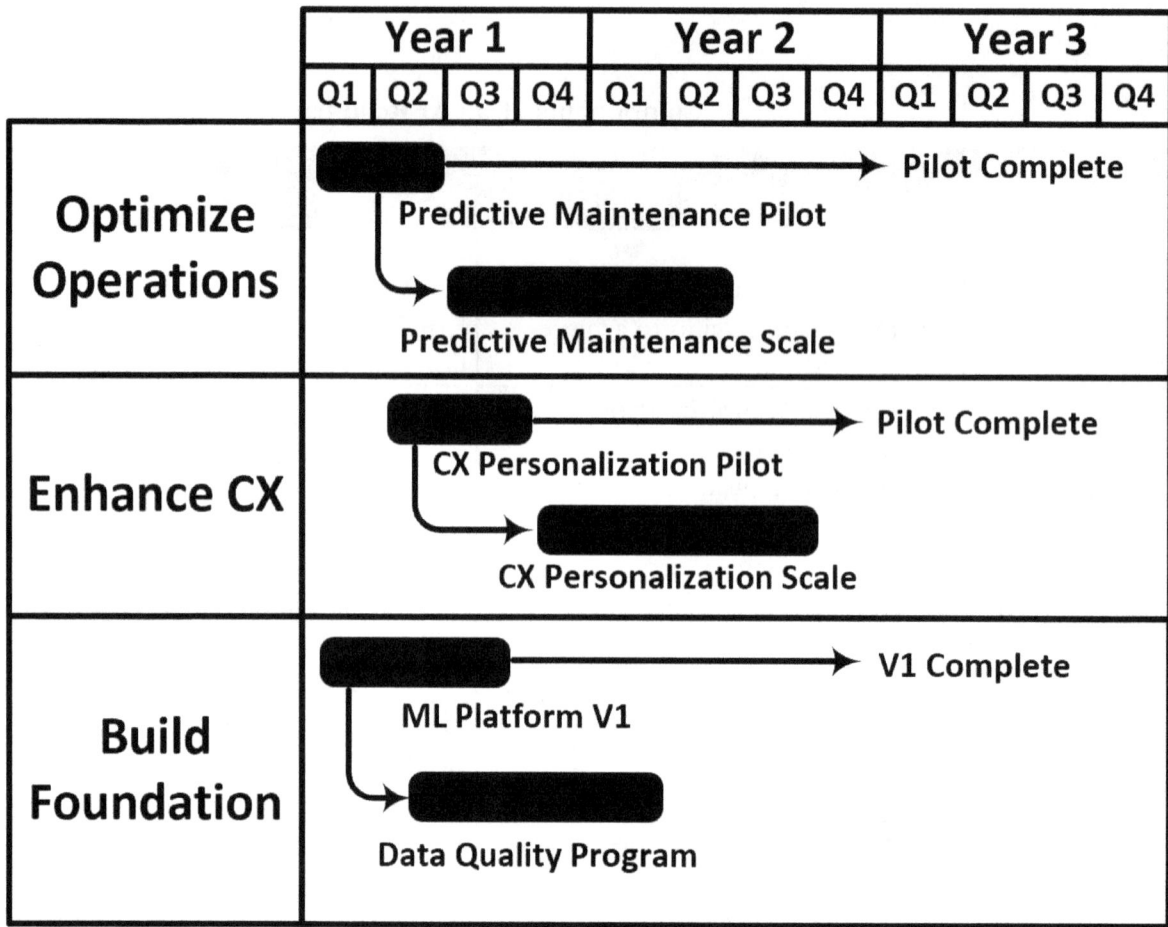

Figure 6.1: Example AI Roadmap Timeline View

6.3 Planning Considerations: Beyond the Timeline

Developing a robust and realistic roadmap involves more than just plotting prioritized projects onto a timeline. Several critical factors must be thoughtfully considered during the planning process:

- **Strategic Goals & EAMI Targets:** Continuously validate that the combination and sequencing of initiatives on the roadmap directly and demonstrably support the achievement of specific, stated business objectives and targeted improvements in overall EAMI maturity levels (or specific EAMI dimension scores) within the planned timeframe. Is the roadmap ambitious enough yet achievable?
- **Organizational Readiness & Capacity:** Honestly assess the organization's current capacity – in terms of available AI/data talent, allocated budget, existing infrastructure maturity, data readiness, and current overall EAMI level (Domain South / Center) – to successfully undertake the planned initiatives within the proposed timelines. Avoid creating a roadmap that significantly outstrips the organization's ability to execute.
- **Phased Approach (Pilot → Scale → Optimize):** For significant or novel AI initiatives, explicitly plan distinct phases. A Pilot phase (often 3-6 months) aims to validate technical feasibility and potential business value on a limited scale with minimal risk. A successful pilot triggers a Scale phase, involving broader rollout, integration with operational systems, and robust MLOps. An Optimize phase often follows scaling, focusing

on continuous monitoring, model retraining, performance improvement, and maximizing value extraction over the long term.

- **Foundational Capabilities:** Critically, ensure the roadmap explicitly includes and appropriately sequences necessary foundational work. This often-overlooked "plumbing" is essential for sustainable success and includes initiatives like establishing core governance processes (Domain East), building required data pipelines or undertaking major data quality improvement programs (Domain South - Data), and implementing essential MLOps infrastructure and practices (Domain South - MLOps). This foundational work often needs to precede or run in parallel with the development of specific AI applications that depend on it. Failing to plan for foundations is a common cause of roadmap failure.

- **Dependency Management:** Proactively identify, document, and plan for critical dependencies – technical (e.g., API availability), data (e.g., access to specific datasets), resource (e.g., availability of a key expert), or organizational (e.g., alignment with another project's timeline). Use the roadmap review process to actively track the status of critical path dependencies.

- **Resource Allocation & Budgeting:** Align the roadmap realistically with corporate budget cycles and projected talent availability (Chapters 16, 19 - Domain South). Can the required specialized roles be hired, trained, or contracted in time? Is the technology budget (cloud spend, software licenses) sufficient for the planned scale?

- **Risk Assessment & Mitigation:** Incorporate risk assessment into roadmap planning (Domain East). Identify potential risks associated with key initiatives (technical hurdles, adoption challenges, changing market conditions) and build in contingency buffers or potential mitigation strategies where appropriate.

> Treat roadmap development as a collaborative, cross-functional exercise (Compass Center activity), not an isolated AI Office activity. Actively involve key leaders and representatives from business units, IT, data management, cybersecurity, finance, legal, and HR in the planning process. This ensures the roadmap is grounded in operational reality, considers diverse perspectives, identifies potential roadblocks early, and crucially, builds shared ownership and commitment for execution.

6.4 Sequencing Initiatives: Principles for Effective Ordering

Determining the optimal order of initiatives on the roadmap is a strategic exercise involving trade-offs. Several guiding principles can help:

- **Value & Quick Wins:** Prioritize initiatives offering significant, measurable business value or visible "quick wins" (often high value, lower effort) early in the roadmap (e.g., within the first 6-12 months). This helps build crucial momentum, demonstrate early ROI (Domain West), secure ongoing stakeholder support, and potentially fund later, more complex initiatives.

- **Dependencies:** Sequence projects logically based on identified technical, data, or capability dependencies. Foundational platform development, critical data integration work, or essential governance policy establishment (Domain South/East) often needs to precede the dependent application development or scaling phases.

- **Risk Management:** Consider tackling higher-risk or technically uncertain projects earlier in the roadmap, but often through smaller, contained pilots or Proofs of Concept (PoCs). This allows for early learning and de-risking before committing significant resources to a potentially flawed approach.

- **Resource Availability & Smoothing:** Sequence initiatives considering the availability of critical, often constrained, resources (e.g., specialized ML engineers, specific datasets, core platform components - Domain South). Avoid scheduling multiple projects that heavily rely on the same scarce resource simultaneously, aiming for a smoother resource utilization profile over time.
- **Learning & Capability Building:** Sequence initiatives strategically to allow the organization and the AI team to build capabilities and confidence incrementally. It may be prudent to start with projects leveraging more mature, well-understood AI techniques before tackling highly complex, novel, or bleeding-edge approaches. Align sequencing with targeted EAMI uplift goals for specific dimensions (e.g., build MLOps capability needed for Level 3 before scaling complex models).

6.5 Portfolio Balancing Revisited

As discussed conceptually in Chapter 5, the resulting roadmap should reflect a deliberately balanced portfolio of AI initiatives when viewed holistically. Regularly review the overall mix of projects planned on the roadmap (e.g., annually or semi-annually) to ensure a healthy balance across key dimensions:

- **Risk vs. Reward:** Blend potentially transformative but higher-risk innovation projects ("explore") with lower-risk, more certain operational efficiency improvements or capability enhancements ("exploit").
- **Time Horizon:** Ensure a mix of short-term deliverables providing value within months and longer-term strategic initiatives building foundational capabilities or enabling major business transformations over multiple years.
- **Business Impact / Strategic Goals:** Verify that the portfolio collectively addresses the organization's key strategic pillars and delivers value across relevant business units, rather than being overly concentrated in one area (unless strategy dictates).
- **Build vs. Buy:** Include a thoughtful mix of initiatives focused on developing custom, proprietary AI solutions versus those focused on implementing and integrating third-party AI tools, platforms, or vendor solutions where appropriate.

6.6 Securing Stakeholder Buy-in and Communication

An AI roadmap, no matter how well-crafted, is only effective if it is understood, accepted, supported, and actively used by key stakeholders across the organization. Securing and maintaining this buy-in (Compass Center/West activity) is a critical, ongoing task for the AI Office:

- **Involve Stakeholders Early and Often:** Engage executives, business sponsors, IT leaders, data owners, finance partners, legal/compliance representatives, and other relevant parties during the roadmap development process, not just presenting a finished plan. Actively solicit their input, address their concerns, and incorporate their feedback to build shared ownership.
- **Develop Clear Communication Materials:** Create clear, concise, and visually compelling presentations and documentation explaining the roadmap's strategic vision, key initiatives, timelines, dependencies, expected business value, and explicit connection to corporate strategy (Chapter 15). Tailor the level of detail and messaging to different audiences.
- **Obtain Formal Endorsement:** Seek formal approval and endorsement of the finalized roadmap from the

executive leadership team and the primary AI governance body (e.g., AI Review Team, AI Steering Committee). This formal sign-off signals enterprise-wide commitment and empowers the AI Office and project teams to execute the plan.

- **Communicate Progress and Changes Proactively:** Treat roadmap communication as an ongoing process. Regularly (e.g., quarterly) communicate progress against milestones, highlight early successes and value delivered, transparently explain any necessary adjustments or changes to the plan (and the rationale behind them), and proactively manage stakeholder expectations. Utilize visual dashboards (Chapter 15) linked to roadmap milestones and EAMI targets where possible.

Establish a clear governance process (Compass East/Center) for managing changes to the approved AI roadmap. Treat the roadmap as a living document, but ensure significant changes (e.g., adding/removing major initiatives, major shifts in timelines or budget) go through a formal review and approval cycle involving key stakeholders and the governance body. This maintains strategic alignment while allowing for necessary agility. Schedule regular roadmap review cycles (e.g., quarterly) as part of the AI Office's operating rhythm.

6.7 The AI Office's Role in Roadmap Management

The AI Office typically plays a central, ongoing role (Compass North/Center focus) throughout the entire lifecycle of the Enterprise AI Roadmap:

- **Facilitation & Development:** Leading the collaborative process for initially developing the roadmap, coordinating stakeholder input, ensuring robust analysis, and confirming alignment with overall strategy and prioritized opportunities.
- **Ownership & Custodianship:** Acting as the official owner and custodian of the Enterprise AI Roadmap document or system of record (e.g., within a dedicated roadmapping tool), ensuring it remains up-to-date and accessible to relevant stakeholders.
- **Monitoring, Tracking & Reporting:** Systematically tracking progress against planned milestones and deliverables, actively monitoring critical dependencies, identifying emerging risks or issues, and providing regular status reporting to stakeholders and governance bodies (Domain West).
- **Change Management & Governance:** Facilitating the established process for requesting, evaluating, and approving changes to the roadmap, ensuring changes are strategically justified, impact-assessed, and communicated.
- **Cross-Functional Integration:** Ensuring the AI roadmap remains aligned and integrated with other relevant corporate strategic roadmaps (e.g., overall IT infrastructure roadmap, data strategy roadmap, digital transformation program plan, product development roadmaps).

6.8 OmnioTech Case Study: Developing the AI Roadmap

Setting the Scene:

Having prioritized the Predictive Maintenance (PredMaint) and Hyper-Personalization (HyperPers) pilots, along-side assessing the AI Support Chatbot feasibility (Chapter 5), Priya Sharma and the OmnioTech AI Office Hub must now create a formal 2-year Enterprise AI roadmap. This roadmap needs to incorporate these initiatives, plan for subsequent phases, include essential foundational work, and align with the goal of reaching EAMI Level 3.

Action & Application:

- **Component Definition & Tooling:** Priya uses a collaborative roadmapping tool (e.g., Aha! or similar) to draft the roadmap, engaging stakeholders to define:
  - » **Themes:** 1. Product Innovation (AI-Enabled Appliances), 2. Customer Engagement (Personalization & Support), 3. Foundational AI Capabilities.
  - » **Initiatives & Phasing:** PredMaint Pilot (Y1Q1-Q2), PredMaint Scale (Y1Q3-Y2Q2); HyperPers Pilot (Y1Q2-Q3), HyperPers Scale (Y1Q4-Y2Q4); AI Support Chatbot Pilot (Decision Y1Q4, Pilot Y2Q1-Q2); ML Platform V1 Build (Y1Q1-Q3 - targeting EAMI Tech L3); Data Quality Program V1 (Y1Q2-Y2Q1 - targeting EAMI DQ L3); AI Governance Framework V1 (Y1Q1-Q2 - targeting EAMI Gov L3); EAMI Assessment Cycle (Y1Q1, Y1Q3, Y2Q1, Y2Q3 - Compass Center activity).
  - » **Milestones:** Pilot Go/No-Go dates, Platform V1 Launch, Governance V1 Policy Approval, Scaling Phase Starts, EAMI Assessment Completions.
  - » **Dependencies:** Scaling phases explicitly depend on ML Platform V1 completion and Data Quality Program V1 milestones. AI Chatbot depends on HyperPers data integration.
  - » **Resources:** High-level FTE estimates (AI Office, IT, BU) and initial budget allocations ($500k for pilots, $1M for Platform V1) documented against initiatives (Domain South).
  - » **Ownership:** Business VPs assigned as Sponsors for pilots; AI Office leads own foundational work and EAMI cycle.
- **Sequencing & Planning Rationale:** Prioritized pilots (PredMaint, HyperPers) start early (Y1Q1/Q2) targeting quick wins and demonstrating value (Domain West). Critical foundational work (ML Platform, Data Quality, Governance V1 - Domain South/East) runs in parallel during Year 1 to enable reliable scaling phases starting late Y1/early Y2. The AI Support Chatbot pilot is deliberately sequenced for Year 2 after foundational elements are more mature. Regular EAMI assessments provide crucial checkpoints for tracking maturity progress.
- **Stakeholder Buy-in & Communication:** The draft roadmap is reviewed iteratively over several sessions with the AI Review Team, key sponsors (Product VP, Marketing VP, Head of Support), and functional leads (IT Platform Lead, Data Architecture Lead). Feedback regarding resource constraints (particularly ML engineering time) and dependency timing risks is incorporated, adjusting some timelines slightly. The final version is formally approved by the COO/CTO sponsors and presented to CEO Julia Weber for final endorsement. A high-level summary is shared with wider leadership, while the detailed view is made accessible to project teams via the roadmapping tool. Priya schedules mandatory quarterly roadmap review meetings for the AI Review Team.

Outcome & Progress:

OmnioTech now has a clear, collaboratively developed, and formally endorsed 2-year Enterprise AI roadmap

(Compass North artifact). This provides a vital shared understanding of priorities, planned activities, timelines, and critical dependencies, effectively guiding resource allocation and execution efforts. By explicitly including foundational capability building targeting specific EAMI Level 3 improvements alongside business-facing initiatives and incorporating regular EAMI assessments, the roadmap positions OmnioTech to systematically build capabilities and track progress towards its EAMI Level 3 maturity target within the defined timeframe.

6.9 Conclusion: The Actionable Blueprint for AI Strategy

The Enterprise AI Roadmap is the indispensable artifact that translates strategic intent (Domain North) and prioritized opportunities into a tangible, actionable, and communicable plan. By clearly defining key initiatives, timelines, milestones, critical dependencies, and high-level resource needs, it provides essential clarity, focus, and direction for the organization's complex AI journey. Developing and maintaining this roadmap requires careful consideration of strategic goals, organizational readiness, thoughtful sequencing principles, portfolio balance, and proactive, continuous stakeholder engagement – all skillfully facilitated by the AI Office. As a dynamic, living document, the roadmap serves not only to guide execution but also to enable effective progress tracking, risk management, and strategic adaptation over time. With a robust roadmap defined, the organization is poised to move forward with focused execution. However, sustainable and successful execution demands strong ethical and operational guardrails, which form the critical focus of Part 3: AI Governance (East).

Key Takeaways:

- An AI Roadmap translates prioritized opportunities into an actionable plan with timelines, resources, and dependencies.
- Key components include strategic themes, initiatives, phasing, milestones, dependencies, resources, ownership, and KPIs.
- Planning requires considering readiness, foundational capabilities, dependencies, and risk mitigation.
- Sequencing balances quick wins, dependencies, risk, resource availability, and capability building.
- Stakeholder buy-in through early involvement and clear communication is crucial for roadmap success.
- The AI Office facilitates development, owns maintenance, monitors progress, and governs changes to the roadmap.

Food for Thought / Application Exercise:

- Does your organization currently have a documented AI roadmap? If so, does it include the key components listed in Section 6.2? If not, what is the biggest barrier to creating one?
- Consider two potential AI initiatives in your area. What are the key dependencies between them or with other organizational factors (data, platforms, skills)? How would these dependencies influence their sequencing on a roadmap?
- Who are the essential stakeholders that would need to be involved in developing and approving an AI roadmap for your business unit or organization? How would you tailor communication about the roadmap to different groups (e.g., executives vs. technical teams)?

PART 3 - AI GOVERNANCE

CHAPTER 7

The Pillars of Trustworthy AI Governance: Building the Foundation

"In the world of AI, trust isn't given, it's earned – through transparent processes, ethical considerations, and robust governance."

7.1 Introduction: Why Governance is Non-Negotiable

Welcome to Part 3, where we navigate the critical Eastern domain of the AI Office Compass™ Framework: Governance. As organizations increasingly harness the profound power of artificial intelligence, they simultaneously take on significant and complex responsibilities. AI systems, particularly sophisticated machine learning models operating on vast datasets, possess the potential to inadvertently perpetuate societal biases, make decisions opaque to human understanding and scrutiny, create new vectors for security vulnerabilities, violate stringent privacy regulations, and deliver unreliable or even harmful outcomes if not developed and deployed within a strong ethical, legal, and operational framework. The potential consequences of neglecting governance are immense, ranging from substantial financial penalties and legal liabilities to severe reputational damage, erosion of customer trust, and negative societal impacts.

Establishing robust AI governance is therefore not merely a "nice-to-have" compliance activity or an administrative burden; it is a fundamental prerequisite for building sustainable, trustworthy, and ultimately value-generating enterprise AI programs. It is about proactively designing and implementing the necessary guardrails – policies, processes, standards, roles, and controls – to ensure AI systems are conceived, developed, deployed, and managed responsibly throughout their entire lifecycle. This governance must align with the organization's core values, comply with applicable legal and regulatory requirements, and meet evolving societal expectations for fairness, transparency, and accountability.

This chapter lays the essential conceptual foundation for effective AI governance by introducing and exploring seven core pillars that underpin trustworthy AI: Accountability, Fairness, Transparency, Security & Safety, Privacy,

Reliability, and Human-Centricity. A clear understanding of these interconnected pillars is crucial before attempting to design the specific governance framework architecture, processes, and tooling detailed in Chapter 8 and subsequent chapters in this part. Furthermore, demonstrating maturity across these pillars is a cornerstone of achieving higher levels within the EAMM framework (introduced in Chapter 1, detailed in Chapter 24), particularly within the critical Governance, Ethics, Security, Compliance, and Risk Management dimensions.

Learning Objectives:

- Understand the critical importance and compelling business case for establishing robust AI governance early in the AI journey.
- **Define and explain the scope and significance of the seven core pillars of trustworthy AI governance:** Accountability, Fairness, Transparency, Security & Safety, Privacy, Reliability, and Human-Centricity.
- Recognize the inherent interplay and potential trade-offs that often exist between these different governance pillars.
- Appreciate how these foundational pillars inform and shape the practical design and implementation of the AI governance framework discussed in subsequent chapters.

7.2 The Business Case for Proactive AI Governance

Investing time and resources in establishing comprehensive AI governance is not just about mitigating risk; it also delivers tangible business benefits and enables sustainable AI adoption:

- **Significant Risk Mitigation:** Proactive governance (Compass East) directly reduces exposure to substantial regulatory fines (e.g., potential multi-million dollar penalties under GDPR or the upcoming EU AI Act), costly lawsuits (e.g., class actions related to discriminatory AI outcomes in hiring, lending, or insurance), damaging security breaches targeting AI models or sensitive training data, and severe, long-lasting reputational damage resulting from public ethical failures or biased AI systems (addresses EAMM Risk Management, Compliance, Ethics, Security).
- **Building and Maintaining Trust:** Demonstrating a clear, consistent commitment to responsible and ethical AI practices is essential for fostering trust with key stakeholders – customers (driving loyalty and willingness to share data), employees (encouraging adoption of internal AI tools), partners (enabling secure data sharing and collaboration), and regulators (facilitating smoother approvals and reducing scrutiny). Trust is the currency of the AI economy (supports EAMM Culture).
- **Ensuring Regulatory Compliance:** Governance provides the necessary structures, processes (like impact assessments), and documentation capabilities to systematically adhere to existing data protection regulations (like GDPR, CCPA) and proactively prepare for and adapt to emerging AI-specific legislation (like the EU AI Act's risk-based requirements or evolving national standards - EAMM Compliance).
- **Improving AI System Quality & Reliability:** Well-defined governance processes typically mandate rigorous standards for data quality management (Chapter 23), comprehensive model validation protocols, continuous performance monitoring in production, and structured model lifecycle management. This inherent discipline leads to more robust, predictable, reliable, and ultimately more effective AI systems (supports EAMM Reliability dimension implicitly, relates to EAMM MLOps).
- **Enhancing Human Decision-Making:** Promoting transparency and explainability (XAI) allows human users and overseers to better understand how AI systems arrive at their recommendations or decisions. This enables humans to appropriately trust, effectively validate, and confidently challenge or override AI outputs

when necessary, leading to better overall decision quality (supports EAMM Ethics - Transparency).

- **Accelerating Responsible Innovation:** Counterintuitively, clear and pragmatic governance guardrails can often accelerate responsible innovation (EAMM Innovation Ecosystem). By providing development teams with clear boundaries, ethical guidelines, and defined processes for risk assessment, governance reduces uncertainty and empowers teams to experiment and build confidently within safe operating parameters, rather than being paralyzed by ambiguity or fear of unintended consequences.
- **Achieving Competitive Differentiation:** Increasingly, organizations demonstrating strong AI ethics and transparent governance practices can leverage this commitment as a source of competitive advantage, attracting ethically conscious customers, top AI talent seeking responsible employers, and impact-focused investors.

> Effective AI Governance cannot be solely delegated to a central office or a compliance checklist. It requires cultivating a pervasive culture of responsibility (EAMM Culture), embedding ethical considerations and governance checkpoints directly into engineering workflows (like MLOps pipelines - EAMM MLOps/Governance), providing continuous training and awareness programs (EAMM Talent/Skills), and ensuring visible commitment and reinforcement from senior leadership across the organization.

7.3 The Seven Pillars of Trustworthy AI Governance

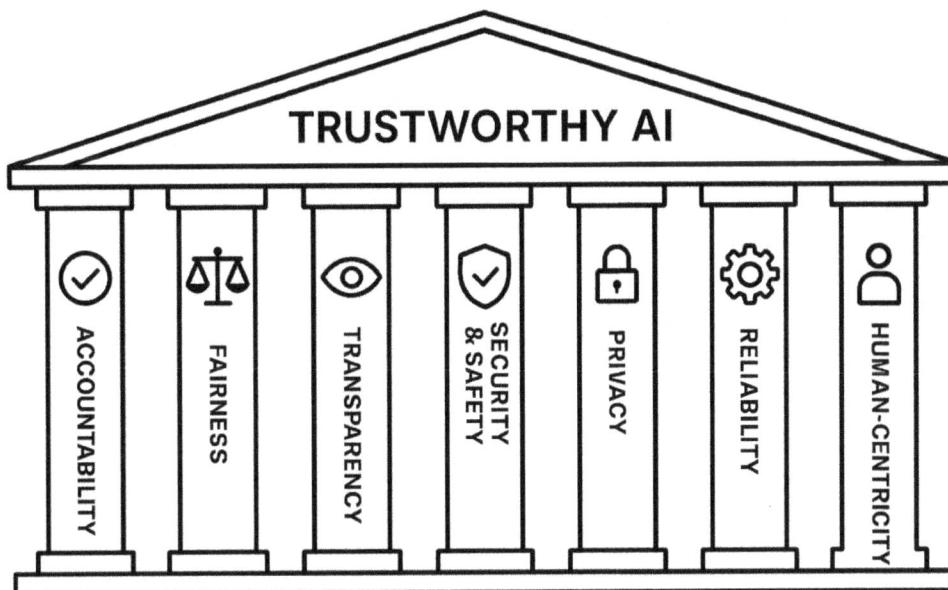

While specific implementations and areas of emphasis will vary based on industry regulations, organizational context, and the risk profile of AI applications, effective and comprehensive AI governance frameworks are typically built upon the following seven foundational and interconnected pillars. These pillars provide a holistic structure for addressing the key dimensions of responsible AI:

Seven Pillars of Trustworthy AI Governance

TRUSTWORTHY AI

ACCOUNTABILITY — FAIRNESS — TRANSPARENCY — SECURITY & SAFETY — PRIVACY — RELIABILITY — HUMAN-CENTRICITY

Figure 7.1: Seven Pillars of AI Governance Diagram

The table below summarizes these foundational pillars:

Table 7.1: Seven Pillars of Trustworthy AI Governance

Pillar	Core Principle	Key Considerations & Activities
1. Accountability	Establishing clear lines of ownership, responsibility, and oversight for AI systems throughout their entire lifecycle.	Defining roles & responsibilities (e.g., using RACI charts for model development, deployment, monitoring); establishing clear governance bodies (e.g., AI Review Team, Ethics Board) with defined decision rights; ensuring clear reporting lines up to executive sponsors; implementing comprehensive audit trails and logging for traceability; defining clear processes for incident management, root cause analysis, and remediation; ensuring accessible mechanisms for redress or appeal exist for affected individuals.
2. Fairness	Proactively identifying, assessing, and mitigating potential sources of unfair bias in data and models to prevent discriminatory outcomes.	Understanding different types of bias (e.g., historical societal bias reflected in data, measurement bias, algorithmic bias, confirmation bias in interpretation); ensuring training datasets are diverse, representative, and carefully evaluated for potential biases; employing quantitative fairness metrics and specialized bias detection tools (e.g., Fairlearn, AIF360, Google's What-If Tool); implementing appropriate mitigation techniques (e.g., data pre-processing like re-sampling, in-processing algorithmic adjustments, post-processing calibration); defining fairness criteria explicitly based on the specific context and potential impact; documenting fairness assessments thoroughly (Chapter 10).
3. Transparency	Ensuring AI system operations and decision-making are understandable to relevant stakeholders to an appropriate degree.	Providing clear documentation of model purpose, design choices, training data characteristics, and known limitations (e.g., via Model Cards or Datasheets for Datasets); utilizing Explainable AI (XAI) techniques (e.g., feature importance via SHAP or LIME, rule-based explanations, counterfactual explanations) to interpret model behavior where appropriate; tailoring the level and type of explanation provided to the specific needs of the audience (e.g., developer vs. end-user vs. regulator); clearly disclosing when users are interacting with an AI system versus a human; managing the inherent trade-offs between model complexity and full interpretability (Chapter 10).

4. Security & Safety	Protecting AI systems (models, data, infrastructure) from malicious attacks and ensuring they operate reliably without causing unintended harm.	Robust data/model security controls; defense against adversarial attacks; security testing (SecMLOps); defining operational boundaries; fail-safe mechanisms; robust monitoring for safety-critical applications; incident response planning (Chapter 11).
5. Privacy	Respecting individual privacy rights and complying with applicable data protection regulations (e.g., GDPR, CCPA).	Data minimization; lawful basis for processing; clear notices & consent management; anonymization/pseudonymization/ differential privacy techniques; Data Protection Impact Assessments (DPIAs); secure data retention/deletion policies (Chapter 12).
6. Reliability	Ensuring AI systems perform accurately, consistently, robustly, and predictably according to their intended function under expected operational conditions.	Rigorous testing & validation (unit, integration, performance); production monitoring for performance/data drift; model retraining/updating strategies; ensuring resilience to noisy/ incomplete data; defining clear performance thresholds (SLOs).
7. Human-Centricity	Designing AI to augment human capabilities, respect autonomy, ensure meaningful human oversight, and align with values.	Intuitive Human-AI interaction design; clear processes for human review/intervention (human-in-the-loop); override capabilities; considering workforce impact (Chapter 26); aligning AI goals with human values; usability testing.

These seven pillars are not independent silos; they are deeply interconnected and often reinforce each other. For example, achieving true Fairness often requires Transparency (to understand and explain potential biases) and depends on Reliability (consistent performance across groups). Strong Security is essential for maintaining Privacy. Accountability provides the framework for ensuring all other pillars are effectively implemented and maintained. A holistic, integrated governance approach (Compass East) is therefore essential.

7.4 Balancing the Pillars: Navigating Trade-offs and Context

Implementing these pillars effectively often involves navigating inherent tensions and making conscious, context-dependent trade-offs. There is rarely a single "perfect" solution that maximizes all pillars simultaneously. Organizations must understand and manage these potential conflicts:

- **Transparency vs. Performance/Accuracy:** Highly complex models (e.g., large ensembles, deep neural networks) might yield better performance but can be harder to fully interpret ("black box") compared to simpler models. The required transparency depends on application risk and regulation.
- **Fairness vs. Accuracy:** Applying specific bias mitigation techniques might slightly reduce overall accuracy to achieve fairness across groups. The acceptable balance requires careful ethical deliberation.
- **Privacy vs. Data Utility:** Strong privacy techniques (like differential privacy) can obscure individual information, sometimes reducing data utility for certain analyses or model training. Evaluation depends on sensitivity, regulation, and task.
- **Governance Rigor vs. Speed-to-Market:** Comprehensive governance reviews and documentation consume time and resources, potentially slowing development compared to less controlled environments. Finding the right, risk-adjusted balance ("just enough" governance) is crucial for responsible innovation.

The appropriate balance between these pillars is highly context-dependent. There is no universal "one-size-fits-all" governance solution. Key factors influencing rigor and trade-offs include:

- **Use Case Risk & Impact:** High-stakes applications (critical infrastructure, medical diagnosis, finance) demand far stricter governance than low-risk internal tools.
- **Regulatory & Industry Context:** Financial services, healthcare, public sector often face specific, strict requirements dictating minimum standards.
- **Data Sensitivity:** Systems processing sensitive personal data require higher focus on Privacy and Security.
- **Organizational Risk Appetite & Culture:** Tolerance for different risk types and existing compliance culture influence choices.

A risk-tiered approach to AI governance (discussed further in Chapter 8), where governance intensity scales with assessed risk level, is often the most pragmatic and effective strategy for balancing rigor with agility.

7.5 OmnioTech Case Study: Embracing the Governance Pillars

Setting the Scene:

As OmnioTech's AI Office Hub begins defining its V1.0 governance framework (Compass East activity), led by the new Governance Lead, David Chen, they use the seven pillars as a guiding structure for initial policy discussions and risk assessments related to their first two pilots (Predictive Maintenance for smart appliances, Hyper-Personalization for marketing).

Action & Application:

- **Accountability:** The AI Review Team charter clearly assigns ownership for model approval (CTO/COO jointly sign off) and designates the AI Office Hub Lead (Priya Sharma) as accountable for ongoing monitoring oversight initially. Pilot project RACIs explicitly define developer responsibilities for documentation and testing. Basic audit logging for model API calls is mandated.
- **Fairness:** For the Hyper-Personalization pilot, David leads a workshop with Marketing and Legal to define specific fairness objectives, explicitly aiming to avoid discriminatory targeting based on protected characteristics. Initial data profiling checks for representation bias. The team plans to evaluate key fairness metrics (e.g., demographic parity, equal opportunity) using the Fairlearn toolkit during pilot validation before scaling (addresses EAMM Ethics).
- **Transparency:** Minimum documentation standards using a Model Card template are mandated, requiring descriptions of intended use, data sources, rationale, limitations, etc. For personalization, generating SHAP values for basic feature importance explanations for internal review is planned before scaling (addresses EAMM Ethics - Transparency).
- **Security & Safety:** Mandatory security reviews by the CISO's team are integrated into the pilot deployment checklist (Azure DevOps). Reviews focus on data access controls (Azure RBAC), API security, and vulnerability scanning (addresses EAMM Security). For predictive maintenance, safety considerations around false positive alerts are documented and addressed via model threshold tuning biased towards precision for high-severity alerts.
- **Privacy:** A formal Data Protection Impact Assessment (DPIA) is completed for Hyper-Personalization, documenting data flows, risks, mitigation, and ensuring alignment with GDPR consent (addresses EAMM Compliance/Privacy). Data minimization principles are reviewed for Predictive Maintenance sensor data collection.
- **Reliability:** Model performance monitoring plans are required, specifying key metrics (accuracy, precision, recall; conversion lift), frequency, drift detection methods (Azure Monitor), and thresholds for alerts or re-training reviews (addresses EAMM Performance Metrics / MLOps).
- **Human-Centricity:** The personalization pilot design includes human-in-the-loop A/B testing where marketing specialists review/approve AI offers. Predictive maintenance alerts provide insights within the existing technician workflow (ServiceNow), augmenting human judgment.

Outcome & Progress:

Using the seven pillars provides a structured framework for OmnioTech to proactively address governance dimensions for their initial pilots (Compass East). It helps systematically identify risks and incorporate controls, ethical considerations, and compliance requirements into their V1.0 governance policies and pilot designs. This establishes a solid foundation aligned with EAMM Governance, Ethics, Compliance, and Security criteria, building confidence for responsible execution and demonstrating progress towards their EAMI Level 3 target in these areas.

7.6 Conclusion: The Foundation of Trust

The seven pillars of Accountability, Fairness, Transparency, Security & Safety, Privacy, Reliability, and Human-Centricity collectively form the essential conceptual foundation for building and maintaining trustworthy AI governance within any enterprise. They provide a comprehensive framework for systematically thinking about the multifaceted ethical, legal, technical, and operational responsibilities associated with deploying powerful AI systems. While effective implementation requires careful context consideration and navigating trade-offs, embedding these principles deeply into the organization's culture, processes, and technology choices (Compass East integration) is non-negotiable for achieving sustainable success, mitigating significant risks, and building lasting stakeholder trust in the age of AI. A solid understanding of these core pillars prepares us for the next crucial step: designing the practical architecture and implementing the specific components of the AI governance framework itself, the focus of Chapter 8.

Key Takeaways:

- Robust AI governance is essential for mitigating risks (fines, reputation), building trust, ensuring compliance, improving AI quality, and accelerating responsible innovation.
- **Trustworthy AI is built on seven key pillars:** Accountability, Fairness, Transparency, Security & Safety, Privacy, Reliability, and Human-Centricity.
- These pillars are interconnected and often involve context-dependent trade-offs (e.g., Transparency vs. Performance).
- A risk-tiered approach often helps balance governance rigor with innovation speed.
- Understanding these pillars provides the conceptual foundation for designing a practical governance framework (Chapter 8).

Food for Thought / Application Exercise:

- Consider an AI system currently used or being developed in your organization. How well does it align with each of the seven governance pillars? Where are the potential gaps or risks?
- Think about a potential trade-off between two pillars (e.g., maximizing model accuracy vs. ensuring fairness or explainability) relevant to your context. How would your organization decide on the appropriate balance? What factors would be considered?
- Which of the seven pillars presents the biggest challenge for your organization to implement effectively today? Why?

CHAPTER 8

Designing the AI Governance Framework: Architecture for Responsibility

"Good governance is the art of making policies that are not only effective but also fair and sustainable."

8.1 Introduction: From Pillars to Practice

Chapter 7 established the foundational importance of AI governance and introduced the seven core pillars – Accountability, Fairness, Transparency, Security & Safety, Privacy, Reliability, and Human-Centricity – that underpin trustworthy AI. Understanding these principles is essential, but principles alone are insufficient without a practical structure to operationalize them. Organizations need a concrete AI Governance Framework: a formally defined system of policies, standards, roles, processes, controls, and tools designed to ensure that AI systems are developed, deployed, and managed consistently in accordance with these guiding principles and organizational objectives.

This chapter transitions from the 'what' and 'why' of AI governance (Chapter 7) to the 'how'. We delve into the architectural components of an effective AI Governance Framework (Compass East responsibility), exploring how to translate the seven pillars into tangible organizational structures and operational workflows. We will discuss the key elements typically included in such a framework, such as tiered policies and standards, specific governance bodies and roles (like an AI Review Board or Ethics Council), critical review processes integrated into the AI lifecycle, necessary controls and monitoring mechanisms, and the supporting documentation and tooling required. Designing this framework architecture is a crucial step guided by the Governance (East) domain of the AI Office Compass™ and is fundamental for achieving higher EAMI maturity levels in Governance, Risk Management, Compliance, and Ethics dimensions.

Learning Objectives:

- Understand the key architectural components of a comprehensive AI Governance Framework.
- Learn how to design risk-tiered policies and standards tailored to different AI application types.
- Identify common governance bodies and roles needed to oversee and execute the framework.
- Recognize critical governance processes that should be integrated into the AI development and deployment lifecycle (MLOps).
- Appreciate the role of documentation, controls, and tooling in supporting effective governance.
- Identify common pitfalls to avoid when designing the governance framework architecture.

8.2 Core Components of the AI Governance Framework

While the specific implementation details will vary based on organizational size, industry, and culture, a robust AI Governance Framework typically comprises several key architectural components working together:

- **Policies & Principles:** High-level statements defining the organization's commitment to responsible AI and outlining core principles (often based on the seven pillars from Chapter 7). Set the tone, endorsed by leadership.
- **Standards & Guidelines:** More detailed, actionable requirements and best practices derived from policies (e.g., data handling, model validation, bias testing, security protocols). May be risk-tiered.
- **Roles & Responsibilities:** Clearly defined roles, responsibilities, and decision rights for individuals/groups (AI Governance Lead, AI Review Board, developers, owners, Legal, Compliance, Security). RACI charts often used.
- **Processes & Workflows:** Defined procedures integrated into the AI lifecycle (risk assessment, ethical review, validation checks, deployment approval, incident response).
- **Controls & Monitoring:** Mechanisms ensuring adherence to standards and monitoring AI systems in production (access controls, automated checks, bias monitoring, security scanning, audits).
- **Documentation & Reporting:** Standardized templates (Model Cards, Data Sheets, risk assessments) and regular reporting on governance activities, compliance, and risk posture.
- **Tooling & Technology:** Supporting technologies enabling/automating governance (governance platforms, risk tools, bias/XAI libraries, MLOps features, data catalogs).

The table below summarizes these core components:

Table 8.1: Core Components of the AI Governance Framework

Framework Component	Purpose	Examples
Policies & Principles	Set high-level ethical tone, values, and organizational commitment to responsible AI.	Enterprise Responsible AI Principles document, AI Ethics Charter, Board-level AI Risk Policy.
Standards & Guidelines	Provide specific, actionable requirements and best practices derived from policies.	Secure AI Development Standard, Data Handling for AI Standard, Model Validation Guideline, Bias Assessment Procedure, Transparency Reporting Standard.

Roles & Responsibilities	Define who does what regarding AI governance design, execution, and oversight.	AI Governance Lead job description, AI Review Board Charter, RACI matrix for model deployment process.
Processes & Workflows	Define standardized procedures for key governance activities integrated into the AI lifecycle.	AI Risk Assessment Process, Ethical Review Workflow, Model Validation Checklist, Deployment Approval Gate Process, AI Incident Response Plan.
Controls & Monitoring	Ensure adherence to standards and monitor ongoing performance and risk.	Access controls on sensitive data/models, automated policy checks in CI/CD, production model monitoring dashboards (drift, bias), regular audits.
Documentation & Reporting	Provide transparency, accountability, and evidence of compliance.	Standardized Model Card template, Data Sheet for Datasets template, Risk Assessment form, Quarterly Governance Compliance Report.
Tooling & Technology	Enable efficiency, automation, and scalability of governance activities.	GRC platforms, MLOps platforms (e.g., Azure ML, SageMaker), bias/explainability libraries (Fairlearn, SHAP), data catalogs (Collibra), monitoring tools (Datadog).

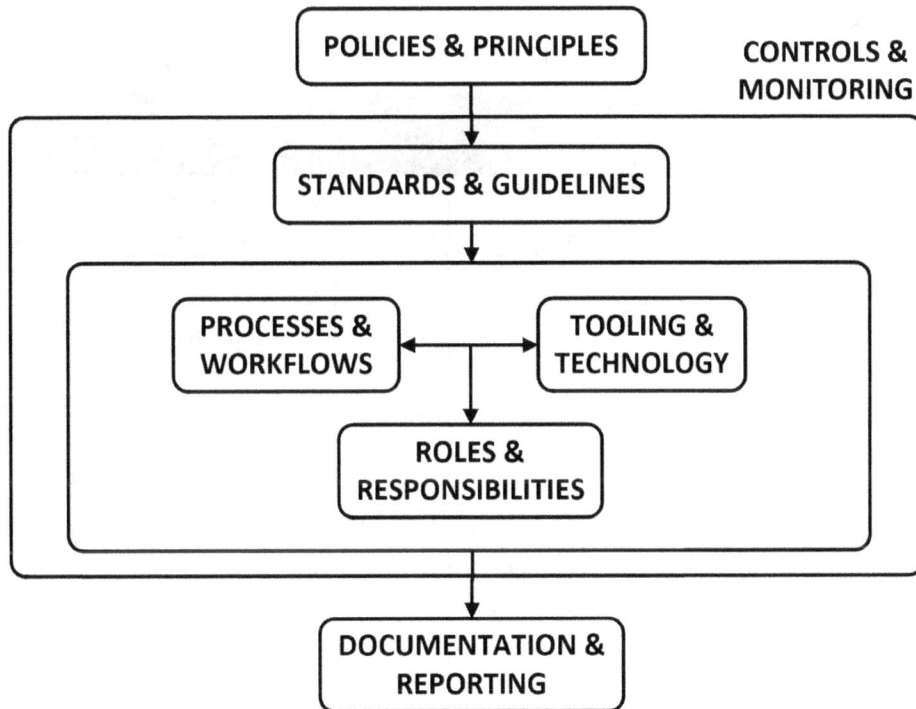

Figure 8.1: AI Governance Framework Components Diagram

8.3 Designing Risk-Tiered Policies and Standards

A "one-size-fits-all" approach to AI governance is often impractical and inefficient. Applying the same scrutiny to a low-risk internal tool as to a high-stakes, customer-facing model can stifle innovation. A risk-tiered approach is generally more effective. This involves:

- **Defining Risk Criteria:** Establish clear criteria for assessing AI application risk based on factors like:
  - » **Impact Domain:** Affects safety, finance, legal rights, essential services, employment?
  - » **Impact Severity:** Potential severity of harm if system fails/is unfair (minor vs. major)?
  - » **Autonomy Level:** Autonomous decisions vs. recommendations for human review?
  - » **Scale of Deployment:** How many people/processes affected?
  - » **Data Sensitivity:** Uses sensitive personal data (health, finance)?
  - » **Regulatory Scrutiny:** Subject to specific regulations (credit scoring, medical devices, EU AI Act high-risk)?
- **Establishing Risk Tiers:** Define distinct risk categories (e.g., Tier 1: Minimal Risk; Tier 2: Low Risk; Tier 3: Medium Risk; Tier 4: High Risk). The EU AI Act provides a potential reference model.
- **Tailoring Requirements:** Define increasingly stringent governance requirements (policies, standards, review processes, documentation, testing, monitoring) for higher risk tiers. Minimal risk applications need basic adherence; high-risk demand extensive testing, formal reviews, robust explainability, continuous monitoring, and executive sign-off.

The table below provides a simplified illustration:

Table 8.2: Example Risk-Tiered Governance Requirements

Risk Tier Example	Example Criteria	Example Tailored Governance Requirements
Tier 1: Minimal	Internal tool, low impact, no sensitive data, human oversight.	Basic documentation (Model Card Lite), adhere to core AI principles, standard code review.
Tier 2: Low	Affects internal processes, moderate impact, non-sensitive data, human review.	Standard Model Card, adherence to data handling standards, basic performance monitoring.
Tier 3: Medium	Customer-facing recommendations, potential financial impact, uses PII, human-in-loop.	Detailed Model Card, bias assessment required, explainability report (e.g., SHAP), DPIA likely needed, robust performance & drift monitoring, periodic governance review by AI Review Board.
Tier 4: High	Autonomous decisions in critical areas (finance, health, safety), high impact, sensitive data, regulated use case.	Extensive documentation (Model Card, Datasheet, etc.), rigorous bias/fairness testing & mitigation, formal ethical review board approval, high degree of explainability, continuous real-time monitoring, enhanced security protocols, executive sign-off.

Clearly document the risk assessment methodology (criteria, scoring, thresholds) and the specific, differentiated governance requirements associated with each tier. Make this framework readily accessible via training and documentation portals. Ensure the AI Office or a designated governance function oversees the risk assessment process for consistency.

8.4 Establishing Governance Bodies and Roles

Effective governance requires clearly defined roles and empowered oversight bodies (Compass East structure):

- **AI Governance Council / Steering Committee:** Highest-level strategic body (senior executives, functional leaders). Sets strategy/risk appetite, approves enterprise policies, oversees aggregate risk, resolves major escalations, champions responsible AI culture.
- **AI Review Board / Ethics Committee:** Cross-functional operational working group (AI Office Gov Lead, ethicists, legal, privacy, security, tech leads, domain experts). Executes governance processes, reviews medium/high-risk projects against standards, provides guidance, approves deployments, recommends policy updates.
- **AI Office Governance Lead(s):** Core AI Office personnel. Develop/maintain framework, facilitate reviews, provide guidance/training, manage tooling, monitor compliance, prepare reports.
- **Model Developers / Data Scientists:** Responsible for understanding and adhering to governance standards (testing, documentation, transparency in reviews, raising concerns).
- **Business Owners / Product Managers:** Accountable for the AI system in its business context. Define use case/value, accept assessed risks, provide domain expertise, ensure process alignment, champion responsible deployment/use.
- **Legal, Compliance, Privacy, Security Officers:** Subject matter experts. Provide input on regulations/standards, review policies, participate in risk assessments, advise on controls, members of governance bodies.

Using RACI (Responsible, Accountable, Consulted, Informed) charts is highly recommended to clarify roles for specific governance processes (e.g., Who is Responsible for bias testing? Accountable for deployment decision? Consulted during risk assessment? Informed of outcome?). Clear RACIs prevent ambiguity.

8.5 Integrating Governance into the AI Lifecycle (MLOps)

Governance must be integrated ("shifted left") into the end-to-end AI development and deployment lifecycle, ideally automated within MLOps practices (Chapter 20):

- **Ideation/Design Phase:** Mandatory initial risk assessment. Early consideration of ethics/privacy (triggering DPIA if needed).

- **Data Acquisition/Preparation:** Automated data quality checks (Chapter 23). Verification of data provenance/rights. Bias assessment. Secure data handling protocols.
- **Model Development/Training:** Use approved/secure libraries/environments. Document choices. Track experiments for reproducibility. Initial bias/fairness tests. Secure coding practices.
- **Model Validation/Testing:** Automated tests in CI pipelines: functional, performance, robustness, fairness/bias, security scans (SAST/DAST/SCA).
- **Pre-Deployment Review:** Formal review gate in CI/CD pipeline (automated evidence collection, manual sign-offs logged). Final risk assessment confirmation.
- **Deployment (CI/CD Pipeline):** Use Infrastructure as Code (IaC) with embedded security/compliance policies (e.g., OPA). Secure deployment environments. Secure secrets management.
- **Production Monitoring:** Continuous automated monitoring: operational metrics, model performance drift, data drift, concept drift, fairness metric drift, security anomalies. Alerts trigger incident response/retraining.
- **Incident Response:** Defined process for handling governance incidents (bias, privacy breach, vulnerability).
- **Decommissioning:** Secure procedures for retiring models, archiving docs/logs, deleting data per retention policies.

> **!** Embedding governance into automated MLOps pipelines (Governance-as-Code) is key to scaling responsible AI. It makes compliance efficient, consistent, verifiable, and the default path, a hallmark of higher EAMI maturity (Level 4/5).

8.6 Documentation, Controls, and Tooling

Robust documentation, verifiable controls, and enabling technology are essential supporting pillars:

- **Standardized Documentation:** Crucial for transparency, accountability, auditability, knowledge transfer, debugging. AI Office should define/mandate templates:
  - » **AI Use Case Registry:** Central inventory (purpose, status, risk tier, owners).
  - » **Risk Assessment Forms:** Consistent template for risk ID, assessment, mitigation.
  - » **Model Cards / AI FactSheets:** Concise summaries (purpose, performance, data, fairness, ethics, limits). Essential for transparency.
  - » **Datasheets for Datasets:** Documenting dataset provenance, characteristics, biases, labeling, usage recommendations.
  - » **Ethical Review & Approval Records:** Formal documentation of governance body deliberations/decisions.
  - » **Validation & Test Reports:** Standardized reports providing evidence of testing results.
  - » **Audit Logs:** Immutable, system-generated logs of critical actions (data access, training, deployment, predictions, approvals).
- **Effective Controls: Mix of technical and procedural controls:**

- » **Technical Controls:** Role-based access controls (RBAC), automated policy checks in CI/CD (e.g., OPA), mandatory security scans (SAST/DAST/SCA), automated monitoring/alerting.
- » **Procedural Controls:** Mandatory review gates/sign-offs, segregation of duties, documented incident response, regular audits.
- **Enabling Tooling: Leverage technology to support, automate, scale governance:**
  - » **MLOps Platforms:** (e.g., Azure ML, SageMaker, Vertex AI, MLflow) Increasingly include registry, tracking, pipeline automation, monitoring, governance features.
  - » **Enterprise GRC Platforms:** (e.g., ServiceNow GRC, OpenPages, Archer) Can sometimes manage AI risks, policies, controls, reporting.
  - » **Specialized AI Governance Platforms:** Growing market for AI model inventory, risk assessment workflows, governance process management, monitoring aggregation, reporting.
  - » **Bias/Fairness/Explainability Libraries/Tools:** (e.g., Fairlearn, AIF360, SHAP, LIME, Captum) Integrated into workflows for specific ethical AI checks.
  - » **Data Catalogs & Lineage Tools:** (e.g., Collibra, Alation, Azure Purview) Provide essential visibility into data usage, supporting governance.

8.7 Common Pitfalls in Framework Design

Avoid these frequent design errors:

- **Overly Complex or Bureaucratic:** Burdensome policies/processes stifle innovation and encourage bypassing ("shadow AI"). Aim for pragmatic, risk-adjusted controls.
- **Lack of Clarity or Ambiguity:** Vague policies, unclear standards, or ambiguous roles lead to inconsistent application and oversight gaps.
- **Insufficient Automation / Over-reliance on Manual Checks:** Manual governance is slow, costly, error-prone, and unscalable. Automate controls in MLOps where feasible.
- **Poor Tooling Integration & Strategy:** Tools not integrating well with development platforms create friction. Lack of coherent tooling strategy leads to fragmentation.
- **"Ivory Tower" Design:** Architecting the framework without sufficient input from those who must operate within it (data scientists, engineers, business users) leads to impractical requirements and resistance.
- **Failure to Communicate, Train & Champion:** Rolling out without explaining the "why," providing role-based training, and securing visible leadership championing results in poor adoption and compliance.

8.8 OmnioTech Case Study: Architecting the V1.0 Governance Framework

Setting the Scene:

With the AI Office Hub established, David Chen (Governance Lead) focuses on designing OmnioTech's V1.0 AI Governance Framework (Compass East), aiming for pragmatic controls suitable for their EAMI Level 2 starting point but enabling progress towards Level 3. He collaborates with Legal, Security, IT, and pilot teams.

Action & Application:

- **Components Defined:** David drafts initial framework components:
  - » **Policy:** Develops high-level "OmnioTech Responsible AI Principles" document based on 7 pillars (Chapter 7), circulated for executive sign-off.
  - » **Standards:** Creates initial mandatory standards (V1.0): 1) AI Risk Assessment & Tiering Methodology, 2) Minimum Data Handling Requirements for AI, 3) Minimum Model Documentation Standard (simplified Model Card template). Stored centrally.
  - » **Roles:** Formalizes AI Review Team charter (membership, scope, decision rights, cadence). Drafts initial RACIs for pilot governance steps.
  - » **Processes:** Defines initial documented workflows: 1) Project Intake & Risk Tiering, 2) Pre-Deployment Review process for AI Review Team (required for Tier 3+), 3) Basic Production Model Monitoring requirements.
  - » **Controls (Initial):** Primarily procedural: mandatory completion/upload of Risk Assessment/Model Card templates; manual review steps as checklist items in Azure DevOps release pipelines; standard Azure RBAC for access.
  - » **Documentation:** Mandates use of Risk Assessment/Model Card templates, stored centrally (SharePoint) linked to project record.
  - » **Tooling:** Leverages existing Azure DevOps for process tracking/checks; Azure ML for core platform/monitoring; plans to evaluate bias/XAI tools (Fairlearn, SHAP) during pilots.
- **Risk-Tiering Implemented:** Simple 3-tier model (Low, Medium, High) defined based on data sensitivity, impact, regulation. Hyper-Personalization pilot classified Tier 3 (Medium) requiring AI Review Team approval; Predictive Maintenance pilot Tier 2 (Low) requiring Head of AI Office & Product VP sign-off based on checklist/docs.
- **Integration Approach:** Focuses on integrating review steps as mandatory approval gates in existing Azure DevOps release pipelines for pilots.
- **Collaboration & Rollout:** David collaborates closely with Legal, Security, IT Platform, pilot leads during drafting. Develops and schedules initial V1.0 framework training for pilot teams and AI Review Team members.

Outcome & Progress:

OmnioTech establishes a foundational V1.0 AI Governance Framework architecture tailored to current needs and maturity. It prioritizes essential policies, standards, roles, processes, adopting a pragmatic risk-tiered approach. While initially reliant on manual controls, it sets the core structure and integrates key checkpoints into workflows. This provides necessary scaffolding for responsible AI deployment, sets the stage for achieving EAMI Level 3 in Governance, and allows future automation/tooling integration as maturity grows.

8.9 Conclusion: Building the Scaffolding for Trust

Designing the AI Governance Framework architecture (Compass East) translates responsible AI principles into a tangible, operational system of policies, standards, roles, processes, controls, and technology. By defining components, adopting risk-tiering, establishing clear roles, integrating governance into the AI lifecycle (MLOps), and avoiding common pitfalls, organizations build the necessary scaffolding for trustworthy AI at scale. With this blueprint established, subsequent chapters explore operationalizing specific aspects, starting with risk management in Chapter 9.

Key Takeaways:

- An AI Governance Framework operationalizes responsible AI principles (policies, standards, roles, processes, controls, tools).
- A risk-tiered approach tailors requirements, balancing rigor and agility.
- Clear roles (e.g., AI Review Board, Gov Lead) and responsibilities (RACI) are essential.
- Integrating governance into MLOps via automation ("Governance-as-Code") is key for scale/efficiency.
- Standardized documentation (Model Cards), controls, and tooling are vital supports.

Food for Thought / Application Exercise:

- Review core components (Section 8.2). Which exist (even informally) for AI governance in your org? Which are missing?
- Consider two different AI applications (varying risk). How would governance requirements differ based on risk-tiering?
- Who holds AI governance accountability? Are roles clear? Would a formal AI Review Board help?

82

AI Office - The Center of AI Excellence

~ CHAPTER 9 ~

Operationalizing AI Governance: Implementing Risk Management and Controls

"The best way to manage risk is to understand it, measure it, and integrate controls proactively, not reactively."

9.1 Introduction: Putting the Framework into Action

Chapter 8 laid out the essential architectural components of an AI Governance Framework – the policies, standards, roles, processes, and tools needed for responsible AI. However, a framework design on paper is only effective when it is actively implemented and integrated into the daily operations of AI development and deployment. Operationalizing governance means translating the architectural blueprint (Compass East) into practical, repeatable actions and embedding necessary controls seamlessly into the workflows where AI systems are built, tested, deployed, and monitored. Without effective operationalization, even the best-designed framework remains purely theoretical, unable to effectively mitigate risks (EAMM Risk Management) or ensure compliance (EAMM Compliance).

This chapter focuses on the practical implementation of the governance framework, a core responsibility within the Governance (East) domain. We will explore how to implement a structured AI risk management lifecycle, discuss strategies for integrating security and compliance checks directly into MLOps pipelines, emphasize the critical role of robust documentation, continuous monitoring, and periodic auditing, and consider challenges and best practices for scaling governance effectively as AI adoption grows. This operational focus is key to ensuring the governance framework moves beyond theory to become a living part of the organization's AI practice, contributing significantly to EAMI maturity in Governance, Risk Management, and Compliance dimensions.

Learning Objectives:

- Understand the key stages of an AI risk management lifecycle (identify, assess, treat, monitor).
- Learn practical strategies for integrating governance controls (security, compliance, ethics checks) into MLOps workflows.
- Recognize the importance of documentation, monitoring, and auditing for operationalizing governance effectively.
- Identify challenges and best practices for scaling AI governance across the enterprise.
- Appreciate the AI Office's role in driving the operationalization and continuous improvement of the governance framework.

9.2 The AI Risk Management Lifecycle

A cornerstone of operationalizing AI governance is implementing a systematic lifecycle approach to managing AI-specific risks. This typically involves several key stages, often executed iteratively:

1. **Identify Risks:** Proactively identify potential risks associated with an AI use case throughout its lifecycle. Consider risks related to data (bias, quality, privacy), model (performance, fairness, explainability, robustness), deployment (security, operations), usage (misinterpretation, unintended consequences, societal impact), and compliance (regulations). Techniques: structured brainstorming, reviewing past incidents, predefined risk checklists (aligned with Chapter 7 pillars), context analysis, "red teaming".

2. **Assess Risks:** Evaluate identified risks based on potential likelihood and impact (financial, reputational, operational, ethical, legal). Utilize the risk-tiering framework (Chapter 8) to categorize risks (Low, Medium, High). Goal: Understand significance to prioritize treatment.

3. **Treat Risks:** Develop strategies to address prioritized risks:
   » **Mitigation:** Implement controls (technical/procedural) to reduce likelihood/impact (e.g., bias mitigation, security protocols, human review). Most common strategy.
   » **Transfer:** Shift risk to a third party (e.g., insurance, outsourcing components with contracts). Accountability often remains.
   » **Acceptance:** Formally accept low-likelihood/low-impact risks where mitigation cost outweighs harm, requiring documented leadership approval.
   » **Avoidance:** Decide not to proceed with the application/feature if risks are too high or mitigation infeasible/costly.

4. **Implement Controls:** Integrate chosen mitigation strategies (controls) into relevant development, deployment, or operational processes and workflows. Translate strategy into concrete actions.

5. **Monitor & Review Risks:** Continuously monitor control effectiveness and system performance for emerging risks, control failures, or unexpected degradation (drift, bias, security issues). Regularly review/update the risk assessment based on monitoring data, incidents, audits, and environmental changes. This feedback loop ensures the process is dynamic.

1. IDENTIFY RISKS
Brainstorming
Checklists
Context Analysis

5. MONITOR & REVIEW
Performance
Monitoring
Audits
Feedback Loop

AI RISK MANAGEMENT LIFECYCLE

2. ASSESS RISKS
Likelihood
Impact
Tiering

4. IMPLEMENT CONTROLS
Technical
Procedural
Training

3. TREAT RISKS
Mitigate
Transfer
Accept
Avoid

Figure 9.1: AI Risk Management Lifecycle Diagram

D **DEFINITION**

AI Risk Management - The systematic and iterative process of identifying, assessing, treating, implementing controls for, and continuously monitoring risks specifically associated with the design, development, deployment, operation, and use of artificial intelligence systems. The primary goal is to minimize potential negative impacts while enabling responsible innovation and achieving strategic AI objectives. Effective risk management is key to the Governance (East) domain and critical for EAMI Risk Management maturity.

9.3 Integrating Governance into MLOps Workflows

To make governance efficient, scalable, and consistently applied, embed controls and checkpoints directly into MLOps pipelines (linking to Chapter 20, EAMM MLOps):

- **Automated Policy & Standard Checks:** Integrate checks into CI/CD pipelines (using OPA, custom scripts) to verify compliance automatically before merge/deploy (e.g., check for required docs, approved libraries, secure configs, schema compliance).

- **Automated Testing Gates for Governance:** Include mandatory, automated tests in CI/CD acting as quality gates:
  - » **Bias/Fairness Tests:** Execute checks (e.g., using Fairlearn, AIF360) against metrics/attributes; fail build if thresholds exceeded (EAMM Ethics).
  - » **Security Scans:** Integrate SAST (code), SCA (dependencies - e.g., Snyk), container image scanning (EAMM Security).
  - » **Robustness & Adversarial Tests:** Check performance against perturbations/known attacks (using ART, CleverHans).
  - » **Compliance & Privacy Checks:** Automate checks for data handling rules (e.g., detect PII), privacy requirements where feasible (EAMM Compliance).
- **Model Registry Integration:** Use the registry (MLflow, Azure ML, SageMaker) to associate governance metadata (risk assessments, Model Cards, validation reports, approval status). Configure registry to enforce workflows (prevent promotion without checks/approvals).
- **Integrated Monitoring & Alerting:** Configure production monitoring (Chapter 14) for operational metrics AND ongoing governance metrics: model performance drift, data drift, concept drift, fairness drift, security anomalies. Automated alerts trigger incident response or retraining pipelines.
- **Infrastructure as Code (IaC) Controls:** Use IaC tools (Terraform, Bicep) combined with policy-as-code (OPA via Checkov) to enforce security/compliance standards for AI infrastructure provisioning.

> Start by automating simpler, clearly defined governance checks in MLOps and increase sophistication gradually with tooling/maturity. Provide fast, actionable feedback to developers within workflows. Supplement automation with well-defined manual review gates for higher-risk applications or complex ethical judgments.

9.4 The Role of Documentation, Monitoring, and Auditing

Operationalizing governance relies heavily on robust, consistent practices:

- **Standardized Documentation:** Essential for transparency, accountability, auditability, knowledge transfer, debugging (EAMM Governance/Compliance). AI Office should define/mandate templates:
  - » **AI Project Charters / Use Case Registry:** Scope, objectives, owners, risk tier.
  - » **Risk Assessment Records:** Document risks, assessment, treatment, controls.
  - » **Model Cards / AI FactSheets:** Standard summaries (purpose, performance, data, fairness, ethics, limits).
  - » **Datasheets for Datasets:** Document dataset provenance, characteristics, biases, labeling, usage.
  - » **Validation & Testing Reports:** Evidence of performance, fairness, security, robustness testing.
  - » **Review & Approval Records:** Formal documentation of governance body decisions/actions.
  - » **Monitoring Plans & Results:** Outline production monitoring (metrics, frequency, thresholds); document performance, drift, incidents.
- **Continuous Monitoring in Production:** Critical governance control as systems interact with real data. Track:

- » **Model Performance:** Key accuracy/business metrics.
- » **Data Drift:** Changes in input data distributions.
- » **Concept Drift:** Changes in underlying input-output relationship.
- » **Fairness Metrics:** Track across subgroups for emerging bias.
- » **Operational Health:** Latency, error rates, throughput.
- » **Security Events:** Anomalous usage, access patterns. Automated alerts feed back into the risk management lifecycle (Stage 5).
- **Periodic Auditing:** Regular internal/external audits provide independent verification that policies/processes are followed and controls are effective. Scope might include:
  - » **Compliance Adherence:** Verify systems meet regulations (GDPR, AI Act).
  - » **Process Adherence:** Confirm governance processes (risk assessments, approvals) were followed.
  - » **Control Effectiveness:** Test if controls (access, pipeline checks, monitoring) function correctly.
  - » **Documentation Completeness & Accuracy:** Review artifacts (Model Cards, risk assessments) against standards. Audit findings drive improvements to the governance framework, processes, controls, training (feeding Compass Center)

9.5 Scaling Governance Across the Enterprise

Scaling governance as AI adoption expands presents challenges:

- **Consistency vs. Context:** Maintaining consistent standards while allowing necessary contextual adaptation.
- **Resource Bottlenecks:** Centralized manual reviews don't scale. Requires automation and distributed responsibility (federated model - Chapter 2).
- **Pace of Change:** Ensuring the framework adapts to new AI techniques, regulations, business needs.
- **Tooling & Integration Complexity:** Implementing and managing an enterprise-wide suite of integrated governance tooling.
- **Culture & Training at Scale:** Embedding consistent understanding and commitment across a large workforce (Chapter 26).

Strategies for Effective Scaling:

- **Embrace Automation:** Heavily invest in automating checks, controls, documentation generation, monitoring within MLOps. Primary lever for scale.
- **Federate Responsibility Appropriately (Hybrid Model):** Empower/train local "AI Governance Champions" or "Spokes" in BUs to handle lower-risk cases using central guidelines, escalating high-risk/complex cases to the central Hub.
- **Risk-Tiering is Crucial for Scale:** Allows central governance to focus deep-review resources on highest-risk applications, streamlining pathways for lower-risk ones.
- **Invest Scalably in Training & Awareness:** Implement modular, role-based online training. Foster CoPs. Ensure ongoing communication.
- **Continuous Improvement of the Framework:** Treat the framework as evolving. Regularly review effectiveness based on operational data/feedback; adjust policies, processes, tools (Compass Center).

9.6 OmnioTech Case Study: Operationalizing V1.0 Governance

Setting the Scene:

With the V1.0 Governance Framework architecture designed (Chapter 8), David Chen (OmnioTech's Governance Lead) focuses on operationalizing it for the initial PredMaint and HyperPers pilots, aiming for repeatable processes suitable for their EAMI Level 2 starting point.

Action & Application

- **Risk Management Lifecycle Implementation:**
  - » **Identify/Assess:** Pilot teams complete standardized Risk Assessment template during initiation, facilitated by David. HyperPers assessed Tier 3 (Medium Risk); PredMaint Tier 2 (Low Risk).
  - » **Treat/Implement:** Mitigation controls (enhanced consent for HyperPers, robust monitoring thresholds for PredMaint) documented as requirements in Azure DevOps backlogs.
  - » **Monitor:** Basic monitoring dashboards in Azure ML configured for pilot models tracking key metrics identified in risk assessments. Alerting thresholds set.
- **MLOps Integration (V1.0):**
  - » **Process Integration:** Mandatory governance checklist items integrated as manual gates in Azure DevOps release pipelines: 1) Link to approved Risk Assessment artifact. 2) Link to completed Model Card artifact. 3) For Tier 3 HyperPers, explicit "AI Review Team Approval" status flag required before production deployment stage.
  - » **Automation (Future):** Plans documented to investigate automated bias checks (Fairlearn) and policy checks (OPA) for Phase 2 MLOps rollout (targeting higher EAMI MLOps maturity).
- **Documentation & Auditing:**
  - » **Templates & Training:** Pilot teams trained on mandatory Risk Assessment/Model Card templates; completed versions stored centrally (SharePoint) linked to Azure DevOps.
  - » **Monitoring Responsibility:** Pilot teams review respective Azure ML monitoring dashboards weekly initially, reporting anomalies to AI Office.
  - » **Auditing Plan:** Internal audit review scheduled for 6 months post-pilot launch to assess adherence to V1.0 processes/standards.
- **Scaling Considerations (Initial):** Risk-tiered approach and AI Review Team process provide initial basis. David outlines role-based governance training modules for Year 2 rollout.

Outcome & Progress:

OmnioTech successfully operationalizes its V1.0 governance framework for pilots. Key risk steps followed/documented. Initial controls integrated as required workflow gates. Documentation enforced. This practical implementation (Compass East) moves governance from theory to tangible operations, ensuring baseline compliance/risk mitigation. It establishes repeatable processes that can be automated/scaled as EAMI maturity grows (Governance and Risk Management dimensions).

9.7 Conclusion: Making Governance Real and Sustainable

Operationalizing the AI Governance Framework (Compass East) translates principles into embedded actions. Implementing a structured risk management lifecycle, integrating controls into MLOps workflows (leveraging automation), enforcing robust documentation, establishing continuous monitoring, and conducting periodic audits are indispensable. While scaling presents challenges, a risk-tiered approach, automation, clear roles, and training enable responsible AI adoption efficiently. Successfully operationalizing governance ensures the framework actively guides responsible innovation, builds trust, and contributes fundamentally to achieving higher EAMI maturity. Having established operational governance, Chapter 10 delves deeper into ethical AI practice (bias, fairness).

Key Takeaways:

- Operationalizing governance means embedding policies, standards, controls into daily AI workflows.
- A structured risk management lifecycle (Identify, Assess, Treat, Implement, Monitor) is crucial.
- Integrating governance checks (security, compliance, ethics) into MLOps pipelines via automation enhances efficiency/consistency.
- Robust documentation (e.g., Model Cards), continuous monitoring (performance, drift, bias), and periodic auditing are essential for operational effectiveness/accountability.
- Scaling governance requires automation, risk-tiering, federated responsibility models, continuous training.

Food for Thought / Application Exercise:

- Consider an AI system in your org. Walk through the AI Risk Management Lifecycle (Section 9.2). Key risks? How assessed/treated?
- How are governance checks performed for AI projects? Manual reviews or automated MLOps integration? Biggest opportunity for automation?
- What level of documentation (like Model Cards) is expected/produced for AI models? How improve standardization/completeness?

~ CHAPTER 10 ~

Ethical AI Practice: Addressing Bias, Fairness, and Accountability

"The true measure of intelligence is not just the ability to solve problems, but the wisdom to solve them ethically."

10.1 Introduction: The Ethical Imperative in AI

Previous chapters established the overall AI governance framework (Chapter 8) and its operationalization (Chapter 9). Within the Governance (East) domain, however, lie some of the most complex and societally critical challenges: ensuring AI systems operate ethically, particularly concerning bias, fairness, transparency, and accountability. While governance provides the structural guardrails, ethics provides the moral compass guiding decisions within those guardrails.

As AI systems increasingly make or influence decisions affecting individuals' lives – from loan applications and job recruitment screening to medical diagnoses, content recommendations, and even judicial sentencing support – the potential for perpetuating or even amplifying existing societal biases and causing significant, unfair harm is substantial and well-documented. This chapter delves into the practical implementation of ethical AI principles, moving beyond abstract ideals to concrete actions. We will explore the common sources of bias in AI systems, discuss techniques and frameworks for detecting and mitigating bias, examine the nuanced concept of fairness and how it can be measured (often involving trade-offs), investigate methods for achieving transparency and explainability (XAI), and outline mechanisms for ensuring accountability when AI systems are deployed. Addressing these ethical dimensions proactively is not only the right thing to do but also essential for building trust, meeting regulatory expectations (like those emerging in the EU AI Act), and achieving genuine AI maturity aligned with EAMM Ethics criteria.

Learning Objectives:

- Understand the ethical imperative for addressing bias, fairness, transparency, and accountability in AI systems.
- Identify common sources of bias throughout the AI lifecycle (data, algorithm, human interaction).
- Learn practical techniques and frameworks for detecting and mitigating bias.
- Explore different definitions and metrics for assessing AI fairness, recognizing inherent trade-offs.
- Understand the importance of transparency and explainability (XAI) and common methods used.
- Recognize key mechanisms for establishing accountability in AI deployment and decision-making.

10.2 Understanding Bias in AI Systems

Bias in AI doesn't necessarily mean the algorithm is intentionally prejudiced; rather, it refers to systematic errors or skewed outcomes that disproportionately disadvantage certain groups. Bias can creep in at various stages:

- **Data Bias:** This is the most common source.
  - » **Historical Bias:** Data reflects existing societal biases or historical discrimination (e.g., loan approval data reflecting past discriminatory lending practices).
  - » **Representation Bias:** Certain groups are underrepresented or overrepresented in the training data (e.g., facial recognition trained primarily on one demographic).
  - » **Measurement Bias:** Features are measured or proxied differently across groups (e.g., using arrest records, which may be biased by policing practices, as a proxy for criminality).
  - » **Label Bias:** Subjective or biased human labeling of training data (e.g., labeling sentiment based on cultural norms).
- **Algorithmic Bias:** The AI algorithm itself might introduce or amplify bias.
  - » **Optimization Bias:** The model optimizes for an objective function that inadvertently favors certain outcomes or groups, even if the data isn't inherently biased.
  - » **Proxy Bias:** The model learns to rely on features that are highly correlated with protected attributes (like race or gender) even if those attributes aren't explicitly included.
- **Human Interaction Bias:** How humans interact with or interpret AI outputs can introduce bias.
  - » **Confirmation Bias:** Users tend to trust AI outputs that confirm their pre-existing beliefs, ignoring potentially fairer but counter-intuitive recommendations.
  - » **Automation Bias:** Over-reliance on AI outputs without critical human oversight, especially when the AI is incorrect.
  - » **Feedback Loop Bias:** If user interactions based on biased AI outputs are fed back into the system as new training data, the bias can be amplified over time (e.g., recommending content only to users who previously clicked similar recommendations).

> Bias is not always obvious. It requires deliberate effort, specialized tools, and diverse perspectives (including domain experts and representatives from potentially affected groups) to proactively identify and address potential biases throughout the AI lifecycle. Simply removing protected attributes (like race or gender) from the data often does not eliminate bias, as other correlated features (proxies) can allow the model to perpetuate discrimination indirectly. Addressing bias is a core component of the EAMM Ethics dimension.

10.3 Bias Detection and Mitigation Frameworks

Addressing bias requires a structured approach involving both detection and mitigation:

- **Bias Detection:**
  - » **Data Analysis:** Thoroughly analyze training data for representation gaps, skewed distributions, missing values across groups, and correlations between features and protected attributes. Use statistical tests and visualization techniques.
  - » **Fairness Metrics:** Quantify potential bias in model outcomes using various statistical fairness metrics (discussed in Section 10.4). Compare model performance across different demographic subgroups.
  - » **Bias Detection Tools:** Leverage specialized open-source libraries (e.g., IBM's AI Fairness 360 (AIF360), Google's What-If Tool, Microsoft's Fairlearn) or commercial platforms providing automated bias detection.
  - » **Qualitative Review:** Involve diverse human reviewers and domain experts to assess potential qualitative biases or harms statistical metrics might miss.
- **Bias Mitigation Techniques:** Mitigation strategies can be applied at different stages of the AI lifecycle:

Table 10.1: Bias Mitigation Techniques by Lifecycle Stage

Mitigation Stage	Goal	Example Techniques
Pre-processing	Modify the training data before model training to reduce inherent biases.	Re-sampling (oversampling minority groups, undersampling majority groups), re-weighting data points, generating synthetic data for underrepresented groups, suppressing biased features (carefully, due to potential proxy issues), dataset balancing techniques.
In-processing	Modify the model training algorithm or objective function to incorporate fairness constraints during training.	Adding fairness regularization terms to the loss function, adversarial debiasing (training a model to predict the target while simultaneously training it not to predict the sensitive attribute), fairness-aware algorithms (e.g., some variants of decision trees or SVMs).
Post-processing	Adjust the model's outputs after training to improve fairness outcomes.	Calibrating prediction thresholds differently for different groups to achieve equal opportunity or other fairness goals, rejecting or modifying certain predictions based on fairness criteria.

Bias mitigation is not a one-time fix. It often requires an iterative approach, combining multiple techniques. Continuously monitor fairness metrics in production (using MLOps - Chapter 20, Compass South) as data distributions can shift and new biases can emerge. Document all detection and mitigation steps thoroughly for accountability and transparency (Compass East).

10.4 Defining and Measuring Fairness

Fairness in AI is a complex, context-dependent, and often contested concept. There is no single, universally accepted definition or metric. Different mathematical formalizations of fairness often conflict (the "impossibility theorems" of fairness). Common categories of fairness metrics include:

Table 10.2: Categories of AI Fairness Metrics

Fairness Category	Focus	Example Metrics	Question Answered
Group Fairness	Ensuring statistical parity or equality across different demographic groups (by protected attributes).	Demographic Parity (Equal selection rates), Equalized Odds (Equal true/false positive rates), Equal Opportunity (Equal true positive rates).	Does the model treat different groups similarly on average?
Individual Fairness	Ensuring similar individuals are treated similarly by the model, irrespective of group membership.	Consistency metrics (measuring output variation for similar inputs), Counterfactual Fairness (checking if output changes if only protected attribute changes).	Does the model make consistent decisions for individuals who are alike in relevant aspects?
Counterfactual Fairness	Ensuring changing a protected attribute would not change the outcome for an individual.	Assessing model output changes when only the sensitive attribute is hypothetically flipped.	Would this individual have received the same outcome if they belonged to a different group?

> Selecting the appropriate fairness definition and metrics depends heavily on the specific use case, societal context, legal requirements, and potential harms associated with unfairness. This selection should involve careful ethical deliberation, stakeholder consultation (including potentially affected groups), and clear documentation of the chosen approach and rationale (Compass East). The AI Office Governance Lead or Ethics Committee often facilitates this critical decision. Be prepared to justify the chosen fairness definition and acknowledge inherent trade-offs.

10.5 Transparency and Explainable AI (XAI)

Transparency refers to understanding how an AI system functions, while Explainable AI (XAI) focuses specifically on providing human-understandable explanations for why a model made a particular prediction or decision. Both are crucial for building trust, enabling debugging, ensuring accountability, and facilitating meaningful human oversight (EAMM Ethics).

- **Levels of Transparency:**
  - » **System Transparency:** Documenting overall design, purpose, data sources, limitations (e.g., via Model Cards).
  - » **Algorithmic Transparency:** Understanding how the model works (e.g., tree structure vs. neural net weights). Challenging for "black boxes".
  - » **Decision Transparency (Explainability):** Providing reasons for specific outputs.
- **Common XAI Techniques:** (See the table below)

Table 10.3: Common Explainable AI (XAI) Technique Categories

XAI Technique Category	Description	Examples	Use Case	Limitations
Model-Specific	Techniques applicable only to certain types of models (often simpler ones).	Coefficients in Linear Regression, Decision Paths in Trees.	Understanding intrinsically interpretable models.	Not applicable to complex black-box models.
Model-Agnostic (Post-hoc)	Techniques applicable to any trained model, analyzing input-output relationships without seeing internals.	LIME (Local Interpretable Model-agnostic Explanations), SHAP (SHapley Additive exPlanations).	Explaining predictions of complex models locally (LIME) or globally (SHAP).	Explanations are approximations; can be computationally expensive; potential instability.
Example-Based	Explaining by showing similar examples from the training data.	Finding prototypes or influential training instances.	Providing intuitive justifications based on past data.	May not reveal underlying model logic; depends heavily on data quality.
Counterfactual Explanations	Showing the minimal changes to input features needed to flip the model's prediction.	"If your income was $X higher, your loan would have been approved."	Providing actionable recourse or understanding decision boundaries.	Can be computationally intensive to find; might not be unique or intuitive.

⭐ Choose XAI methods appropriate for the audience and required detail level (developer vs. customer). Document methods and limitations. Explainability does not guarantee fairness or accuracy.

10.6 Establishing Accountability Mechanisms

Accountability ensures clear responsibility and mechanisms for redress (EAMM Governance). Key mechanisms include:

- **Clear Roles & Responsibilities:** Defined in the governance framework (Chapter 8), ensuring individuals/ teams know their accountability for ethical AI practices (RACI charts).
- **Documentation Standards:** Mandating comprehensive documentation (Model Cards, Datasheets, Risk Assessments, Fairness Reports) creates an essential evidence trail.
- **Governance Oversight:** Empowering bodies (AI Review Board, Ethics Committee) to review high-risk projects, enforce standards, investigate concerns/incidents, mandate corrections.
- **Audit Trails & Logging:** Implementing robust, immutable logging of critical events (data usage, training runs, predictions, monitoring results, approvals) enables post-hoc analysis/investigation.
- **Impact Assessments:** Requiring formal assessments (DPIAs, EIAs) before deploying high-impact systems helps proactively identify, evaluate, mitigate risks, assigning mitigation responsibility.
- **Performance Monitoring & Alerting:** Continuously monitoring deployed models for performance and fairness degradation detects issues promptly, triggering accountability for investigation/remediation.
- **Redress Mechanisms:** Establishing clear channels for affected individuals to seek explanations, challenge outcomes, or request human review/corrections.

10.7 OmnioTech Case Study: Implementing Ethical Practices

Setting the Scene:

As OmnioTech develops its PredMaint and HyperPers pilots, David Chen (Governance Lead) works proactively with the teams to implement practical ethical controls based on the V1.0 governance framework (Compass East), aiming to meet EAMI Ethics dimension targets for Level 3.

Action & Application

- **Bias/Fairness (HyperPers):**
  - » **Detection:** Team uses AIF360 toolkit in Azure ML pipeline to analyze customer data for representation bias (age, location). Evaluates initial model predictions using Demographic Parity and Equal Opportunity metrics. Finds slight under-representation for rural customers and lower offer relevance for younger demographics.
  - » **Mitigation:** Pre-processing uses data augmentation (synthetic data) for rural segment. Plans to explore post-processing threshold adjustments for younger segment fairness metric during pilot validation if needed after retraining. Mitigation steps documented in Model Card.
- **Transparency (HyperPers):**
  - » **Documentation:** Team completes standardized Model Card template, including fairness testing details, limitations, data sources.
  - » **Explainability:** Implements SHAP analysis during validation. Generates feature importance reports (e.g., past purchase influence) for Marketing team review, increasing trust.
- **Accountability (Both Pilots):**
  - » **Roles:** Updated RACIs clarify developer accountability for fairness tests/docs, and assign Business Own-

er (VP Product for PredMaint, VP Marketing for HyperPers) accountability for final deployment approval based on governance review.

» **Oversight:** Medium Risk (Tier 3) HyperPers pilot undergoes formal documented review by AI Review Team before deployment authorization. Lower Risk (Tier 2) PredMaint pilot requires documented sign-off from Head of AI Office and VP Product based on checklist/docs.

» **Monitoring:** Production monitoring plans explicitly include checks for fairness metric drift (HyperPers, quarterly) and alert accuracy across appliance types (PredMaint, monthly), with assigned owners for investigating alerts.

Outcome & Progress:

By actively implementing bias detection tools (AIF360), defining fairness metrics, requiring standardized documentation (Model Cards), exploring XAI (SHAP), and assigning explicit accountability via governance reviews, OmnioTech embeds ethical considerations into its pilot lifecycle (Compass East practices). This strengthens their governance posture (improving EAMI Ethics score towards Level 3) and builds internal confidence for responsible AI deployment.

10.8 Conclusion: Embedding Ethics into AI Operations

Ensuring ethical AI is an integral part of robust AI governance (Compass East) and responsible development. By proactively addressing biases, defining/measuring fairness contextually, striving for appropriate transparency/explainability (XAI), and establishing clear accountability mechanisms throughout the lifecycle, organizations reduce ethical risks and build trustworthy AI systems. While challenges exist, embedding these principles into operational practices, supported by tools and training, is essential for sustainable success, meeting regulatory expectations, maintaining trust, and realizing AI's positive potential. Having considered these critical ethical dimensions, Chapter 11 focuses on AI security strategy.

Key Takeaways:

- Ethical AI requires proactively addressing bias, fairness, transparency, and accountability.
- Bias can originate from data, algorithms, or human interaction; detection/mitigation need deliberate effort.
- Fairness is context-dependent; select appropriate metrics and document trade-offs.
- Transparency (documentation) and Explainability (XAI tools like SHAP/LIME) build trust and enable oversight.
- Accountability is established via roles, documentation, governance oversight, audit trails, impact assessments, monitoring, and redress mechanisms.

Food for Thought / Application Exercise:

- Consider an AI system your org uses/develops. Primary sources of potential bias? How detected?
- For that system, what's the most relevant "fairness" definition? Which metrics apply?
- What transparency/explainability level is needed for different stakeholders? Which XAI techniques might suit?

~ CHAPTER 11 ~

Architecting AI Security Strategy: Protecting Models, Data and Infrastructure

"Security in AI is not just about protecting data.
It's about protecting the integrity and trustworthiness of automated decisions."

11.1 Introduction: The Unique Security Challenges of AI

Chapters 7 through 10 explored critical aspects of AI governance, focusing heavily on ethics, fairness, transparency, and accountability. This chapter addresses another vital pillar within the Governance (East) domain: Security & Safety. While traditional cybersecurity principles remain essential, AI systems introduce unique vulnerabilities and significantly expand the potential attack surface, requiring a specialized, proactive, and integrated security strategy (EAMM Security).

AI models themselves can be targets: manipulated via adversarial inputs, poisoned during training, or stolen through extraction techniques. The vast datasets used are prime targets for theft or compromise. The complex infrastructure underpinning AI (cloud platforms, distributed frameworks, specialized hardware, MLOps pipelines) presents its own security challenges.

Crucially, the impact of a compromised AI system – controlling critical decisions or physical processes – can be far more severe than traditional software breaches, potentially leading to incorrect medical diagnoses, biased decisions causing lawsuits, financial losses, or physical safety incidents. This chapter delves into AI-specific security challenges, outlines principles for a robust AI security strategy, discusses Secure Machine Learning (SecML / MLSecOps) practices, explores common attack vectors and defenses, and emphasizes tailored incident response planning. Establishing a dedicated AI security strategy is crucial for protecting assets, maintaining trust, ensuring compliance, and achieving higher EAMI Security maturity.

Learning Objectives:

- Understand unique security challenges and attack surfaces introduced by AI systems.
- Learn core principles for architecting AI security strategy, extending traditional cybersecurity.
- Explore key SecML / MLSecOps practices for integration throughout the AI lifecycle.
- Identify common AI-specific attack vectors (data poisoning, adversarial attacks, model theft) and defense strategies.
- Appreciate the need for AI-specific incident response plans.
- Recognize the collaborative role of the AI Office with Cybersecurity (CISO) teams in defining and enforcing AI security standards.

11.2 Unique AI Security Challenges

AI systems introduce security challenges needing specific attention beyond standard practices:

- **Expanded and Dynamic Attack Surface:** The complex AI lifecycle (data collection to monitoring) with diverse tools/platforms creates numerous potential vulnerabilities. Models evolve via retraining, changing the surface.
- **Data Poisoning Vulnerabilities:** Malicious actors can inject manipulated data into training sets, subtly skewing model behavior, embedding biases, or creating backdoors. Detection is difficult, especially with large datasets or federated learning.
- **Adversarial Attacks at Inference Time:** Deployed models can be targeted:
  - » **Evasion Attacks:** Subtle, often imperceptible input perturbations cause misclassification (e.g., fooling object detectors, bypassing content filters).
  - » **Extraction Attacks (Model Stealing):** Repeated API queries infer model parameters or train a functional equivalent, stealing IP.
  - » **Membership Inference Attacks:** Determine if specific individual data was in the training set by analyzing model outputs (privacy breach).
- **Model Integrity and Confidentiality Risks:** Trained models (valuable IP) must be protected from unauthorized access, theft, or tampering.
- **Infrastructure and Dependency Complexity:** AI relies on complex distributed infrastructure (Kubernetes, Spark) and numerous open-source libraries. Misconfigurations or dependency vulnerabilities pose significant risks. Managing the software supply chain securely is challenging.
- **Potential Security vs. Explainability Trade-offs:** XAI techniques making models transparent might inadvertently reveal information exploitable for adversarial attacks.
- **Lack of AI-Specific Security Expertise and Standards:** Many organizations lack personnel skilled in both AI/ML and cybersecurity. Established security standards are still evolving for AI risks.

Traditional cybersecurity measures (firewalls, IDS, endpoint security, AppSec testing) are necessary but insufficient alone to address unique AI threats. A dedicated, AI-aware security focus (Compass East) is mandatory.

11.3 Core Principles of AI Security Strategy

A robust AI security strategy extends classic principles (CIA Triad) and integrates AI-specific concepts:

- **Confidentiality:** Protecting sensitive training/inference data, proprietary models/parameters, and confidential predictions from unauthorized disclosure/access.
- **Integrity:** Ensuring trustworthiness/accuracy of data throughout the lifecycle (guarding against poisoning/corruption) and guaranteeing the model functions as intended without tampering.
- **Availability:** Ensuring AI systems and supporting infrastructure are available and performant when needed, resilient to disruptions.
- **Robustness:** Designing AI systems resilient against adversarial perturbations, noisy inputs, distribution shifts, maintaining acceptable performance/safety under non-ideal/malicious conditions.
- **Accountability & Traceability:** Maintaining clear, comprehensive logs/audit trails for critical activities (data access, training, deployment, predictions, security events) enabling investigations and compliance demonstration.
- **Privacy Preservation:** Integrating privacy-enhancing techniques (PETs) proactively to protect sensitive personal information (aligns with Privacy pillar - Chapters 7, 12).
- **Secure Lifecycle Management (SecML / MLSecOps):** Systematically embedding security considerations, controls, testing throughout the AI/ML lifecycle (MLOps - Chapter 20), making security integral.

11.4 Secure Machine Learning (SecML / MLSecOps) Practices

Integrating security into the MLOps lifecycle (Compass South / East) is key for secure AI by default:

- **Secure Data Acquisition & Handling:** Strong access controls (RBAC) on data stores/feature stores. Encrypt data (rest/transit). Data validation pipelines for anomalies/poisoning indicators. Track provenance/lineage. Anonymize/pseudonymize where appropriate.
- **Secure Development Environment & Practices:** Secure coding standards (OWASP adapted). Mandate approved, vetted libraries (SCA in CI). Robust input validation (prevent injection). Secure development environments (access controls).
- **Secure Model Training:** Train in secure, isolated environments (cloud VNETs). Strict access controls (infra, tracking platforms, artifact storage). Log parameters, code versions, lineage meticulously. Consider differential privacy or robust algorithms for high-risk cases.
- **Model Hardening & Validation:** Rigorously test models for accuracy AND security/robustness against relevant adversarial attacks (using ART, CleverHans). Perform security validation (pen testing endpoints) alongside performance validation. Consider model signing/watermarking for integrity.
- **Secure Deployment (CI/CD Integration):** Integrate automated security checks in MLOps CI/CD: SAST (code), SCA (dependencies), container image scanning, IaC security policy checks (Checkov). Mandatory security review gates (automated evidence, manual sign-offs).
- **Secure Inference Endpoints:** Protect deployed APIs (REST, managed endpoints) with strong authentication, authorization (RBAC), strict input validation, rate limiting (prevent abuse/extraction), potentially WAFs with AI-specific rules.
- **Continuous Monitoring & Logging:** Comprehensive, immutable logging (API calls, predictions, data access, health, security events). Continuously monitor deployed models/infra for anomalous behavior (performance drops, drift, suspicious queries), security alerts. Feed alerts to incident response.
- **Supply Chain Security:** Vet third-party data, pre-trained models, libraries, tools. Maintain Software Bill of

Materials (SBOM) for dependencies.

INTEGRATING SECURITY INTO THE MLOps LIFECYCLE (MLSecOps)

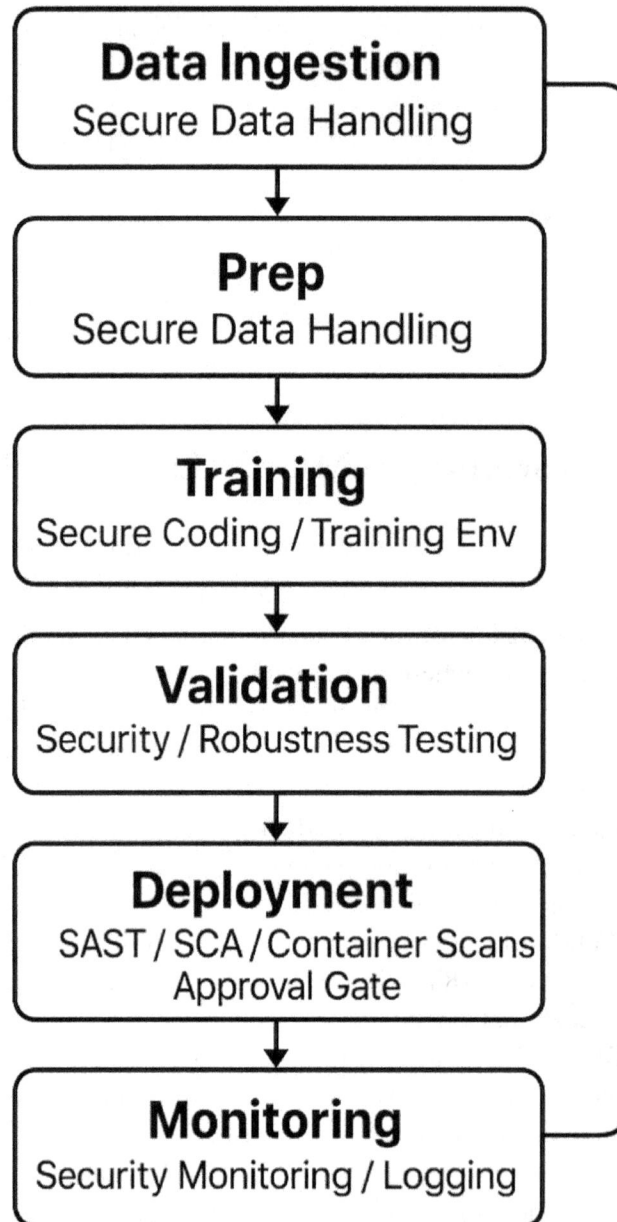

Data Ingestion
Secure Data Handling

Prep
Secure Data Handling

Training
Secure Coding / Training Env

Validation
Security / Robustness Testing

Deployment
SAST / SCA / Container Scans
Approval Gate

Monitoring
Security Monitoring / Logging

Figure 11.1: MLSecOps Lifecycle Diagram

Foster close collaboration and shared responsibility between AI/ML engineers, central Cybersecurity (CISO), and IT operations. Leverage cybersecurity expertise for AI threat modeling, pen testing, control selection, SIEM integration, incident response planning. Define clear AI security roles/responsibilities in RACIs.

11.5 Common AI Attack Vectors and Defenses

Understanding specific attack types helps design targeted defenses:

Table 11.1: Common AI Attack Vectors and Defenses

Attack Vector	Description	Example Defenses
Data Poisoning	Attacker injects malicious data into the training set to compromise model behavior or create backdoors.	Input data validation & outlier/anomaly detection; data provenance tracking; differential privacy during training; robust training algorithms; monitoring training data sources; secure data ingestion channels.
Evasion Attack	Attacker crafts slightly perturbed inputs at inference time (often imperceptible) to cause misclassification or bypass detection systems.	Adversarial training; input sanitization/ validation/transformation (smoothing, feature squeezing); defensive distillation; using more robust model architectures; detecting adversarial inputs; rate limiting queries.
Model Extraction / Stealing	Attacker queries deployed API repeatedly to infer parameters or train a functional substitute model (IP theft).	API rate limiting/throttling; monitoring query patterns for suspicious behavior; output perturbation (noise); differential privacy on outputs; watermarking models; deploying less informative APIs (e.g., top class vs. probabilities).
Membership Inference	Attacker tries to determine if specific individual data was in the training set by analyzing model outputs (privacy violation).	Differential privacy during training; techniques to reduce overfitting (regularization, dropout); using aggregated/federated data; robust access controls.
Model Inversion / Attribute Inference	Attacker attempts to reconstruct sensitive training data features/ attributes by analyzing model outputs or accessing parameters.	Differential privacy; limiting output granularity/ information; secure aggregation in federated learning; robust access controls on trained models.

Traditional Software & Infra Attacks	Exploiting standard vulnerabilities (OS, cloud config, network, libraries, APIs - SQLi, XSS, insecure keys, unpatched OS).	Apply standard cybersecurity best practices: patching, secure config management, vulnerability scanning (SAST, DAST, SCA, container), network security, strong auth/authZ, secure coding.

11.6 Incident Response for AI Security Events

Organizations need an AI-specific Incident Response (IR) plan integrated with the corporate plan:

- **Detection:** How will AI-specific attacks (poisoning, evasion, extraction) or failures (biased outputs, data leakage) be detected? Requires specialized monitoring beyond traditional tools (model behavior, data stats, API usage).
- **Containment:** Immediate steps to isolate compromised AI system/endpoint? (Revoke keys, scale down, disable features).
- **Analysis & Forensics:** Who has combined AI/ML + cybersecurity expertise to analyze attack vector, impact (bias introduced? IP stolen? data accessed? incorrect predictions?), root cause, compromise extent? Requires detailed logs, lineage.
- **Remediation:** Secure procedures? (Retrain from clean snapshot, patch vulnerabilities, update configs, deploy robust defenses, rollback model).
- **Communication & Reporting:** Who notified internally (Legal, Compliance, Privacy, PR, Leadership) and externally (Regulators, customers) per policies/law?.
- **Post-Mortem & Learning:** Conduct blameless reviews to capture lessons learned (attack effectiveness, defense/response effectiveness), drive improvements (AI security strategy, governance, MLOps, IR plan).

> ★ Regularly conduct tabletop exercises or simulated drills for plausible AI security scenarios (adversarial attack, poisoning, extraction attempt). Tests IR plan effectiveness, clarifies roles, identifies gaps before real incidents.

11.7 OmnioTech Case Study: Implementing Initial AI Security Measures

Setting the Scene:

As OmnioTech pilots PredMaint and HyperPers, Priya Sharma (AI Office Head) and David Chen (Gov Lead) collaborate with CISO Ravi Singh to implement baseline AI security controls (V1.0 governance, Compass East), aiming for EAMI Security Level 3.

Action & Application:

- **Risk Assessment Integration:** Standard Risk Assessment template updated with AI security threat prompts (poisoning - Low risk initially; evasion - Medium risk for HyperPers; extraction - Low risk). CISO team reviews assessments for Tier 3 HyperPers pilot.
- **Secure Development Standards (Initial):** Basic secure coding guidelines (Python/OWASP) distributed. Mandated use of approved libraries in Azure ML (enforced via code review initially). Strong Azure RBAC for sensitive data/workspaces (least privilege).
- **MLOps Integration (V1.0):**
  - » **Pre-Deployment Checks:** Azure DevOps pipeline checklist includes mandatory CISO team sign-off on inference endpoint config (network, auth) and data access permissions.
  - » **Monitoring: Basic** Azure Monitor alerts configured for unusual API errors/latency. CISO ops team added to alert distribution. Plans noted for more sophisticated anomaly detection later.
- **Pilot-Specific Controls:**
  - » **HyperPers:** Input validation logic added to API endpoint. API rate limiting configured via Azure API Management.
  - » **PredMaint:** Access to raw sensor data tightly restricted via Azure RBAC. Model output (probability score) deemed less sensitive.
- **Incident Response (Initial):** Simple procedure drafted: Monitoring alerts trigger notification to AI Office/ MLOps Lead, Gov Lead, CISO via emergency channel. CISO ops team leads initial investigation/containment, requesting AI/ML engineer support as needed.

Outcome & Progress:

OmnioTech implements foundational AI security controls integrated with V1.0 governance. Lacking advanced features (automated adversarial testing), these measures establish security as a key consideration, integrate CISO team, leverage cloud controls (RBAC, monitoring), and provide baseline protection/IR awareness. This contributes positively to their EAMI Security dimension score (towards Level 3), acknowledging further maturity needs more specialized SecML/MLSecOps tooling and automation.

11.8 Conclusion: Building Secure and Resilient AI

Securing AI requires a dedicated strategy extending traditional cybersecurity (Compass East). Understanding unique vulnerabilities (data, models, pipelines, infra) allows targeted defenses. Integrating security throughout the MLOps lifecycle (SecMLOps), robust secure data handling, adversarial testing, diligent endpoint security, and tailored incident response plans are critical for mature AI security (EAMM Security). Strong collaboration between AI Office, developers, MLOps, and Cybersecurity is essential. Prioritizing security alongside ethics/compliance builds secure, resilient, trustworthy AI systems capable of delivering value without unacceptable risk. Next, Chapter 12 examines navigating AI regulations and compliance.

Key Takeaways:

- AI introduces unique security risks (data poisoning, adversarial attacks, model theft) beyond traditional cybersecurity.
- AI security strategy extends CIA principles to include Robustness and Secure Lifecycle Management (SecML/MLSecOps).

- Key SecML practices involve secure data handling, secure development/training, model hardening/validation, secure deployment (CI/CD checks), and continuous monitoring.
- Defenses must target specific attack vectors (e.g., adversarial training for evasion, rate limiting for extraction).
- AI-specific incident response plans and close collaboration between AI teams and Cybersecurity are essential.

Food for Thought / Application Exercise:

- Consider an AI system in your org. Top 2-3 potential AI-specific security vulnerabilities?
- How are security considerations currently integrated into your AI development lifecycle? Specific checks for data security, model robustness, secure deployment?
- How prepared is your org to detect/respond to an AI security incident (adversarial attack, model extraction)?

CHAPTER 12

Navigating the Regulatory Landscape: AI Laws and Compliance Management

> "Compliance is not just about following the rules.
> It's about building systems that inherently operate within ethical and legal boundaries."

12.1 Introduction: The Evolving Legal Landscape for AI

Previous chapters in Part 3 explored core AI governance pillars like ethics (Chapter 10) and security (Chapter 11). This chapter addresses another critical aspect of the Governance (East) domain: navigating the complex and rapidly evolving landscape of laws, regulations, and standards governing artificial intelligence. As AI becomes more pervasive and impactful, governments and regulatory bodies worldwide are increasingly focused on establishing legal frameworks to mitigate risks, protect citizens, and ensure responsible innovation. Operating AI systems, especially those handling personal data or making significant decisions, requires organizations to understand and comply with a growing patchwork of requirements (EAMM Compliance).

This includes established data protection laws like GDPR and CCPA, as well as emerging AI-specific regulations like the EU AI Act and evolving standards from bodies like NIST and ISO. Failure to comply can result in severe financial penalties, legal action, operational disruptions, and significant reputational damage. This chapter provides an overview of key regulatory trends and major legal frameworks impacting enterprise AI. We will discuss strategies for operationalizing compliance within the AI governance structure, address challenges like data residency and conflicting jurisdictions, and outline the essential role of the AI Office in proactively managing regulatory risk. Staying abreast of and ensuring compliance with this dynamic legal landscape is crucial for sustainable AI deployment and achieving higher EAMI maturity in the Compliance dimension.

Learning Objectives:

- Understand key trends and drivers behind the evolving regulatory landscape for AI.
- Recognize major existing and emerging AI-relevant legal frameworks (e.g., GDPR, CCPA, EU AI Act, NIST AI RMF, ISO 42001).
- Appreciate challenges posed by data residency and conflicting international regulations.
- Learn strategies for operationalizing AI compliance within the governance framework and MLOps lifecycle.
- Understand the AI Office's role collaborating with Legal/Compliance to manage regulatory risk.

12.2 Key Regulatory Trends and Drivers

The increasing focus on AI regulation stems from several key drivers:

- **Potential for Harm:** Recognition of AI's potential to cause harm through bias, discrimination, privacy violations, safety failures, or opacity.
- **Protection of Fundamental Rights:** Desire to safeguard rights (privacy, non-discrimination, due process, safety).
- **Building Public Trust:** Establishing rules aims to foster public trust and confidence.
- **Economic Competitiveness:** Regulations can aim to create level playing fields or position regions as trustworthy AI leaders.
- **Geopolitical Factors:** Concerns about national security, data sovereignty, and influence shape approaches.

Key trends include:

- **Risk-Based Approaches:** Many regulations (EU AI Act) categorize AI based on risk, applying stricter requirements to higher-risk applications.
- **Focus on High-Risk Applications:** Intense scrutiny on critical infrastructure, medical devices, employment, law enforcement, rights administration.
- **Emphasis on Transparency & Documentation:** Common requirements for documenting data, design, testing, risk assessments.
- **Data Governance Linkage:** Strong links to data protection laws (GDPR) regarding data quality, minimization, lawful basis.
- **Global Divergence:** Different jurisdictions (EU, US, China) taking distinct approaches, creating complexity.
- **Rise of Standards:** Increased development of voluntary technical standards (ISO/IEC, NIST) for practical guidance.

12.3 Major Regulatory Frameworks and Standards

Organizations deploying AI globally need awareness of key frameworks:

Table 12.1: Major AI-Relevant Regulatory Frameworks and Standards

Framework	Scope / Focus	Key AI-Relevant Aspects	Status / Applicability
EU GDPR (General Data Protection Regulation)	Comprehensive data protection law for EU residents' personal data.	Strict rules on lawful basis, consent, minimization, data subject rights (access, rectification, erasure), DPIAs for high-risk processing, restrictions on automated decision-making.	Enforceable since 2018
CCPA/CPRA (California Privacy Rights Act)	Data privacy law for California residents, granting rights over personal information.	Rights to know, delete, opt-out of sale/sharing, limit use of sensitive personal info. Impacts AI personalization data use, requires transparency.	CPRA amendments effective 2023
EU AI Act	Landmark comprehensive AI regulation establishing a risk-based framework (Unacceptable, High, Limited, Minimal Risk).	Prohibits certain AI (unacceptable risk). Strict requirements for High-Risk AI (data governance, docs, transparency, human oversight, accuracy, robustness, cybersecurity). Requires registration & conformity assessment.	Adopted 2024, phased rollout
NIST AI RMF (AI Risk Management Framework)	Voluntary US framework providing guidance on managing AI risks.	Structured process (Govern, Map, Measure, Manage) & guidance on trustworthiness characteristics (valid/reliable, safe, secure/resilient, accountable/transparent, explainable/interpretable, privacy-enhanced, fair).	Voluntary framework (V1.0 2023)
ISO/IEC 42001	International standard specifying requirements for an AI Management System (AIMS).	Certifiable management system standard focusing on organizational processes, policies, risk assessment, objectives for responsible AI, integrating with other ISO standards (e.g., 27001).	Published standard (2023)
Other Jurisdictions	Canada (AIDA - proposed), China (sector-specific rules), UK (pro-innovation approach), Brazil (LGPD), various US state laws.	Organizations must track relevant laws in all operating jurisdictions or where AI impacts individuals.	Varies / Evolving

> (!) This landscape is highly dynamic. New regulations, interpretations, and standards emerge constantly. Maintaining regulatory awareness requires dedicated resources (Legal, Compliance, AI Office) and potentially external monitoring services.

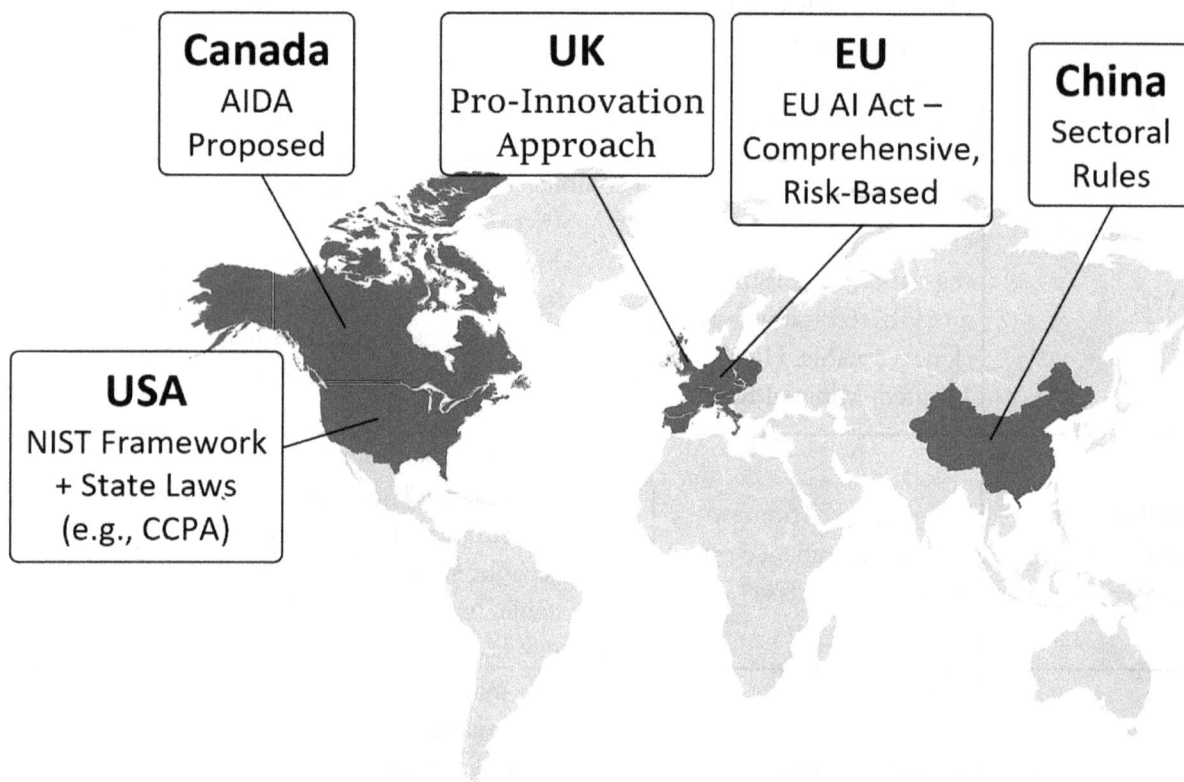

Figure 12.1: Global AI Regulatory Landscape Map

12.4 Challenges: Data Residency and Conflicting Requirements

Operating globally introduces specific compliance challenges:

- **Data Residency & Sovereignty:** Some countries mandate data remain within borders, complicating global AI systems (requiring infrastructure replication or limiting training data).
- **Cross-Border Data Transfers:** Transferring personal data across borders (e.g., EU to US) is heavily regulated (e.g., GDPR adequacy, SCCs). Requires appropriate legal/technical safeguards.
- **Conflicting Requirements:** Different regulations may impose varying requirements (transparency, data access, risk assessment). Harmonization requires careful legal analysis, often adopting the strictest standard

globally ("highest watermark") where feasible.

12.5 Operationalizing AI Compliance

Ensuring compliance requires embedding requirements into the AI governance framework and operational workflows (Compass East integration):

- **Policy & Standard Integration:** Incorporate requirements from relevant regulations (GDPR, AI Act risk levels) into internal AI policies, standards, guidelines. Ensure risk-tiering (Chapter 8) aligns with regulatory definitions.
- **Mandatory Assessments:** Integrate required assessments (DPIAs under GDPR, AI Impact Assessments for high-risk systems) into project initiation or pre-deployment phases.
- **Compliance Checks in MLOps:** Automate checks in CI/CD pipelines (approved libraries, required docs, data type checks) where possible.
- **Documentation & Record Keeping:** Maintain meticulous records (risk assessments, DPIAs, reviews, agreements, validation reports, audit trails) as required by regulations (often for years). Crucial for audits.
- **Training & Awareness:** Ensure relevant personnel are trained on applicable regulations and internal compliance procedures.
- **Vendor Risk Management:** Include AI compliance requirements/clauses in third-party contracts. Conduct due diligence on vendor practices.
- **Cross-Functional Collaboration:** Foster strong collaboration between AI Office, Legal, Compliance, Privacy, Cybersecurity to interpret requirements, implement controls, manage incidents.

> Treat compliance as a continuous process. Regularly review/update policies, standards, controls based on new regulations, interpretations, audits, incidents. Embed compliance from the start ("Compliance-by-Design").

12.6 The AI Office's Role in Compliance Management

While Legal/Compliance hold ultimate accountability, the AI Office plays a crucial coordinating/enabling role (Compass East support):

- **Translating Requirements:** Help translate legal/regulatory requirements into practical technical standards/ guidelines for AI teams.
- **Facilitating Assessments:** Coordinate execution of required DPIAs/AI Impact Assessments, ensuring expert input.
- **Developing Standards & Templates:** Create standardized templates (docs, risk assessments) incorporating compliance requirements.

- **Promoting Awareness & Training:** Collaborate with HR/Compliance on AI-specific regulatory training.
- **Monitoring & Reporting:** Track AI project compliance status, report metrics to governance bodies.
- **Staying Informed:** Actively monitor regulatory landscape, advise leadership/teams on changes.
- **Tooling Evaluation:** Assess/recommend tools to automate/streamline compliance checks/documentation.

12.7 OmnioTech Case Study: Addressing Initial Compliance Needs

Setting the Scene:

As OmnioTech prepares pilots, David Chen (Governance Lead) works with Legal Counsel/Privacy Officer to ensure the V1.0 governance framework addresses immediate compliance needs, particularly GDPR given their European customer base (Compass East activity).

Action & Application:

- **GDPR Focus:** Confirm both pilots process personal data subject to GDPR.
  - » **HyperPers:** Processing customer data requires consent management (via existing preference center, verified) and data minimization. Formal DPIA completed (documenting risks, mitigation - anonymization explored, clear opt-out).
  - » **PredMaint:** Processing user-linked appliance data also under GDPR. DPIA focuses on data minimization, stream security, clear privacy policy communication.
- **AI Act Awareness:** Team reviews EU AI Act risk categories. Tentatively classify HyperPers as potentially "Limited Risk" (transparency needed), PredMaint as potentially "High Risk" (if failure prediction impacts safety). Flag need for monitoring Act requirements as pilots scale.
- **Policy Integration:** V1.0 Data Handling for AI Standard explicitly incorporates GDPR principles (lawful basis, minimization, rights access). Risk Assessment template includes privacy/regulatory prompts.
- **Documentation:** Model Card template includes fields for data sources, purpose, links to DPIAs/privacy notices.
- **Training:** Initial AI awareness training includes module on GDPR basics and data handling policies.

Outcome & Progress:

Proactively addressing GDPR via DPIAs, policy integration, and documentation standards mitigates immediate compliance risks for pilots (EAMM Compliance). Early AI Act consideration prepares for future obligations. This structured approach enhances their EAMI Compliance score and demonstrates responsible governance.

12.8 Conclusion: Building Compliance into the AI Fabric

Navigating the complex, evolving regulatory landscape is indispensable for responsible AI governance (Compass East). Organizations must proactively understand laws (GDPR, CCPA, AI Act), integrate requirements into internal policies, standards, MLOps processes, and implement robust documentation/monitoring (EAMM Compliance). Challenges like data residency/conflicting rules require careful planning. The AI Office, collaborating with Legal/Compliance, translates requirements, facilitates assessments, and ensures compliance is continuous and embedded. Building compliance in mitigates risks, builds trust, and enables responsible AI operation globally. Having addressed core governance (ethics, security, compliance), Part 4 focuses on Performance & Value Delivery (West).

Key Takeaways:

- Navigating AI regulations (GDPR, AI Act, etc.) is critical for compliance and trust.
- Operationalize compliance by integrating requirements into policies, standards, MLOps, assessments (DPIAs), documentation, training, vendor management.
- Challenges include data residency and conflicting international rules.
- Compliance is continuous; use a "Compliance-by-Design" approach.
- AI Office collaborates with Legal/Compliance to manage regulatory risk.

Food for Thought / Application Exercise:

- What are the 1-2 most significant AI-related regulations impacting your organization today or in the near future?
- How are compliance requirements currently incorporated into your AI development or procurement processes? Is it proactive or reactive?
- How does your organization stay informed about the rapidly evolving global AI regulatory landscape?

PART 4 - AI PERFORMANCE & VALUE DELIVERY

~ CHAPTER 13 ~

Defining and Measuring AI-Driven Business Value: Focusing on Impact

"What gets measured gets managed. And what gets managed gets improved."
— Peter Drucker (adapted)

13.1 Introduction: Beyond Technical Metrics – The Pursuit of Value

Welcome to Part 4, where we turn our attention to the Western point of the AI Office Compass™ Framework: Performance & Value Delivery. Parts 2 and 3 focused on setting the right strategic direction (North) and establishing the necessary governance guardrails (East). Now, we address the critical question that ultimately determines the success and sustainability of any enterprise AI program: "Is it actually delivering meaningful business value?"

While technical performance metrics like model accuracy, precision, recall, or latency are important for development and operational monitoring, they are often insufficient on their own to demonstrate strategic worth. Stakeholders, particularly executives and budget holders, need to understand the tangible impact AI initiatives have on core business outcomes – revenue growth, cost reduction, risk mitigation, efficiency gains, or improvements in customer and employee experience. Failing to define, measure, and communicate this business value effectively can lead to skepticism, challenges in securing funding, and the perception of AI as a costly experiment rather than a strategic investment capable of driving transformation. This directly impacts the EAMM Performance Metrics and Value Realization dimensions.

This chapter focuses on establishing a clear framework for defining and measuring the business value derived from AI initiatives. We will explore the limitations of relying solely on technical metrics, discuss various dimensions of AI-driven value, provide guidance on selecting relevant Key Performance Indicators (KPIs) linked directly to business goals (referencing examples in Appendix B), emphasize the importance of establishing accurate baselines before implementation, and introduce core financial concepts like Return on Investment (ROI) and Total Cost of Ownership (TCO) as applied specifically in the AI context. Mastering value definition and measurement is essential for demonstrating impact, justifying continued investment, optimizing the AI portfolio (Domain North), and achieving higher EAMI maturity in the critical Value Realization and Performance Metrics dimensions.

Learning Objectives:

- Understand the limitations of relying solely on technical AI metrics and recognize the critical importance of focusing on measurable business value.
- Identify and define various dimensions of potential AI-driven business value (e.g., revenue, cost, risk, experience, sustainability, strategic).
- Learn how to select relevant, specific, measurable, achievable, relevant, and time-bound (SMART) Key Performance Indicators (KPIs) that directly link AI initiatives to prioritized business outcomes.
- Appreciate the crucial role of establishing accurate and reliable baseline measurements before implementing AI solutions to enable objective comparison.
- Understand the core concepts of Return on Investment (ROI) and Total Cost of Ownership (TCO) and how to apply them practically to AI initiatives.
- Recognize the AI Office's vital role in establishing, promoting, and standardizing value measurement practices across the enterprise.

13.2 The Limits of Technical Metrics

Data science and engineering teams naturally focus heavily on technical metrics during model development, validation, and operational monitoring. These metrics are indeed vital for assessing the technical quality, robustness, and operational health of AI systems:

- **Classification Metrics:** Model Accuracy, Precision, Recall, F1-Score, AUC (Area Under the Curve) – measure how well a classification model performs its predictive task against labeled data.
- **Regression Metrics:** Mean Absolute Error (MAE), Root Mean Squared Error (RMSE), R-squared – measure the magnitude and fit of errors for regression models (e.g., forecasting).
- **Operational Metrics:** Inference Latency (prediction speed), Throughput (predictions per second), System Uptime/Availability – measure the operational performance and reliability of deployed models.
- **Drift Metrics:** Statistical measures indicating changes in input data distributions (data drift) or the relationship between inputs and outputs (concept drift) over time compared to the training data.

However, these technical metrics, while essential for ML practitioners, often fail to resonate with business leaders or directly translate into strategic impact. They represent how well the AI is doing its technical task, not necessarily what business value that task is creating.

> A model can exhibit excellent technical accuracy (e.g., 98% accuracy in predicting a minor event) yet deliver zero net business value if it addresses the wrong problem, if its insights aren't integrated into decision-making processes, if user adoption is low, or if the total cost of developing and operating the system significantly outweighs the benefits it generates. Conversely, a model with slightly lower technical accuracy (e.g., 85% accuracy in predicting a high-cost failure event) might deliver immense business value if it effectively solves a critical, high-impact business problem in a cost-efficient manner. The primary focus for justifying AI investment (Compass West) must shift from purely technical performance to demonstrable business impact reflected in relevant EAMM Value Realization KPIs.

13.3 Dimensions of AI-Driven Business Value

AI can create value across multiple facets of the business. When assessing the potential or actual impact of an AI initiative, it's crucial to consider a broad spectrum of value dimensions beyond just direct cost savings or revenue uplift. The table below outlines key categories:

Table 13.1: Dimensions of AI-Driven Business Value

Value Dimension	Description	Example AI Applications & Potential Metrics
1. Revenue Enhancement	Directly increasing top-line growth through improved sales effectiveness, marketing reach, pricing strategies, or new AI-powered product/service offerings.	Personalized product/content recommendations ($\rightarrow$ Increased cross-sell/upsell revenue, higher Average Order Value - AOV, improved conversion rates), dynamic pricing optimization engines ($\rightarrow$ Increased profit margins), AI-driven lead scoring & qualification ($\rightarrow$ Improved sales pipeline conversion rates), generative AI for creating new digital product features ($\rightarrow$ New revenue streams). (See Appendix B for related KPIs, e.g., Market Positioning)
2. Cost Reduction / Efficiency	Decreasing operational expenses, reducing resource consumption, or improving the efficiency and speed of core business processes through automation or optimization.	Intelligent Process Automation (IPA) of manual back-office tasks ($\rightarrow$ Reduced Full-Time Equivalent (FTE) costs, reduced error rates, faster processing times), predictive maintenance for assets ($\rightarrow$ Reduced unplanned downtime, lower repair/replacement costs), supply chain route/ inventory optimization ($\rightarrow$ Lower logistics and holding costs), AI-powered resource scheduling ($\rightarrow$ Improved asset/ personnel utilization). (See Appendix B for related KPIs, e.g., Operational Efficiency, Financial Performance [cost reduction])

3. Risk Mitigation	Proactively identifying, preventing, or reducing exposure to various types of risks: financial, operational, security, compliance, or reputational.	AI-powered fraud detection systems ($\rightarrow$ Reduced direct fraud losses), anomaly detection for cybersecurity threats ($\rightarrow$ Reduced breach likelihood/impact), automated compliance monitoring tools ($\rightarrow$ Reduced penalty risk, improved audit readiness), predictive safety alerts in industrial settings ($\rightarrow$ Reduced accident rates), bias mitigation tools applied during hiring/lending ($\rightarrow$ Reduced discrimination lawsuit risk). (See Appendix B for related KPIs, e.g., Risk Management, Compliance & Ethics [violations reduction])
4. Customer Experience (CX)	Improving overall customer satisfaction, loyalty, retention, engagement, and the ease/quality of interactions across touchpoints.	AI-powered chatbots/virtual assistants for instant 24/7 support ($\rightarrow$ Faster issue resolution times, higher Customer Satisfaction (CSAT) scores), personalized website content/journeys ($\rightarrow$ Higher engagement rates, improved Net Promoter Score (NPS)), sentiment analysis of customer feedback ($\rightarrow$ Faster identification of pain points, improved product design). (See Appendix B for related KPIs, e.g., Customer Experience [CSAT, resolution time, engagement])
5. Employee Experience (EX)	Improving employee productivity, job satisfaction, skill development opportunities, or reducing tedious and repetitive tasks.	AI assistants automating routine administrative tasks ($\rightarrow$ Increased employee focus on higher-value strategic work), intelligent knowledge management systems providing faster access to information, personalized learning recommendations for upskilling ($\rightarrow$ Improved skill development velocity), AI tools for code generation/debugging ($\rightarrow$ Increased developer productivity and satisfaction). (See Appendix B for related KPIs, e.g., Workforce Productivity [task time reduction, output improvement])
6. Sustainability / ESG	Contributing positively to Environmental, Social, and Governance (ESG) goals and metrics, enhancing corporate responsibility and reputation.	AI for optimizing energy consumption in buildings or data centers ($\rightarrow$ Reduced carbon footprint), AI optimizing logistics routes for lower fuel consumption/emissions, AI enabling supply chain transparency for ethical sourcing verification (Social/Governance), AI identifying environmental risks from satellite imagery. (See Appendix B for related KPIs, e.g., Sustainability [waste reduction, goal attainment speed])
7. Innovation / Strategic Enablement	Enabling entirely new business models, significantly accelerating research and development cycles, providing novel strategic insights, or fundamentally enhancing the quality/speed of critical decisions.	AI accelerating drug discovery or materials science simulations, generative AI rapidly prototyping new product designs or marketing campaigns, advanced market trend prediction informing long-range strategic planning, AI-driven "digital twin" simulations for complex strategic decision support (e.g., M&A analysis), enhanced competitive intelligence gathering and analysis. (See Appendix B for related KPIs, e.g., Innovation & Product Dev [cycle time, success rate], Strategic Alignment [decision accuracy])

> **D** **DEFINITION**
>
> **Business Value (in AI Context):** The measurable improvement (quantitative or rigorously assessed qualitative) in key organizational outcomes – spanning financial performance, operational efficiency, risk posture, customer/employee experience, sustainability, and strategic positioning – directly or indirectly attributable to the implementation and effective use of AI technologies and capabilities, assessed relative to predefined strategic objectives and baselines. Reflects performance within the Compass West domain and contributes to the EAMM Value Realization dimension.

13.4 Selecting Relevant KPIs and Establishing Baselines

Defining and demonstrating value requires selecting the right Key Performance Indicators (KPIs) and measuring t hem correctly against a reliable starting point.

- **Link Directly to Business Objectives:** Ensure selected KPIs directly reflect the specific business objective the AI initiative aims to impact (as defined during Strategic Alignment - Chapter 4, Compass North). If the goal is cost reduction, the primary KPI must be a cost metric, not just model speed. (Appendix B provides many examples linking KPIs to strategic criteria).
- **Measurable & Attributable:** Choose KPIs reliably measurable via available data/systems. Consider how to plausibly attribute changes to the AI intervention versus other factors. Careful experimental design (A/B testing) or statistical controls may be needed to isolate AI's impact. Simple before-and-after can be misleading.
- **Leading vs. Lagging Indicators:** Track a balanced set. Lagging indicators measure final outcomes (e.g., quarterly revenue, annual savings). Leading indicators measure intermediate progress predictive of outcomes (e.g., daily tool adoption, weekly prediction accuracy improvement, CTR on recommendations). Leading indicators provide earlier feedback.
- **Establish Accurate Baselines:** This is critical but often overlooked. Rigorously measure and document the state of key KPI(s) before AI implementation. Provides the objective reference point for measuring improvement and calculating value/ROI. Requires planning data collection well in advance.
- **Consider Counter Metrics / Unintended Consequences:** Be mindful of potential negative side effects. E.g., an automation AI might reduce costs (positive KPI) but negatively impact CSAT (critical counter metric). Tracking counter metrics provides a balanced view.

> Collaborate closely and early with business sponsors and finance partners when defining value metrics, KPIs, baselines, and attribution methodologies. Business sponsors link to strategic importance; Finance validates calculations, ensures alignment with corporate reporting, and adds credibility to ROI assessments. Define primary value KPIs during project initiation (Compass West planning).

13.5 Measuring Customer Experience (CX) and Trust

Quantifying impact on "softer" dimensions like CX or trust is challenging but critical for customer-facing AI. Strategies include:

- **Standard CX Metrics:** Leverage industry-standard metrics, tracking consistently pre- and post-AI:
  - » **Net Promoter Score (NPS):** Measures loyalty ("How likely to recommend...?").
  - » **Customer Satisfaction (CSAT):** Measures satisfaction with interaction/service ("How satisfied...?").
  - » **Customer Effort Score (CES):** Measures ease of interaction ("How easy...?").
  - » **Churn Rate / Retention Rate:** Tracks loyalty over time.
  - » **Task Completion Rates:** Measures user success with AI tools (e.g., chatbot resolution rate). (See Appendix B for example KPIs like % increase in CSAT scores, % faster customer support resolution)
- **Behavioral Metrics (Digital Interaction):** Track observable user behavior changes indicating improved experience/engagement from AI features:
  - » Increased usage/adoption frequency of AI feature.
  - » Higher click-through rates (CTR)/conversion rates on AI recommendations/offers.
  - » Reduced task completion time with AI assistance vs. manual.
  - » Lower bounce rates / higher time-on-page for AI-personalized content.
- **Qualitative Feedback & Sentiment Analysis:** Supplement quantitative metrics with direct feedback:
  - » **User Surveys & Interviews:** Ask about experience, usability, benefits, trust.
  - » **Focus Groups:** Gather in-depth feedback from target users.
  - » **Sentiment Analysis:** Use NLP on feedback (tickets, social media, reviews) to identify sentiment and themes related to AI interaction. (See Appendix B KPI: % improvement in brand sentiment)
- **Trust Proxies:** While directly measuring trust is hard, proxy indicators help:
  - » **Adoption Rates:** How quickly/widely is the AI tool adopted?
  - » **User Reliance:** Do users accept AI recommendations or frequently override?
  - » **Willingness to Engage/Share Data:** Are users comfortable interacting or providing data?
  - » **Sentiment Expressed in Feedback:** Do comments reflect confidence or skepticism?

13.6 Introduction to ROI and TCO for AI

Justifying significant AI investments often requires demonstrating positive financial return using standard metrics adapted for AI (Compass West deliverable, key for EAMM Financial & Value Realization):

- **Return on Investment (ROI):** Measures profitability relative to cost.

- » **Simplified formula: ROI (%) = [(Total Gain from Investment - Total Cost of Investment) / Total Cost of Investment] * 100**
- » **Gain from Investment:** Quantified, incremental business value (revenue gain, cost savings, risk avoidance) attributable to AI over a specific period (e.g., 1-3 years). Based on KPIs and attribution methods.
- » **Cost of Investment:** Must encompass the Total Cost of Ownership (TCO).
- **Total Cost of Ownership (TCO):** Comprehensive assessment of all direct/indirect costs over the AI system's lifecycle (concept to decommission). Crucial for realistic ROI. Key categories (detailed further in Chapter 19):
  - » **Development Costs:** Personnel, software/hardware, data prep labor, training compute (can be high).
  - » **Deployment Costs:** Infra setup, integration, MLOps pipeline/tooling costs.
  - » **Ongoing Operational Costs:** Cloud inference/storage, license renewals, platform maintenance, monitoring, retraining compute, API fees. (Often largest component, variable).
  - » **Governance & Compliance Costs:** Personnel time (reviews, audits, docs), specialized tooling licenses.
  - » **Change Management & Training Costs:** User training, communication, process redesign support.

Establish a standardized, documented methodology, developed with Finance, for calculating ROI and TCO for AI projects. Be transparent about assumptions (timeframe, attribution, discount rates). Focus on demonstrating positive ROI relative to the full TCO estimate for a realistic view of financial viability.

13.7 The AI Office's Role in Value Measurement

The AI Office (Compass West domain lead) plays a key leadership and enabling role:

- **Developing Frameworks & Standards:** Creating/disseminating standardized frameworks, methodologies, KPIs, templates (ROI/TCO), guidelines for baselines, tracking, attribution.
- **Providing Expertise & Guidance:** Offering expert advice/coaching to teams/sponsors on selecting metrics, designing measurement approaches (A/B tests), interpreting results.
- **Ensuring Consistency & Comparability:** Promoting consistent application of value measurement practices across initiatives for meaningful comparison and portfolio analysis.
- **Consolidating & Reporting Overall Value:** Aggregating validated value metrics/ROI results for an enterprise view. Reporting aggregate value compellingly to leadership.
- **Linking Value to EAMI:** Ensuring value realization metrics and ROI results are tracked and used as inputs for EAMI assessment (Chapter 24) to measure progress in the EAMM Value Realization dimension.

13.8 OmnioTech Case Study: Defining Value for Pilots

Setting the Scene:

As PredMaint and HyperPers pilots launch (Chapter 6), Priya Sharma (AI Office Head) collaborates proactively with sponsors (VP Product, VP Marketing) and the Finance rep on the AI Review Team to establish clear, measurable value metrics (Compass West activity) from the outset.

Action & Application

- **KPI Definition Workshop:** Priya facilitates workshops using the value dimensions framework (Section 13.3). Primary/secondary KPIs defined:
  - » **PredMaint:** Primary KPI: % reduction in emergency/unscheduled maintenance calls vs. control group (linking to Appendix B KPI). Secondary KPIs: Average reduction in repair cost/incident, impact on relevant CSAT scores.
  - » **HyperPers:** Primary KPI: % increase in conversion rate for AI-personalized offers vs. standard offers (via A/B testing) (linking to Appendix B KPI). Secondary KPIs: Increase in Average Order Value (AOV), impact on email open/CTR.
- **Baseline Establishment Process:** Mandated upfront effort. Product Analytics provides 12 months historical maintenance data. Marketing Ops provides baseline campaign metrics. Finance validates cost data.
- **Value Tracking Process & Tools:** Standardized "AI Value Tracking Template" (Excel initially) created by AI Office/Finance. Pilot teams populate quarterly with measured KPIs vs. baseline, validated by sponsor. Finance provides TCO estimation templates/guidance, including cloud costs via Azure tagging (Chapter 19).
- **ROI Calculation (Projected & Actual):** Initial ROI projections in pilot business cases. Quarterly Value Review meetings (Chapter 14) track actual pilot TCO against measured benefits (e.g., calculated savings from reduced calls; incremental revenue from conversion lift) to refine ROI before scaling decisions.

Outcome & Progress:

By collaborating proactively, OmnioTech establishes clear, business-focused KPIs, reliable baselines, and a defined process for tracking pilot value from the start. Success is defined by business impact, not just technical metrics. This structured approach (Compass West) lays groundwork for systematic value tracking and credible ROI, aligning with EAMI Value Realization principles (targeting Level 3) and building confidence.

13.9 Conclusion: Measuring What Truly Matters

Defining and rigorously measuring business value delivered by AI (Compass West) is fundamental for demonstrating strategic impact, securing investment, guiding prioritization, and ensuring contribution to goals. Moving beyond technical metrics requires understanding diverse value dimensions, selecting relevant KPIs linked to strategy, establishing accurate baselines, and a disciplined approach to tracking performance and calculating ROI against full TCO. The AI Office leads in establishing these practices. By relentlessly focusing on measuring tangible business impact, organizations manage their AI portfolio effectively, communicate success compellingly, and build a sustainable, credible, value-driven AI program. Having defined what value to measure, Chapter 14 explores frameworks for managing AI performance over time.

Key Takeaways:

- Relying solely on technical AI metrics is insufficient; focus must be on measurable business value.
- **AI value spans multiple dimensions:** revenue, cost/efficiency, risk, CX, EX, sustainability, strategic enablement.
- Select specific, measurable KPIs linked to business goals and establish accurate baselines before AI implementation.
- Calculate ROI based on attributed gains vs. the full Total Cost of Ownership (TCO) across the AI lifecycle.
- The AI Office should lead in standardizing value measurement frameworks and collaborating with Finance.

Food for Thought / Application Exercise:

- Think of an AI initiative. Which value dimensions (Table 13.1) are most relevant? What specific, measurable business KPIs could track impact?
- For that initiative, what data is needed for baseline KPIs? Is it available?
- Beyond development costs, what are likely major ongoing TCO elements (cloud inference, monitoring, retraining)? How estimated?

~ CHAPTER 14 ~

Implementing AI Performance Management Frameworks: Tracking Success

> "Continuous improvement is better than delayed perfection."
> — Mark Twain

14.1 Introduction: From Measurement to Management

Chapter 13 focused on the critical task of defining and measuring the business value derived from AI initiatives, shifting the focus from purely technical metrics to tangible business impact aligned with strategic goals (Compass West). However, measurement alone is insufficient; organizations need systematic Performance Management Frameworks to actively monitor, analyze, report on, and manage the performance and value delivery of their AI systems over time. This ensures accountability, facilitates continuous improvement (Compass Center), enables proactive risk management (Compass East), and provides the data needed for ongoing strategic decision-making regarding the AI portfolio (Compass North).

This chapter, situated within the Performance & Value Delivery (West) domain of the AI Office Compass™ framework, delves into the practical implementation of these management frameworks. We will explore the key components involved, such as establishing processes for regular Key Performance Indicator (KPI) tracking (using KPIs like those in Appendix B), implementing value realization methodologies throughout the AI lifecycle, defining cadences for performance reviews, linking AI performance to broader concepts like system observability and financial operations (FinOps), and ensuring effective reporting mechanisms are in place. The AI Office plays a central role in designing, implementing, and overseeing these frameworks, which are essential for demonstrating ongoing value, justifying investments, and driving continuous improvement aligned with EAMI maturity goals (specifically in the Performance Metrics and Value Realization dimensions).

Learning Objectives:

- Understand the necessity of moving beyond one-time value measurement to ongoing AI performance management.
- Identify the key components of an effective AI Performance Management Framework.
- Learn how to establish processes for regular KPI tracking and value realization monitoring.
- Recognize the importance of defined review cadences and reporting mechanisms.
- Understand the links between AI performance management, system observability, and FinOps.
- Appreciate the AI Office's role in implementing and overseeing performance management practices.

14.2 Key Components of an AI Performance Management Framework

A comprehensive framework for managing AI performance typically integrates several key elements:

- **Defined KPIs & Metrics:** Clear, agreed-upon KPIs (as defined in Chapter 13) covering both business value (e.g., revenue impact, cost savings) and relevant operational/technical performance (e.g., model accuracy, latency, drift metrics, uptime). These must be measurable and linked to strategic objectives (Domain North).
- **Data Collection & Monitoring Systems:** Reliable mechanisms for collecting data related to the defined KPIs. This often involves integrating with operational systems, logging platforms, MLOps monitoring tools (Chapter 20), data quality monitoring systems (Chapter 23), and potentially business intelligence (BI) platforms.
- **Value Realization Tracking Process:** A defined process for tracking the realization of expected business value throughout the lifecycle of an AI initiative – from initial business case projections, through pilot validation results, to ongoing performance monitoring after full-scale deployment. This often involves comparing actual results against initial targets and baselines.
- **Performance Dashboards & Reporting:** Clear, audience-tailored dashboards and regular reports (Chapter 15) visualizing KPI trends, value realization progress, ROI calculations, operational health, and highlighting key insights or areas needing attention.
- **Review Cadences & Governance:** Scheduled meetings (e.g., monthly operational reviews, quarterly business reviews with sponsors) involving relevant stakeholders (AI Office, business owners, technical leads, finance) to review performance data, discuss insights, make decisions (e.g., optimize, retrain, decommission), and ensure accountability.
- **Feedback Loops & Continuous Improvement:** Processes for feeding performance insights back into strategic planning (Domain North), roadmap prioritization (Chapter 6), model development practices, MLOps processes (Domain South), and governance standards (Domain East) to drive continuous improvement (Domain Center).

STRATEGY/GOALS

DEFINE KPIs & TARGETS

COLLECT & MONITOR DATA

DECIDE & ACT (Optimize/ Improve)

ANALYZE PERFORMANCE & VALUE

REPORT & REVIEW (stakeholders)

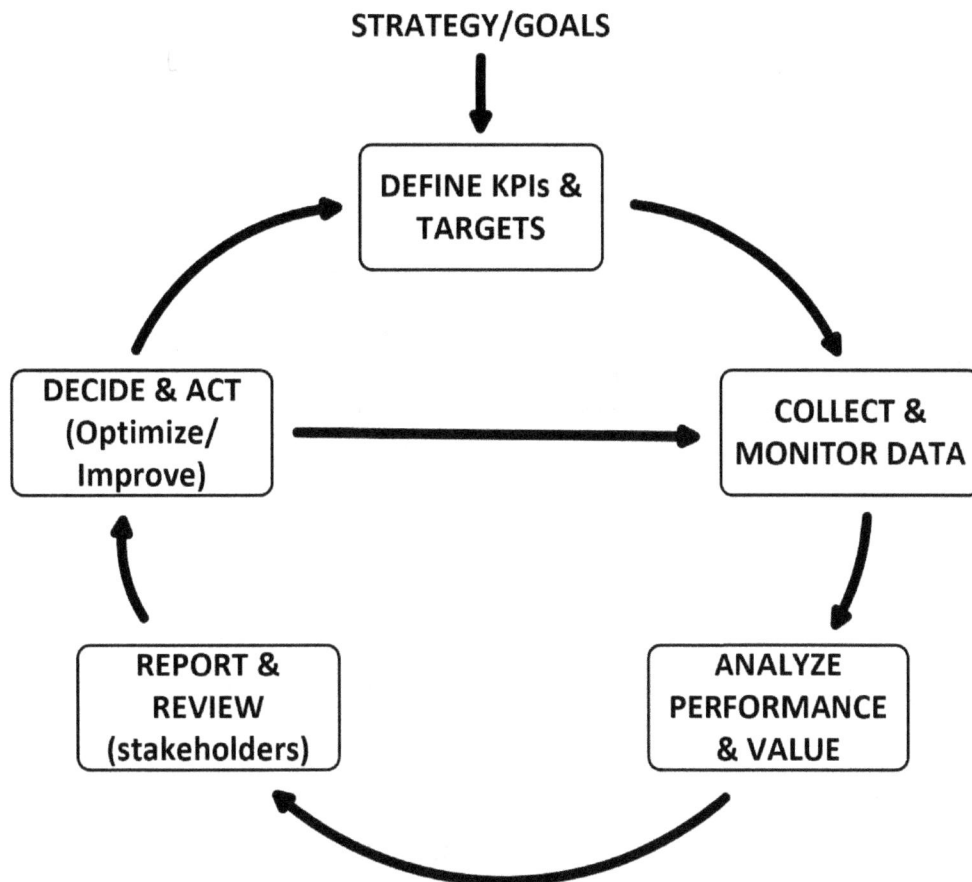

Figure 14.1: AI Performance Management Cycle Diagram

14.3 Establishing KPI Tracking and Value Realization Processes

Operationalizing performance management requires structured processes (Compass West):

- **KPI Definition & Ownership:** For each significant AI initiative, clearly define the primary business value KPIs and supporting operational metrics during the planning phase. Assign clear ownership for tracking and reporting each KPI (e.g., business owner tracks revenue impact, MLOps team tracks latency). Ensure KPIs are SMART (Specific, Measurable, Achievable, Relevant, Time-bound).
- **Baseline Measurement:** As emphasized in Chapter 13, rigorously establish baseline measurements for all key KPIs before the AI solution is implemented. Document the baseline period and methodology clearly.
- **Data Collection Automation:** Automate the collection of KPI data wherever possible by integrating with source systems, logging platforms (like Splunk or ELK stack), cloud monitoring services (like AWS Cloud-Watch, Azure Monitor), and MLOps monitoring tools. Avoid reliance on manual data gathering, which is error-prone and not scalable.
- **Value Realization Checkpoints:** Integrate value realization checkpoints into the AI project lifecycle milestones. For example:
  - » **Business Case:** Document expected value and KPIs.
  - » **Pilot Completion Review:** Compare pilot results against expected value and baseline; validate initial ROI estimates.
  - » **Post-Deployment (e.g., 3/6/12 months):** Conduct formal reviews to measure actual value realized

against the business case and track ongoing performance trends.

- **ROI & TCO Calculation Cadence:** Define how often ROI and TCO calculations will be updated based on actual performance data and operational costs (e.g., quarterly for major systems). Ensure alignment with Finance on methodologies (Chapter 19).

> ★ Start simple but be consistent. For early-stage AI programs, begin with tracking a few critical business value KPIs and operational health metrics for key initiatives. Focus on establishing reliable data collection and baseline measurement processes first. The sophistication of the framework can grow with EAMI maturity (EAMI level).

14.4 Performance Reviews and Reporting Cadences

Regular reviews are essential for translating performance data into insights and actions:

- **Operational Reviews (e.g., Weekly/Monthly):** Focus on technical performance, system health, data drift, model accuracy, latency, error rates. Involve technical leads, MLOps engineers, data scientists. Goal: Identify and address operational issues quickly, trigger necessary model retraining or technical fixes.
- **Project/Initiative Reviews (e.g., Monthly/Quarterly):** Focus on progress against milestones for specific AI initiatives, value realization tracking against the business case, key risks and issues, and resource utilization. Involve project managers, business sponsors, technical leads, AI Office representatives. Goal: Ensure projects are on track to deliver expected value, make go/no-go decisions for scaling, adjust plans as needed.
- **Business/Portfolio Reviews (e.g., Quarterly/Semi-Annually):** Focus on the overall business impact and ROI of the AI portfolio, alignment with strategic goals, progress against EAMI targets, major cross-initiative risks or opportunities, and budget allocation decisions. Involve senior business leaders, executive sponsors, AI Office leadership, Finance representatives. Goal: Demonstrate overall program value, secure ongoing strategic commitment and funding, make portfolio-level adjustments.

Reporting should be tailored to these review cadences and audiences (Chapter 15), often leveraging visual dashboards for clarity.

14.5 Links to Observability and FinOps

Effective AI performance management increasingly connects with two related disciplines:

- **Observability:** Goes beyond traditional monitoring (tracking predefined metrics) to enable deeper exploration and understanding of system behavior ("Why is this happening?"). For AI, this involves tools and techniques to gain insights into model predictions (XAI - Chapter 10), data pipeline health, infrastructure performance, and complex interactions within distributed AI systems. Mature performance management leverages observability principles for faster root cause analysis of performance issues or unexpected model behavior.

- **FinOps (Financial Operations):** A cultural practice and operational framework bringing financial accountability to the variable spend model of cloud, enabling organizations to make informed trade-offs between speed, cost, and quality. As AI/ML workloads (especially training large models or running inference at scale) can be major drivers of cloud spend, integrating FinOps principles is crucial for AI performance management. This involves cost visibility, optimization (e.g., right-sizing, spot instances, auto-scaling), and governance (budgeting, cost awareness). The AI Office should collaborate closely with the central FinOps team or champion FinOps practices within the AI community (Compass South, Chapter 19).

14.6 The AI Office's Role in Performance Management

The AI Office typically plays a central coordinating and enabling role in AI performance management (Compass West):

- **Framework Design & Standardization:** Developing and promoting standardized frameworks, methodologies, KPIs, templates, and processes for value measurement, performance tracking, and ROI calculation across the enterprise.
- **Tooling & Platform Enablement:** Providing or recommending tools and platforms for monitoring, dashboarding, and reporting AI performance and value (often collaborating with IT/BI teams).
- **Consolidation & Enterprise Reporting:** Aggregating performance data and value realization results from across the AI portfolio to provide an enterprise-level view for executive leadership and governance bodies. Linking performance to EAMI progress.
- **Facilitation & Coaching:** Guiding project teams and business sponsors on how to effectively define KPIs, establish baselines, track value, and interpret performance data. Facilitating key performance review meetings.
- **Driving Continuous Improvement:** Analyzing performance trends across the portfolio to identify systemic issues or best practices, feeding insights back into governance standards, MLOps practices, and strategic planning (Compass Center).

> Automate reporting and dashboarding as much as possible. Manual report generation is time-consuming and prone to errors. Leverage BI tools (Tableau, Power BI) connected to underlying monitoring and data systems to provide near real-time, self-service performance visibility tailored to different stakeholder needs.

14.7 OmnioTech Case Study: Implementing Performance Tracking

Setting the Scene:

With the PredMaint and HyperPers pilots approved for scaling (based on initial results communicated in Chapter 15), Priya Sharma (Head of AI Office) and her team, collaborating with Finance and the pilot sponsors, implement a more structured V1.0 performance management framework (Compass West) to ensure ongoing value delivery and track progress towards EAMI Level 3.

Action & Application:

- **KPI Tracking System:**
  - » **Data Collection:** Automated scripts pull key operational metrics (model predictions, latency, system uptime) from Azure Monitor into a central Azure SQL database weekly. Business outcome data (maintenance calls, conversion rates, AOV) is extracted monthly from ServiceNow and the Marketing data mart via scheduled Azure Data Factory pipelines.
  - » **Dashboarding:** Initial dashboards are built using Power BI connected to the Azure SQL database. One dashboard focuses on operational health for the AI Office/MLOps team. Separate dashboards tailored for Product (PredMaint KPIs vs. baseline) and Marketing (HyperPers KPIs vs. baseline/control) are created for business sponsors.
- **Value Realization Process:** A quarterly "AI Value Review" meeting cadence is established for each scaled initiative, involving the AI Office, business sponsor, Finance representative, and technical lead. The agenda includes: review of KPI trends vs. baseline/targets, updated ROI calculation based on actual costs (provided by Finance using FinOps tagging) and measured benefits, discussion of issues/opportunities, and confirmation of continued value delivery.
- **Reporting Cadence:**
  - » **Monthly:** Operational health dashboards reviewed by AI Office/MLOps.
  - » **Quarterly:** Value Realization reports and dashboards reviewed in dedicated meetings for each major initiative. High-level summary included in AI Office update to AI Review Team.
  - » **Annually:** Aggregated ROI and overall program value contribution reported to executive leadership as part of the strategic planning cycle. EAMI Value Realization score updated based on documented results.
- **EAMI Calculation Context:** The quarterly value reviews explicitly track progress against the initial EAMI assessment baseline for relevant KPIs (e.g., % cost reduction from PredMaint contributes to the Value Realization dimension score). A simplified ROI calculation, showing the tangible benefit ($ savings/revenue) exceeding the documented TCO (including development, cloud compute, monitoring costs), is used as a key input for validating improvement in the EAMI Value Realization dimension. For instance, if PredMaint achieves $500k savings against a $200k TCO in Year 1, this strong positive ROI (ROI = [($500k - $200k) / $200k] * 100 = 150%) contributes significantly to justifying a higher score for that dimension during the next EAMI assessment cycle (Chapter 24). Similarly, achieving the target 18% conversion lift with HyperPers against its TCO would bolster the Value Realization score.

Outcome & Progress:

OmnioTech implements a structured process for ongoing AI performance management (Compass West). Automated data collection and tailored dashboards provide visibility. Quarterly value reviews ensure accountability and link performance directly to business outcomes and EAMI goals, including using ROI evidence to support EAMI scoring. This framework allows OmnioTech to move beyond one-off pilot validation to continuously manage and demonstrate the value of its scaling AI initiatives, solidifying support and enabling data-driven decisions for future optimization and investment.

14.8 Conclusion: Sustaining Value Through Continuous Management

Defining and measuring AI value (Chapter 13) is the starting point; implementing robust performance management frameworks (Compass West) is essential for sustaining that value over the long term. By establishing clear KPIs linked to business goals, implementing reliable data collection and monitoring systems, instituting regular review cadences involving key stakeholders, and fostering a culture of accountability for results, organizations can ensure their AI investments deliver ongoing returns. Integrating performance management with broader observability and FinOps practices further enhances efficiency and insight. The AI Office plays a critical role in designing, implementing, and championing these frameworks, ensuring that AI performance is not just measured, but actively managed to drive continuous improvement, optimize value delivery, and contribute demonstrably to achieving higher levels of enterprise EAMI maturity. The final piece of the value puzzle, communicating these results effectively to build trust, is explored in Chapter 15.

Key Takeaways:

- Ongoing performance management is needed beyond initial value measurement for sustained AI impact.
- Key framework components include defined KPIs, monitoring systems, value tracking process, dashboards/ reports, review cadences, and feedback loops.
- Automate KPI data collection and reporting where possible; establish clear ownership and baselines.
- Regular operational, project, and portfolio reviews ensure accountability and drive decisions.
- Link performance management to Observability (deeper insights) and FinOps (cost optimization).
- The AI Office standardizes frameworks, enables tooling, consolidates reporting, and drives continuous improvement.

Food for Thought / Application Exercise:

- Select a deployed AI system in your organization. What KPIs (business and operational) are currently being tracked? How frequently are they reviewed, and by whom?
- How is the value realization of AI projects tracked over time in your organization after deployment? Is there a formal process to compare actual value against the initial business case?
- How visible are the operational costs (e.g., cloud compute/storage for inference and monitoring) of deployed AI systems? How could FinOps practices be better integrated with AI performance management?

~ CHAPTER 15 ~

Communicating AI Value and Building Organizational Trust

"The single biggest problem in communication is the illusion that it has taken place."
— George Bernard Shaw

15.1 Introduction: Beyond Measurement – Making Value Visible and Trusted

Chapter 13 established the critical importance of defining and measuring the business value delivered by AI initiatives, while Chapter 14 focused on the frameworks for managing AI performance over time (Compass West activities). However, simply measuring value is insufficient if it isn't effectively communicated and understood by key stakeholders across the organization. Furthermore, realizing the full potential of AI requires building broad organizational trust – trust in the technology itself, trust in the processes governing its development and deployment (Domain East), and trust in the teams leading the charge.

This chapter, concluding Part 4 on Performance & Value Delivery (West) within the AI Office Compass™ framework, focuses on the crucial arts of communication and trust-building specifically in the context of enterprise AI. Effective communication translates complex performance data and quantified value metrics into clear, concise, and compelling narratives that resonate with vastly different audiences – from executives demanding high-level ROI summaries and strategic implications to end-users needing practical guidance and confidence in the AI tools integrated into their daily workflows. Building enduring organizational trust requires unwavering transparency (even when discussing challenges or setbacks), consistent messaging from leadership, active engagement with skeptics to address their valid concerns, and demonstrating a steadfast, visible commitment to the responsible AI principles outlined in the governance framework (Domain East). The AI Office plays a pivotal, central role in orchestrating these communication and trust-building efforts, ensuring that the value generated by AI is not only meticulously measured but also widely recognized, clearly understood, and actively leveraged to foster broader adoption, continuous improvement, and advancement along the EAMI maturity scale (particularly in the Value Realization and Culture/Change Management dimensions).

Learning Objectives:

- Understand the critical role of strategic communication in demonstrating AI value, securing ongoing support, and driving adoption.
- Learn practical strategies for tailoring communication about AI performance, value, risks, and benefits to different stakeholder audiences.
- Explore effective formats (dashboards, reports, stories) and channels (meetings, newsletters, CoPs) for communicating AI results and progress.
- Recognize the power of data storytelling techniques to create compelling and memorable value narratives for AI initiatives.
- Appreciate the fundamental importance of transparency, including acknowledging challenges and failures, in building and maintaining organizational trust around AI.
- Identify constructive techniques for engaging with and addressing concerns from individuals or groups skeptical of AI.
- Understand the AI Office's leadership role in developing and executing communication plans and trust-building initiatives.

15.2 The Importance of Communicating AI Value

Systematically and proactively communicating the value derived from AI initiatives is not merely good practice; it is essential for the long-term success and sustainability of any enterprise AI program. Key reasons include:

- **Justifying Investment & Securing Resources:** Clear communication of delivered value (ROI, cost savings, revenue uplift - Compass West output) provides the hard evidence needed to justify past investments and, more importantly, secure ongoing funding, budget, and critical resources (talent, technology - Domain South) for future AI initiatives and the continued operation of the AI Office itself.
- **Maintaining Executive Sponsorship:** Keeping senior leaders consistently informed and engaged with tangible results reinforces their support, advocacy, and willingness to champion AI initiatives across the organization. Demonstrating measurable value maintains crucial momentum, especially for long-term programs.
- **Driving User Adoption & Engagement:** Building confidence among business users and functional teams through clear communication about benefits, reliability, and responsible use encourages them to adopt AI-powered tools, integrate AI insights into their decision-making processes, and provide valuable feedback for improvement (supports EAMM Culture).
- **Managing Expectations Realistically:** Setting clear, realistic expectations about what specific AI systems can (and importantly, cannot) achieve, based on transparent communication of capabilities and limitations, helps prevent disillusionment resulting from initial hype cycles or misunderstandings.
- **Facilitating Organizational Learning & Reuse:** Actively sharing successes, challenges overcome, technical breakthroughs, effective data strategies, and lessons learned from both successful and unsuccessful AI projects helps disseminate valuable knowledge and best practices across the organization, accelerating learning and preventing redundant efforts (supports Compass Center).
- **Building AI Office Credibility:** Consistent, transparent, and data-driven communication about performance, value delivered, risks managed, and progress against the roadmap establishes the AI Office as a credible, accountable, and strategically vital function within the organization.
- **Reinforcing Strategic Alignment:** Continuously linking reported value metrics and project outcomes back to the overarching corporate strategic objectives (Domain North) reinforces the AI program's purpose, relevance, and direct contribution to the organization's success.

> Communication about AI value and performance should not be treated as a separate, periodic activity conducted only at the end of a project or during annual reviews. It must be an ongoing, integrated process throughout the entire AI lifecycle – from clearly articulating the expected value in the initial business case, to reporting interim pilot results, tracking scaled performance metrics, sharing user feedback, and disseminating lessons learned post-deployment.

15.3 Tailoring Communication to Different Audiences

A "one-size-fits-all" communication approach rarely works for complex topics like AI. Effective communication requires understanding the distinct information needs, technical literacy levels, primary concerns, and preferred communication styles of various stakeholder groups within the organization:

- **C-Suite Executives (CEO, CFO, COO, etc.):** Require highly concise, strategic summaries focused on bottom-line impact (ROI, market share, competitive advantage), overall program health, alignment with corporate goals, major risks and mitigation strategies, and key decisions needed. Prefer executive dashboards, brief presentations, and clear "so what?" takeaways.
- **Business Unit Leaders & Project Sponsors:** Need information directly relevant to their specific operational or financial KPIs, detailed performance results for use cases within their domain, quantified value realized, resource implications for their teams, potential operational benefits or disruptions, and clear updates on project timelines and deliverables affecting their area. Require regular, context-specific updates.
- **IT & Core Technical Teams (Data Engineering, MLOps, Infrastructure, Security):** Need technical details regarding model performance metrics (accuracy, latency), platform utilization and stability, data pipeline health, integration requirements and challenges, MLOps pipeline efficiency, security posture, resource consumption (e.g., cloud costs), and identification of technical roadblocks or dependencies. Communication often involves technical dashboards, detailed logs, system alerts, and technical review meetings.
- **AI/Data Science Teams:** Interested in detailed model performance metrics, algorithmic insights and innovations, data quality discoveries, challenges in feature engineering or model training, MLOps pipeline effectiveness, access to compute resources and new tools, opportunities for experimentation, and forums for technical knowledge sharing (e.g., CoPs, internal seminars).
- **End-Users of AI Tools/Systems:** Require clear, simple explanations of what the AI tool does for them, how it benefits their specific workflow, how to use it effectively and reliably, what its known limitations are, and clear channels for providing feedback or getting support. Focus must be on practical utility, ease of use, and building trust in the tool's outputs.
- **Legal, Compliance, Risk, Audit & HR:** Need specific information related to governance adherence (policy compliance, risk assessment completion), ethical considerations addressed (bias testing results, fairness metrics), privacy compliance documentation (DPIAs), security assessment outcomes, audit trail availability, impacts on workforce roles or required skills, and details of relevant training programs. Often require formal reports and participation in governance review processes (Domain East).
- **Wider Organization / All Employees:** Benefit from general awareness communications reinforcing the

company's overall AI strategy, celebrating key successes and their impact, highlighting the commitment to responsible and ethical AI practices, promoting available AI literacy resources, and demystifying AI concepts to reduce fear or misinformation.

> Develop a formal Stakeholder Communication Plan as part of the overall AI program governance (Compass West/Center task). This plan should explicitly identify key stakeholder groups, analyze their specific information needs and concerns regarding AI, define the key messages for each group, determine the most effective communication channels and formats (see below), establish an appropriate frequency for updates, and assign clear responsibility for delivering each communication component. Regularly review and update this plan.

15.4 Effective Formats and Channels for AI Communication

Choosing the right format and channel is crucial for ensuring messages are received, understood, and impactful:

- **Executive Dashboards:** Highly visual, interactive summaries (e.g., built in Tableau, Power BI, Qlik Sense) displaying critical strategic KPIs, overall program ROI trends, key milestone status (from the roadmap), top risks heatmaps, and potentially EAMI maturity progress. Designed for quick consumption by senior leadership.
- **Business Review Presentations:** Structured presentations (e.g., quarterly) delivered to leadership teams and key business sponsors, summarizing performance against goals, value delivered in specific areas, progress against the roadmap (Chapter 6), key challenges encountered and mitigation plans, upcoming priorities, and required decisions. Often includes data visualizations and concise narratives.
- **Project Status Reports & Updates:** More frequent, often weekly or bi-weekly, updates for project teams and immediate stakeholders, covering technical progress, tasks completed/pending, milestone tracking, risk/issue logs, and resource status. Frequently managed via project management software (e.g., Jira dashboards, Asana updates) or brief stand-up meetings.
- **Value Realization Case Studies / Reports:** Dedicated documents or presentations detailing the quantified business value achieved by specific, significant AI deployments against their initial business case and baseline measurements. Often co-developed with the Finance department to ensure rigor and credibility. Powerful tools for justifying future investments.
- **Internal Communication Channels:**
  - » **Newsletters / Blogs / Intranet Portals:** Used for broader communication of AI success stories (written as compelling narratives), announcements of new AI tools or capabilities available, promoting AI training opportunities, sharing high-level strategy updates, and reinforcing the organization's commitment to responsible AI.
  - » **Town Halls / All-Hands Meetings:** Provide opportunities for senior leaders (e.g., CEO, CDO, Head of AI Office) to communicate the strategic importance of AI, celebrate major milestones or successes, and address employee questions or concerns directly, fostering transparency and alignment.
  - » **Communities of Practice (CoPs) & Technical Forums:** Dedicated meetings, workshops, or online

channels (e.g., Slack/Teams channels) for technical practitioners (data scientists, ML engineers, MLOps engineers) to share detailed technical learnings, discuss specific challenges, demo innovative work, and foster peer-to-peer knowledge transfer (Compass Center).

    » **Training Materials & User Documentation:** Clear, concise, accessible guides, FAQs, video tutorials, and embedded help features specifically designed for end-users of particular AI-powered tools or systems, focusing on practical usage and building confidence.

- **Formal Governance & Compliance Reporting:** Standardized reports detailing risk assessment outcomes, compliance audit results, ethical review decisions, privacy impact assessments (DPIAs), and security posture updates, distributed to relevant oversight bodies (e.g., AI Review Team, Risk Committee, Audit Committee, regulators if required).

Table 15.1: Communication Channels & Audiences Matrix

Stakeholder Group	Exec Dashboards	Business Reviews	Status Reports	Newsletters/ Intranet	Town Halls	CoPs	User Guides
Executives	High	High	Medium	Low	Low	Low	Low
BU Leaders	Medium	High	High	Medium	Medium	Low	Low
Tech Teams	Low	Medium	High	Medium	Low	High	Low
End Users	Low	Low	Low	Medium	Medium	Low	High
Wider Org.	Low	Low	Low	High	High	Low	Low

15.5 Storytelling with Data: Crafting Compelling Value Narratives

Raw data and metrics, while important, often fail to capture attention or drive understanding on their own. To make the value of AI truly resonate, especially with non-technical audiences, the AI Office and project teams must become adept at storytelling with data (Compass West skill). This involves weaving the quantitative results into a clear, compelling narrative structure:

- **Establish the Context (The "Why"):** Start by clearly reminding the audience of the specific business problem, strategic objective, or user pain point that the AI initiative was designed to address. What was the situation before AI? (Link back to Domain North).
- **Introduce the Solution (The "How"):** Briefly explain, in understandable terms, the AI approach or solution that was implemented. Avoid deep technical jargon; focus on what it does and how it addresses the problem.
- **Present the Results (The "What"):** Showcase the key performance metrics and quantified value delivered in a clear, visual, and easily digestible format (e.g., using impactful charts or graphs). Highlight the comparison against the baseline or control group. Use specific numbers ($ saved, % improvement, time reduced). Refer to Appendix B for relevant KPI examples.
- **Explain the Impact (The "So What?"):** This is crucial. Explicitly connect the results back to the initial business problem and strategic goals. Clearly articulate the tangible benefits realized for the organization, its customers, or its employees. Translate technical results into business implications.
- **Add the Human Element:** Where possible and appropriate, incorporate brief anecdotes, quotes, or tes-

timonials from users or stakeholders illustrating how the AI solution has positively impacted their work or experience. This makes the value more relatable and memorable.

- **Acknowledge the Journey (Challenges & Learnings):** Briefly and transparently mention any significant hurdles overcome during development or deployment, or key lessons learned. This builds credibility and demonstrates a learning mindset.
- **Look Ahead (The "Now What?"):** Conclude by outlining next steps, potential future enhancements or optimization opportunities, or how the learnings from this initiative will be applied to other areas or inform future roadmap decisions.

Develop a standardized template or internal framework for crafting "AI Value Stories." Encourage project teams to proactively capture the narrative elements alongside the quantitative metrics throughout the project lifecycle. Invest in training key personnel (e.g., AI Translators, Program Managers, AI Office staff) in effective data visualization and storytelling techniques.

15.6 Transparency and Trust-Building

Trust is foundational for successful AI adoption. Communication (Compass West) plays a vital role in building and maintaining that trust:

- **Be Transparent:** Openly communicate about AI initiatives, including their goals, how they work (at an appropriate level), the data they use, and their known limitations. Avoid over-promising or hype.
- **Acknowledge Challenges and Failures:** Don't just report successes. Transparently sharing challenges faced, pilots that didn't scale, or lessons learned from failures builds credibility and fosters a culture of learning rather than blame.
- **Explain Governance Processes:** Communicate clearly about the governance framework, ethical guidelines, and review processes in place to ensure responsible AI (Domain East). This reassures stakeholders that risks are being managed proactively.
- **Provide Channels for Feedback:** Create mechanisms for users and stakeholders to ask questions, raise concerns, or provide feedback about AI systems. Actively listen and respond to this feedback.
- **Consistency is Key:** Ensure consistent messaging about AI strategy, value, and governance principles across all communications and from all relevant leaders.

15.7 Engaging with Skeptics

Not everyone in the organization will immediately embrace AI. The AI Office must proactively engage with skeptics (Compass Center/West activity):

- **Listen Actively:** Understand the root causes of their skepticism – is it fear of job loss, lack of understanding, past negative experiences with technology, valid concerns about risks, or something else?

- **Address Concerns Directly:** Provide clear, evidence-based responses to specific concerns. Tailor explanations to their level of understanding.
- **Highlight Benefits (for Them):** Focus on how AI can help them or their teams achieve their goals, reduce tedious work, or make better decisions.
- **Show, Don't Just Tell:** Demonstrate working prototypes or pilot results that showcase tangible benefits relevant to the skeptics' area.
- **Involve Them:** Where appropriate, involve skeptical stakeholders in pilot projects or feedback sessions to give them a sense of ownership and allow them to see the process firsthand.
- **Leverage Champions:** Use enthusiastic early adopters and respected leaders within their peer group to advocate for AI benefits.

15.8 The AI Office's Role in Communication and Trust

The AI Office typically leads or coordinates many aspects of AI communication and trust-building (Compass West/Center):

- **Developing the Communication Strategy:** Creating the overall plan for communicating AI strategy, progress, value, and governance.
- **Creating Standardized Materials:** Developing templates, dashboards, and reporting formats.
- **Consolidating and Reporting Value:** Aggregating results from across the AI portfolio for executive reporting.
- **Managing Internal & External Communications:** Crafting newsletters, blog posts, intranet content, and potentially external communications about the organization's AI efforts and commitment to responsibility.
- **Facilitating Stakeholder Engagement:** Organizing review meetings, workshops, and feedback sessions.
- **Promoting Transparency:** Ensuring governance processes and AI system documentation are accessible and understandable.
- **Supporting Change Management:** Working with HR and internal communications to manage the human aspects of AI adoption (Chapter 26).

15.9 OmnioTech Case Study: Communicating Pilot Value

Setting the Scene:

The initial pilots (PredMaint, HyperPers) at OmnioTech have completed their first three months. Priya Sharma (Head of AI Office) needs to communicate the early results (Compass West report) to the AI Review Team and executive sponsors (COO, CTO, CEO) to maintain momentum and justify continued investment in the scaling phases.

Action & Application:

- **Audience Tailoring:**
  - » **Executives (CEO, COO, CTO):** Priya prepares a concise presentation focusing on: 1) Linkage to strategic goals (warranty cost reduction, cross-sell revenue). 2) Key pilot results vs. baseline (e.g., "Pred-Maint pilot identified 7% of potential critical failures proactively, projecting $300k annualized savings if

scaled";"HyperPers pilot achieved 18% conversion lift vs. control group in A/B test"). 3) High-level ROI projection update. 4) Key risks/learnings. 5) Recommendation to proceed with scaling. A one-page dashboard summary is also created using Power BI.

> » **AI Review Team & Sponsors (VPs Product/Marketing):** A more detailed review includes specific KPI trends, model performance metrics (briefly), user feedback summaries, challenges encountered (e.g., data integration delays for HyperPers), detailed TCO refinement, and discussion of scaling plan risks.

- **Storytelling with Data:** Priya frames the results as demonstrating progress towards the strategic goals set in Chapter 4. She highlights how the predictive maintenance alerts helped specific technicians prevent costly failures (using anonymized examples) and how personalized offers drove measurable engagement uplift.
- **Transparency:** She openly discusses the data integration challenges faced by the HyperPers pilot and the mitigation plan involving closer collaboration with the central Data team, demonstrating proactive problem-solving. She also shares feedback about initial user interface awkwardness in the technician app for Pred-Maint alerts and the plan to address this in the scaling phase based on user input.
- **Channels:** Uses formal presentations for review meetings, shares the executive Power BI dashboard via email link, and drafts an article for the next company-wide internal newsletter highlighting the pilot progress and responsible AI approach.

Outcome & Progress:

The clear, tailored communication, focused on business value and transparent about challenges, successfully builds confidence among executives and sponsors. The AI Review Team approves proceeding to the scaling phases for both pilots based on the demonstrated value and clear plans. This effective communication (Compass West) reinforces the AI Office's credibility and secures continued support for the roadmap, aligning with EAMI Value Realization best practices.

15.10 Conclusion: Making AI Value Resonate and Trusted

Measuring AI performance and quantifying its business value, as explored in the preceding chapters, is essential groundwork (Compass West). However, realizing the full benefits of these efforts requires mastering the art of communication and actively cultivating organizational trust. Effectively communicating AI value involves translating complex data into compelling narratives tailored to diverse stakeholder audiences, utilizing appropriate formats and channels, and demonstrating clear linkage to strategic objectives. Building enduring trust necessitates unwavering transparency – openly discussing capabilities, limitations, risks, and governance processes, and actively engaging with concerns and feedback. By leading these communication and trust-building initiatives, the AI Office ensures that the impact of AI is not just achieved but also recognized, celebrated, and leveraged to fuel the ongoing journey towards greater AI maturity and enterprise transformation. Having established how to demonstrate AI's value, Part 5 shifts focus to the foundational inputs required to generate that value: Resource Management (South).

Key Takeaways:

- Effective communication is vital to demonstrate AI value, justify investment, drive adoption, and build trust.
- Tailor communication content, format, and channels to specific stakeholder audiences (executives, tech teams, users, etc.).
- Use data storytelling techniques to create compelling narratives that explain the why, how, what, and so what

of AI initiatives.

- Transparency, including acknowledging challenges, is crucial for building and maintaining organizational trust in AI.
- Proactively engage with AI skeptics by listening, addressing concerns directly, and demonstrating relevant value.

The AI Office plays a key role in orchestrating communication strategy and trust-building efforts.

Food for Thought / Application Exercise:

- Identify the key stakeholder groups for a specific AI initiative in your organization. What are their primary interests and concerns regarding this initiative? How would you tailor communication to each group?
- Think about a recent AI success (or challenge) in your organization. How could you frame this using the data storytelling elements (Why, How, What, So What, Human Element, Learnings, Now What)?
- What are the biggest sources of skepticism or lack of trust regarding AI in your organization? What specific communication or engagement strategies could help address these?

PART 5 - AI RESOURCE

MANAGEMENT

~ CHAPTER 16 ~

Building High-Performing AI Teams: Talent and Collaboration

"The strength of the team is each individual member. The strength of each member is the team."
— Phil Jackson

16.1 Introduction: The People Imperative in AI

Welcome to Part 5, where we explore the crucial Resource Management (South) domain of the AI Office Compass™ Framework. While previous parts focused on strategy (North), governance (East), and value delivery (West), successful execution ultimately hinges on having the right people with the right skills, organized effectively, and collaborating seamlessly. AI is not just about algorithms and data; it's fundamentally conceived, built, deployed, governed, and managed by human talent.

Building and nurturing high-performing, diverse AI teams is arguably one of the most critical challenges – and key differentiators – for organizations seeking to achieve significant and sustainable AI maturity (EAMI Levels 4 and 5). The rapid evolution of AI technologies creates intense demand for highly specialized skills (like ML engineering, MLOps, AI ethics, prompt engineering), often outstripping the available supply in the global market and leading to fierce competition for top talent. Organizations cannot rely solely on traditional recruitment methods; they must adopt a strategic, multi-faceted approach to acquiring, developing, organizing, and, crucially, retaining the diverse expertise needed to succeed with enterprise AI.

This chapter delves into the key roles and essential skillsets required for enterprise AI, explores different strategies for sourcing talent (Build, Buy, Borrow, Retain), discusses various team structures aligned with AI operating models (Chapter 2), and emphasizes the importance of fostering cross-functional collaboration and diversity within AI teams. Effectively managing the "people" dimension of AI, a core component of the Resource Management (South) domain, is paramount for sustained success and directly impacts the EAMM Talent Acquisition and Skill Development criteria.

Learning Objectives:

- Understand the critical importance of a strategic approach to talent for successful enterprise AI adoption and maturity.
- Identify the key roles (e.g., Data Scientist, ML Engineer, AI Ethicist, AI Translator) and essential skillsets (technical, analytical, domain, ethical, soft) required for high-performing AI teams.
- Explore different talent sourcing strategies (Build, Buy, Borrow, Retain) and their respective advantages and disadvantages in the context of AI skills.
- Learn about different organizational structures for AI teams (e.g., centralized pool, embedded "spokes," matrixed structures) and their alignment with different AI operating models and maturity levels.
- Appreciate the crucial importance of fostering effective cross-functional collaboration and diversity within AI teams.
- Recognize the AI Office's pivotal role in shaping, enabling, and supporting the enterprise AI talent strategy in partnership with HR.

16.2 The AI Talent Landscape: Key Roles and Skills

Building effective AI systems requires a synergistic blend of diverse expertise, extending far beyond the traditional perception of needing only data scientists. High-performing enterprise AI teams typically incorporate individuals (or individuals wearing multiple hats in smaller organizations) proficient in several key areas, as summarized below:

Table 16.1: Key Roles and Skills in Enterprise AI Teams

Key Role Category	Example Titles	Primary Focus & Responsibilities	Essential Skills
Data Science & Modeling	Data Scientist, ML Scientist, AI Researcher, Quantitative Analyst	Applying statistical/ML techniques to explore data, develop predictive/ prescriptive models, conduct feature engineering, select/ tune algorithms, design/run experiments, stay abreast of research advancements.	Strong statistics & math foundation, deep understanding of ML algorithms (supervised, unsupervised, deep learning, reinforcement learning), proficiency in programming (Python/R) and relevant libraries (Scikit-learn, TensorFlow, PyTorch), experimental design rigor, often significant domain knowledge.

ML Engineering & MLOps	ML Engineer, MLOps Engineer, AI Platform Engineer	Building robust, scalable, and reliable ML pipelines; deploying models into production environments; managing ML infrastructure (cloud/on-prem); implementing CI/CD/CT & monitoring practices (Chapter 20); optimizing model serving performance and cost.	Strong software engineering principles (testing, modularity, version control), cloud platforms (AWS/Azure/GCP), containerization (Docker/Kubernetes), MLOps tools (MLflow, Kubeflow, Airflow), infrastructure-as-code (Terraform), scripting (Python, Bash), systems design.
Data Engineering	Data Engineer, Analytics Engineer	Designing, building, and maintaining scalable data pipelines (ETL/ELT - Chapter 22); ensuring data quality, availability, and accessibility for AI use cases (Chapter 23); managing data stores (lakes/warehouses/lakehouses - Chapter 21); collaborating closely with data scientists on data requirements.	Advanced SQL, data modeling, expertise in big data technologies (Spark, Hadoop, Flink), cloud data services (BigQuery, Redshift, Snowflake, Databricks), data pipeline orchestration tools (Airflow, Prefect), programming (Python/Scala/Java), data quality and testing frameworks.
AI Governance & Ethics	AI Ethicist, Responsible AI Lead, Governance Lead, Risk Analyst	Developing and implementing AI policies/standards (ethics, fairness, transparency, compliance); conducting ethical reviews and AI risk assessments; advising teams on bias detection/mitigation techniques; ensuring compliance with regulations (Chapters 7-12); promoting an ethical AI culture.	Deep understanding of AI ethics principles & frameworks, familiarity with fairness/bias/XAI concepts & tools, knowledge of relevant regulations (GDPR, AI Act), risk assessment methodologies, policy development & communication skills, strong analytical and cross-functional collaboration abilities.

AI Product Management	AI Product Manager, Technical Product Manager	Defining the vision, strategy, and roadmap for specific AI-powered products or features; translating business needs and user requirements into technical specifications; prioritizing features and managing the product backlog; driving user adoption and measuring business value delivered (Domain West).	Strong product management fundamentals (market analysis, user research, roadmap planning), solid understanding of AI/ML capabilities and limitations, user-centered design thinking, proficiency in Agile methodologies, business acumen, excellent technical communication skills.
AI Translator / Business Analyst	AI Translator, Business Analyst (AI Focus), Domain Liaison	Bridging the communication and understanding gap between business stakeholders and technical AI teams; eliciting and clarifying business requirements for AI solutions; explaining complex AI concepts and implications in clear business terms; facilitating communication and stakeholder management; supporting value tracking and reporting.	Strong business domain knowledge specific to the application area, exceptional analytical thinking and problem-solving skills, excellent communication (verbal, written, presentation) skills, foundational AI/data literacy, requirements gathering techniques, stakeholder engagement expertise.
Domain Expertise	Subject Matter Expert (SME) from Business Units	Providing deep, contextual knowledge of the specific business process, operational environment, data nuances, user needs, and validation criteria relevant to the AI application. Essential input for building relevant, accurate, and usable AI solutions.	Deep expertise and credibility within their specific functional area (e.g., marketing campaign management, financial fraud investigation, supply chain logistics, clinical research protocols), ability to clearly articulate operational needs and constraints, capacity to critically validate AI outputs against real-world scenarios.

Emerging Roles	Prompt Engineer, AI Ops Specialist, Data/ML Architect, AI UX Researcher, AI Auditor	Focusing on optimizing interactions with large language models (LLMs), managing the operational health and performance of deployed AI systems, designing overarching AI system/data architectures, ensuring user-centric design for AI interfaces, providing independent assurance on AI governance/controls.	Specialized skills evolving rapidly: LLM interaction techniques, advanced operational monitoring & automation, complex systems architecture design for AI, user research methods adapted specifically for AI systems, AI auditing frameworks and techniques.

> **D**
> **DEFINITION**
>
> **AI Skillsets:** Beyond specific job titles, building effective AI capabilities requires cultivating a blend of skill-sets within the team or through collaboration:
> - **Technical Skills:** Foundational programming (Python/R), robust statistics, diverse ML algorithms knowledge, data engineering techniques (SQL, Spark), cloud platform proficiency, MLOps tooling expertise.
> - **Analytical Skills:** Strong problem-solving aptitude, critical thinking, experimental design methodology, data interpretation and visualization skills, business process analysis.
> - **Domain Expertise:** Deep understanding of the specific business function, industry context, or scientific field where AI is being applied.
> - **Ethical & Governance Skills:** Understanding of responsible AI principles, awareness of bias types and mitigation strategies, knowledge of fairness concepts and metrics, familiarity with relevant privacy regulations and security best practices.
> - **Soft Skills:** Exceptional communication (translating between technical and business audiences), effective collaboration and teamwork, data storytelling ability, stakeholder management and influence, strong change leadership capabilities, and critically, a continuous learning mindset to keep pace with the field.

Successfully acquiring and nurturing this blend of roles and skills (Compass South) is fundamental for advancing across multiple EAMM dimensions, particularly Talent Acquisition and Skill Development, but also impacts areas like Collaboration, Innovation, and overall execution capability.

16.3 Strategic Talent Sourcing: Build, Buy, Borrow, Retain

Given the high demand and limited supply for many specialized AI roles (like experienced ML Engineers or AI Ethicists), organizations typically need a multi-pronged, strategic approach to sourcing the necessary talent (Compass South). Relying solely on external hiring ("Buy") is often insufficient and unsustainable. A balanced strategy usually incorporates:

Table 16.2: AI Talent Sourcing Strategies

Strategy	Description	Pros	Cons	EAMM Alignment Focus
Build	Investing significantly in targeted training, upskilling, and reskilling programs to develop AI capabilities within the existing internal workforce.	Leverages valuable existing domain knowledge, institutional memory, and cultural fit; potentially lower long-term cost; improves employee retention, morale, and internal career pathways.	Takes considerable time (months/years) to develop deep expertise; requires substantial investment in high-quality training programs; may not address immediate needs for highly specialized or senior expertise.	Skill Development, Culture, Collaboration
Buy	Actively recruiting and hiring experienced AI talent (from entry-level to senior experts) directly from the external market.	Can bring in cutting-edge skills, specific niche expertise, and fresh external perspectives relatively quickly; essential for filling immediate critical gaps or acquiring leadership roles.	Highly competitive and often expensive talent market; onboarding takes time for new hires to become productive/ understand context; potential challenges with cultural integration or retention if internal environment isn't supportive.	Talent Acquisition
Borrow	Engaging external resources like specialized consulting firms, individual contractors, or staff augmentation partners for specific projects, expertise, or temporary capacity needs.	Provides flexible, on-demand access to highly specialized skills; can significantly accelerate specific projects or overcome temporary resource bottlenecks; useful for evaluating new technologies or getting expert advice.	Can be very expensive; risk of limited knowledge transfer back to internal teams; potential for creating long-term dependencies; requires careful vendor management and contract definition.	Collaboration, Innovation, Project Management
Retain	Implementing proactive strategies focused on keeping high-performing, valuable AI talent engaged, motivated, and committed to staying long-term.	Protects critical institutional knowledge and prior investment in talent development; maintains team stability, cohesion, and productivity; significantly reduces high costs/disruption of constant recruitment/ onboarding in competitive market.	Requires continuous, proactive effort: competitive total compensation, challenging projects, clear career paths, ongoing learning opportunities, strong team culture, effective leadership.	Talent Acquisition, Culture, Leadership

Employ a blended talent strategy that dynamically combines elements of Build, Buy, Borrow, and Retain based on evolving needs, strategic roadmap priorities, budget, urgency, and long-term workforce goals. Don't rely solely on one approach. Strategically "Buy" critical leadership/niche roles, heavily invest in "Build" for broader literacy/core skills, selectively "Borrow" for acceleration/advice, and always prioritize "Retain" for valuable contributors. Align this blended strategy explicitly with targeted EAMI Talent Acquisition and Skill Development maturity goals.

16.4 Organizing AI Teams: Structures and Collaboration

How AI talent is organized (Compass South decision) impacts efficiency, collaboration, knowledge sharing, business alignment, and the ability to execute the chosen operating model (Chapter 2). Common structures include:

- **Centralized Pool (Often aligns with Centralized Operating Model):** Most AI specialists reside within a single, central team (AI Office/CoE Hub), allocated to projects across BUs based on central prioritization.
  - » **Pros:** Facilitates strong knowledge sharing, consistent standards, peer learning; efficient allocation of scarce resources; builds deep technical expertise centrally. Good for initial capability building (EAMI Level 2-3).
  - » **Cons:** Can disconnect from BU needs/context; potential prioritization conflicts; risk of bottleneck; perceived "ivory tower" if weak translator roles.
- **Decentralized/Embedded Teams (Aligns with Decentralized Operating Model):** AI specialists fully embedded within BUs/product teams, reporting solely through that unit.
  - » **Pros:** Deep domain expertise, close alignment with BU needs/priorities; strong BU ownership.
  - » **Cons:** High risk of silos, duplication, inconsistent standards, poor governance oversight, difficult knowledge sharing, inefficient talent use. Often reflects low EAMI maturity.
- **Hybrid/Federated Structure (Aligns with Hybrid Operating Model):** Combines central coordination with distributed execution. Widely adopted.
  - » **Central "Hub" (AI Office/CoE):** Provides strategy, governance, core platforms/tools, shared services, deep expertise, enablement programs.
  - » **Distributed "Spokes":** Dedicated AI specialists/teams embedded within BUs/product lines. Focus on domain-specific solutions using Hub resources/standards, provide business context, drive local adoption, feed learnings back. (Enabling Spokes key for EAMI Level 4/5).
  - » **Pros:** Balances central coordination/standards with BU agility/ownership; enables scalable deployment; fosters collaboration; optimizes expertise use.
  - » **Cons:** Requires clear roles/interfaces; complex initial setup; needs strong coordination; potential matrix reporting complexity.
- **Matrixed Structures:** Often used within Hybrid/Centralized models. Individuals report functionally (e.g., central Data Science) but have dotted-line/assignment to projects/domains. Fosters deep expertise and project alignment.

CENTRALIZED POOL

DECENTRALIZED / EMBEDDED

HYBRID / FEDERATED

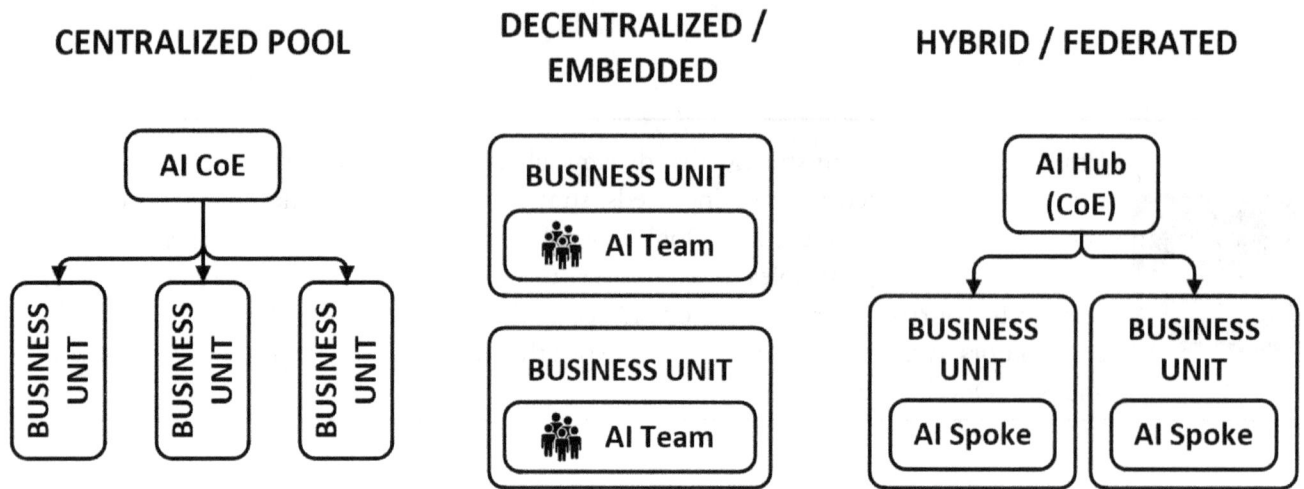

Figure 16.1: AI Team Structures Diagram

Fostering Collaboration and Diversity: Regardless of structure, actively fostering collaboration and diversity is crucial:

- **Mandate Cross-Functional Project Teams:** Ensure core project teams include representation from all necessary disciplines (business SME, DS, MLE, DE, MLOps, Gov, UX, Change) from the start. Avoid technical isolation.
- **Establish and Nurture Communities of Practice (CoPs):** Create/support CoPs for key AI roles (DS, MLOps, Responsible AI). Vital for informal knowledge sharing, best practices, peer problem-solving, community building across silos (Compass Center activity). AI Office often sponsors/facilitates.
- **Utilize Shared Tools & Platforms:** Standardize on common dev platforms, code repos (Git), project tools (Jira/Azure DevOps), communication channels (Slack/Teams) to enhance collaboration, visibility, reuse.
- **Actively Promote Diversity & Inclusion:** Consciously build diverse teams (skills, backgrounds, perspectives, demographics). Diverse teams are more innovative, less prone to groupthink, better at identifying/ mitigating bias. Requires inclusive hiring/culture.

> Collaboration extends externally. Encourage interaction with the broader AI ecosystem (industry consortia, conferences, open-source, partners, research). Helps stay current, benchmark, bring in new ideas (supports EAMM Innovation Ecosystem).

16.5 The AI Office's Role in Talent Strategy

The AI Office (Compass South), partnering closely with HR, plays a crucial strategic role in the enterprise AI talent ecosystem:

- **Defining AI Roles, Skills & Career Paths:** Collaborating with HR on clear, standardized role definitions, competency levels, attractive career paths for AI professionals.
- **Developing the Blended Talent Strategy:** Providing input on Build/Buy/Borrow/Retain strategy based on roadmap needs, current capabilities (EAMI assessments), market, budget.
- **Supporting Strategic Recruitment ("Buy"):** Partnering with HR/TA on job descriptions, technical screening/interviewing, attracting top candidates.
- **Championing Upskilling & Training ("Build"):** Identifying skill gaps, advocating for training investments, collaborating with HR/L&D on effective internal/external AI training (technical, tools, MLOps, responsible AI, literacy).
- **Facilitating Communities & Knowledge Sharing:** Establishing, sponsoring, nurturing AI CoPs for peer learning and best practice dissemination.
- **Promoting Collaboration & Best Practices:** Encouraging shared tools, standard processes (MLOps), collaborative work methods.
- **Advising on Team Structure & Operating Model:** Providing expert guidance to BUs (esp. in Hybrid models) on structuring embedded "Spoke" teams and establishing productive collaboration with the central "Hub".

16.6 OmnioTech Case Study: Building the Initial AI Team

Setting the Scene:

OmnioTech needs to staff its new AI Office Hub (Chapter 2) and provide expertise for initial pilots (PredMaint, HyperPers). Priya Sharma (AI Office Head) collaborates with HR Lead Maria Garcia, noting baseline EAMI assessment showed gaps in Talent Acquisition and Skill Development (Level 1-2).

Action & Application:

- **Talent Strategy (Initial Blended Approach):**
  - » **Buy:** Prioritize immediate external hiring for critical missing roles: AI Governance Lead (ethics/regulatory expertise), Lead Platform/MLOps Engineer (Azure ML/pipeline skills). Develop attractive JDs.
  - » **Borrow/Leverage:** Secure 75% FTE allocation from experienced Lead Data Engineer (central IT Data team). Assign existing capable PMs (central PMO) to pilots, with Priya providing AI coaching (initial AI Translator).
  - » **Build (Future Focus):** Plan internal development. Propose internal "AI Catalyst" upskilling program (Y1) for high-potential analysts/engineers (Product/Marketing) targeting future "Spoke" roles. Identify initial online AI literacy modules for broader rollout (Ch 26).
  - » **Retain:** Priya focuses on creating exciting vision/projects to retain self/attract hires. Longer-term retention strategies (career paths, training budgets, recognition) flagged as Y1 AI Office/HR deliverable.
- **Team Structure (Initial Hub):** Core AI Office Hub starts small, functionally oriented (Priya + Gov Lead + Platform Lead + Data Eng Liaison). Pilot teams operate matrixed: Hub members, IT Data team member,

BU domain experts/sponsors (Product for PredMaint, Marketing for HyperPers).
- **Collaboration Mechanisms:** Mandatory weekly Hub syncs. Bi-weekly "AI Pilot Sync" meetings (wider teams incl. BU reps). Dedicated MS Teams channel ("Enterprise AI Hub") for discussions/docs. Plans noted for formal Data Science CoP in Y2.

Outcome & Progress:

OmnioTech adopts pragmatic, blended talent strategy (Compass South) addressing immediate needs ("Buy"), leveraging internal resources ("Borrow"), planning for future ("Build"). Initial matrix structure supports Hub-heavy hybrid model. Focused approach provides expertise for pilots, establishes longer-term plan to improve EAMI Talent Acquisition and Skill Development maturity towards Level 3/4 targets.

16.7 Conclusion: People Powering AI Progress

Enterprise AI success rests fundamentally on the capabilities, motivation, organization, and collaboration of people (Compass South). Building high-performing teams requires deliberate, strategic talent management: defining roles/skills, executing blended sourcing (Build/Buy/Borrow/Retain), organizing effectively (often evolving to hybrid), fostering collaboration/diversity/learning. The AI Office, partnering with HR/leadership, orchestrates this talent strategy, ensuring the vital human capital needed to execute the AI vision. With the right people/teams, Chapter 17 examines the enabling technology ecosystem.

Key Takeaways:

- High-performing AI teams require diverse roles (DS, MLE, DE, Ethicist, Translator, PM, SME) and skills (technical, analytical, domain, ethical, soft).
- **Strategic talent approach blends sourcing:** Build (upskill), Buy (hire), Borrow (contractors), Retain (keep top talent).
- Team structures (Centralized, Decentralized, Hybrid) align with operating model/maturity; Hybrid often balances central expertise/BU agility.
- Cross-functional collaboration (project teams, CoPs, tools) and diversity are crucial.
- AI Office, with HR, drives talent strategy, roles, recruitment, training, collaboration.

Food for Thought / Application Exercise:

- Review key AI roles (Table 16.1). Which exist (formally/informally) in your org? Biggest gaps?
- Considering Build/Buy/Borrow/Retain, what's the most appropriate initial mix for your org's AI skills? Why?
- Which team structure (Centralized, Decentralized, Hybrid) best describes your current AI approach? Effective? How evolve for scale?

CHAPTER 17

Designing the Enabling Technology Ecosystem: Platforms and Tools

"The best tools are the ones that amplify human capability, reliably and at scale."

17.1 Introduction: The Technology Foundation for AI

Chapter 16 focused on building high-performing AI teams. This chapter shifts attention to another essential pillar within the Resource Management (South) domain: the Enabling Technology Ecosystem. AI initiatives require a robust, scalable, secure, and well-integrated set of technologies: infrastructure, data platforms, development environments, MLOps tooling, and potentially specialized hardware (GPUs/TPUs). Designing and managing this landscape strategically is crucial for enabling efficient AI development, reliable deployment, and effective operations at scale, forming the technical backbone for achieving higher EAMI maturity (Technology Selection, Infrastructure, MLOps dimensions).

Making suboptimal technology choices early can lead to technical debt, vendor lock-in, poor performance/reliability, security vulnerabilities, integration nightmares, and spiraling costs. Conversely, a well-architected AI technology ecosystem, aligned with needs, strategy, and current EAMI level, significantly accelerates adoption, improves developer productivity, enhances model reliability/trustworthiness, and optimizes resource use.

This chapter explores key components of a modern enterprise AI tech stack. We discuss infrastructure choices (cloud, on-prem, hybrid, edge), essential data platform elements for AI, criteria for selecting AI/ML development platforms and MLOps tooling (linking to Chapter 20), "build vs. buy" considerations, vendor management importance, interoperability, and managing technical debt. The AI Office, collaborating with IT Architecture/ Infrastructure, guides this technology strategy (Compass South) ensuring the ecosystem supports AI ambitions and EAMI Technology goals.

Learning Objectives:

- Understand key layers/components of a comprehensive enterprise AI technology ecosystem.
- Compare infrastructure options (cloud, on-prem, hybrid, edge) for AI workloads considering pros/cons and EAMI level suitability.
- Identify essential elements/capabilities of modern data platforms supporting AI at scale.
- Learn key criteria for evaluating/selecting AI/ML development platforms and MLOps toolchains (relevant to EAMM Technology/MLOps).
- Understand strategic trade-offs in "build vs. buy" decisions for AI tools/platforms/models.
- Recognize importance of vendor risk management, integration/interoperability, managing technical debt.
- Appreciate AI Office's collaborative role guiding AI technology strategy, standards, selection.

17.2 The AI Technology Imperative

Selecting/managing the right technology is a strategic imperative for AI success. The ecosystem must:

- **Enable Scalability:** Support processing massive datasets and handle high inference volumes without degradation.
- **Ensure Reliability & Performance:** Provide stable, resilient, performant infrastructure/platforms for training/inference.
- **Facilitate Developer Productivity:** Offer integrated tools, familiar environments (notebooks), efficient data access, streamlined workflows.
- **Support Mature MLOps:** Provide integrated tooling for automating the ML lifecycle (versioning, testing, CI/CD/CT, monitoring, registry - Chapter 20, EAMM MLOps).
- **Maintain Security & Compliance:** Adhere to security standards and regulations (Domain East, EAMM Security/Compliance).
- **Manage Costs Effectively:** Allow efficient resource use, cost visibility, optimization (FinOps - Chapter 19, EAMM Financial).
- **Promote Interoperability:** Enable seamless data flow/integration between AI stack components and core enterprise systems.

A poorly designed/fragmented ecosystem creates friction, increases costs, slows innovation, hinders scaling, limits potential EAMI maturity across multiple dimensions (Technology, MLOps, Data Management, Value Realization).

17.3 Infrastructure Choices: Cloud, On-Premise, Hybrid, Edge

Where AI workloads run impacts cost, scalability, control, flexibility. Choice depends on needs, existing infra, regulations, AI maturity.

Table 17.1: Comparison of AI Infrastructure Options

Infrastructure Option	Description	Key Pros	Key Cons	Best Suited For
Public Cloud	Utilizing IaaS/ PaaS from providers (AWS, Azure, GCP), including managed AI/ ML platforms, scalable storage, specialized compute (GPUs/ TPUs).	High scalability & elasticity (pay-as-you-go); rapid access to cutting-edge managed AI services & hardware; reduced internal infra management overhead; global reach.	Potential for high/ unpredictable costs (FinOps crucial); data residency/ sovereignty complex; potential vendor lock-in; security is shared responsibility requiring careful customer configuration.	Most orgs pursuing significant AI, needing scalability, latest services/ hardware, or minimizing internal infra management.
On-Premise	Running AI workloads entirely on company-owned/managed hardware in private data centers.	Complete control over hardware, software, security; potentially lower long-term OpEx for predictable workloads; simplified compliance with strict data residency/security mandates.	Significant upfront CapEx; requires substantial internal expertise for setup, management, maintenance, upgrades; slower scalability/ provisioning; limited access to latest hardware/managed PaaS AI services.	Orgs with extreme security needs, strict data residency mandates, very large/predictable workloads justifying CapEx, or existing massive private data center investments.
Hybrid Cloud	Strategically combining public cloud resources/ services with private cloud or on-premise infrastructure; run workloads where most appropriate.	Flexibility ("best of both worlds"); leverage cloud scale/services for less sensitive tasks, keep sensitive data/low-latency inference on-premise; enables phased cloud migration.	Increased architectural complexity (connectivity, security, orchestration); requires robust networking & hybrid management tools; challenges in consistent policy/ cost management across platforms.	Orgs needing flexibility for diverse workloads, undergoing gradual cloud migration, or with specific data residency/security needs dictating placement.

Edge Computing	Deploying/executing AI models directly on endpoint devices near data source/action point (IoT, factory, cameras, vehicles, smartphones).	Enables very low-latency inference; reduces bandwidth/costs (local processing); enhances privacy (minimizes data transmission); allows offline operation.	Limited compute/memory on devices restricts model complexity; challenges managing monitoring/updating distributed models; physical security of devices.	Applications needing real-time decisions (industrial IoT, autonomous vehicles, video analytics, on-device mobile ML), offline operation scenarios, or strong privacy constraints.

> ★ Infrastructure choice is rarely enterprise-wide. Many adopt hybrid/multi-cloud, selecting the best environment (public cloud A/B, private, edge) per workload based on technical needs (performance, latency), cost, security, data governance, regulations.

17.4 Data Platform Essentials for AI

AI is fueled by data (Chapter 18). The underlying data platform (Compass South) needs specific capabilities:

- **Scalable & Flexible Data Storage:** Cost-effectively store/manage large volumes (TBs/PBs) of diverse data (structured, semi-structured, unstructured). Paradigms:
  - » **Data Lakes:** (e.g., S3, ADLS Gen2, GCS) Store raw data flexibly/cost-effectively (schema-on-read). Need governance to avoid "swamps".
  - » **Data Warehouses:** (e.g., Snowflake, BigQuery, Redshift) Optimized for structured data analytics/BI (SQL, schema-on-write). Less ideal for raw/unstructured ML data.
  - » **Lakehouse Architecture:** (e.g., Delta Lake/Iceberg/Hudi on lakes via Databricks/Snowflake) Combines lake flexibility with warehouse management (ACID, schema enforcement/evolution, versioning). Unified platform for BI and ML on same data copies. Increasingly favored for high EAMI Data Management maturity. (Discussed Chapter 21).
- **Efficient Data Pipelines & Integration:** Robust tools/frameworks for reliable data ingestion, transformation (ETL/ELT - Chapter 22), quality assurance (Chapter 23), efficient data movement. (e.g., Cloud services like ADF/Glue/Dataflow, orchestrators like Airflow, transformation tools like dbt).
- **Data Governance & Cataloging:** Integrated capabilities/tools for metadata management, lineage tracking, data discovery (catalog), quality rules enforcement, access control, compliance (e.g., Collibra, Alation, Purview, Glue Data Catalog). Essential for responsible AI (Compass East/South).
- **Feature Stores:** Centralized platforms managing ML feature lifecycle (definition, storage, discovery, sharing, serving, versioning, monitoring). Promote consistency, reduce redundant engineering, bridge data prep/deployment, crucial for mature MLOps (Chapter 20) and higher EAMI maturity. (Examples: Feast, Tecton,

cloud services). (Discussed Chapter 21).

17.5 AI/ML Development Platforms

Core environment where teams build, train, evaluate, prepare models:

- **Cloud AI Platforms (PaaS):** Comprehensive suites (AWS SageMaker, Azure Machine Learning, Google Vertex AI). Provide managed notebooks, AutoML, scalable training, registries, MLOps features, data labeling, foundation models. Abstract infrastructure complexity, preferred for enterprise scale/integration, support higher EAMI Technology maturity.
- **Open Source Frameworks & Libraries:** Foundational blocks (TensorFlow, PyTorch, Keras, Scikit-learn, XGBoost, Hugging Face Transformers, spaCy, OpenCV). Offer flexibility/control but require more internal effort for infra/MLOps management. Typically used within managed/commercial platforms.
- **Commercial AI/ML Platforms:** Specialized end-to-end platforms (DataRobot, H2O.ai, C3 AI). Offer automated workflows, industry solutions, aim for faster time-to-value (esp. for less mature teams). Involve licensing costs, potential lock-in risks.
- **Key Selection Criteria:** Consider ease of use, breadth/depth of capabilities, scalability/performance, integration (data/MLOps), security/governance controls, cost model, vendor support/docs, team skill alignment.

17.6 MLOps Tooling (Integration Point)

As detailed in Chapter 20, a mature, integrated MLOps tool-chain is essential for reliable, scalable, governed AI (Compass South, EAMM MLOps). It's a critical layer in the tech ecosystem. Key tool categories needing integration:

- **Experiment Tracking & Management:** (MLflow Tracking, W&B, Comet, Neptune.ai, cloud platform equivalents)
- **Version Control:** (Git; DVC for data/models)
- **Feature Stores:** (Feast, Tecton, cloud provider services)
- **CI/CD/CT Pipeline Orchestration:** (Jenkins, GitLab CI, GitHub Actions, Azure Pipelines, Kubeflow Pipelines, Argo)
- **Model Registries:** (MLflow Registry, cloud platform registries)
- **Model Serving & Deployment Frameworks:** (KServe, Seldon, BentoML, TF Serving, cloud managed endpoints, Kubernetes)
- **Production Monitoring & Observability:** (Prometheus/Grafana, Datadog, specialized AI monitoring tools, cloud provider services)
- **Infrastructure as Code (IaC):** (Terraform, Bicep, CloudFormation, Pulumi)

The AI Office (Platform/MLOps Eng - Domain South) typically guides selection, standardization, integration, adoption of these tools.

17.7 Build vs. Buy Decisions and Vendor Management

Whether to build custom solutions or buy COTS/use PaaS applies to the AI ecosystem:

- **Build:** Develop custom AI platforms/frameworks/models internally.
  - » **Pros:** Full customization for unique needs; potential competitive advantage; full IP/roadmap control.
  - » **Cons:** Requires substantial/scarce internal expertise; long dev time, high upfront investment; higher failure risk; ongoing internal maintenance/support. Often only feasible for large tech firms or core IP components.
- **Buy/Use Services (COTS or PaaS):** Purchase licenses or use managed cloud AI services (AWS/Azure/GCP).
  - » **Pros:** Faster time-to-market; leverage vendor expertise/R&D; lower upfront costs (esp. PaaS); vendor handles infra/platform upgrades.
  - » **Cons:** Ongoing license/subscription costs (can be high); potential vendor lock-in; less customization flexibility; requires careful due diligence (security, compliance, SLAs, data handling, roadmap alignment).

> Adopt a pragmatic, hybrid approach. Leverage established open-source frameworks and robust managed cloud PaaS (SageMaker, Azure ML, Vertex AI) for core infra, data platforms, standard ML dev, foundational MLOps to accelerate/reduce burden. Focus internal "build" highly selectively on unique, strategic differentiators where COTS/PaaS is inadequate. Implement rigorous Vendor Risk Management for all third-party AI tools/platforms/data (involve Security, Legal, Compliance, Procurement).

17.8 Integration, Interoperability, and Technical Debt

Disconnected tools aren't an effective ecosystem. Integration is critical:

- **Integration & Interoperability:** Design ecosystem for integration. Favor platforms/tools with open APIs, standard support (e.g., ONNX), pre-built connectors. Lack of interoperability creates brittle pipelines, manual handoffs, complexity, hinders automation.
- **Managing Technical Debt:** Expedient but suboptimal tech choices (non-standard libraries, poorly engineered custom solutions, lack of maintenance) lead to technical debt. Manifests as fragility, higher maintenance costs, slowed innovation. AI Office should advocate for strategic platform investment, sound engineering, periodic refactoring/decommissioning to manage debt proactively.

17.9 The AI Office's Role in Technology Strategy

The AI Office (Compass South), collaborating with Enterprise Architecture, IT Infra, Data Mgmt, Cybersecurity, influences/leads AI tech strategy:

- **Developing Technology Strategy & Principles:** Defining architectural vision, principles, standards for AI

162

platforms/infra/data/tools, aligning with business strategy (North) and maturity goals (Center).

- **Platform/Tool Selection & Standardization:** Leading/participating in evaluation, selection, procurement of core enterprise AI platforms/tools. Promoting standardization for efficiency/support.
- **Architecture Guidance & Best Practices:** Providing expert architectural guidance, reference architectures, best practices for building scalable, reliable, secure, cost-effective solutions.
- **Vendor Assessment & Management:** Assisting in technical evaluation, due diligence, ongoing relationship management for third-party AI tech vendors/data providers.
- **Promoting Foundational Practices:** Championing modern data engineering, robust MLOps (Chapter 20), disciplined FinOps (Chapter 19) adoption.
- **Staying Abreast of Technology Trends:** Monitoring evolving AI tech landscape (linking to Chapter 27 - Innovation), advising on impact/adoption path.

17.10 OmnioTech Case Study: Defining the Initial Tech Stack

Setting the Scene

As OmnioTech launches its AI Office Hub and prepares pilots, the new Platform/MLOps Lead (reporting to CTO, collaborating with IT Arch) defines the foundational V1.0 tech stack (Compass South activity), supporting immediate needs and future scaling, aiming for EAMI Technology Level 3.

Action & Application

- **Infrastructure Choice:** Standardize on Microsoft Azure (existing enterprise agreement, skills, PaaS AI offerings). Azure Kubernetes Service (AKS) for custom inference endpoints.
- **Data Platform:** Utilize existing Azure Data Lake Storage (ADLS) Gen2 (raw/processed data). Implement Azure Data Factory (ADF) for initial batch pipelines. Plan documented to evaluate Databricks on Azure (Year 2) for advanced processing/Feature Store. Azure Purview initiated for data cataloging (prioritizing pilot sources).
- **AI/ML Platform:** Select Azure Machine Learning (Azure ML) as primary, standardized platform (managed compute, notebooks, experiments, registry, MLOps integration, endpoints). Supports baseline MLOps goals (Chapter 20). Data scientists use Python with standard libraries within Azure ML.
- **MLOps Tooling (V1.0):** Leverage integrated Azure capabilities: Azure Repos (Git), Azure ML Model Registry, Azure Pipelines (CI/CD), Azure Monitor (basic monitoring). Defer specialized third-party MLOps tools initially.
- **Build vs. Buy:** Strategy favors "Buy/Use Services" (Azure PaaS). Internal "Build" focuses solely on custom ML models/logic within Azure ML platform. No specialized commercial end-to-end AI platform bought initially.
- **Integration Plan:** Initial focus on secure/efficient data flow (source systems → ADLS → Azure ML via Azure networking best practices). Plan for future output integration (ServiceNow, Marketing tool).

Outcome & Progress

OmnioTech defines pragmatic, cloud-native V1.0 AI tech stack (Azure-centric). Provides standardized, scalable foundation for pilots, leverages existing relationships/skills, reduces initial infra burden, establishes core components for EAMI Technology Level 3. While lacking advanced features initially (Feature Store, sophisticated MLOps), architecture is extensible, providing base for future capability growth.

17.11 Conclusion: The Engine for AI Execution

The technology ecosystem (Compass South) is the engine enabling enterprise AI. Designing strategically requires considering infra options, robust data platforms, integrated ML dev environments, mature MLOps tooling. Informed choices (leveraging cloud PaaS, pragmatic build-vs-buy, ensuring integration) facilitate efficient development, reliable deployment, security, cost management, scaling. The AI Office, collaborating with IT/Data, ensures the ecosystem is an accelerator, not bottleneck, towards higher AI maturity and business impact. Next, Chapter 18 examines the critical fuel: strategic data management.

Key Takeaways:

- Robust tech ecosystem (infra, data/ML platforms, MLOps tools) essential for scalable, reliable, efficient AI.
- Infra choices (Cloud, On-Prem, Hybrid, Edge) depend on workload needs (scalability, cost, control, latency).
- Modern data platforms for AI often use lakehouses, need scalable storage, pipelines, governance, Feature Stores.
- Cloud AI PaaS platforms offer integrated environments; MLOps tools critical for automation.
- Balance "Build vs. Buy" pragmatically; prioritize integration, vendor management, managing tech debt.
- AI Office collaborates with IT/Data to guide tech strategy, standards, platform selection.

Food for Thought / Application Exercise:

- Evaluate your org's current infra for analytics/AI. Cloud, On-Prem, Hybrid? Pros/cons for AI goals?
- Standardized AI/ML dev platform? If yes, how well integrated? If no, challenges created?
- Consider recent "Build vs. Buy" decision (analytics/AI tools). Factors? TCO considered?

~ CHAPTER 18 ~

Data as the Lifeblood of AI: Strategic Data Enablement

"Data is the new oil. It's valuable, but if unrefined, it cannot really be used."
— Clive Humby

18.1 Introduction: Fueling the AI Engine

Previous chapters within Part 5 (Resource Management - South) addressed talent (Chapter 16) and technology (Chapter 17). Now, we address arguably the most critical resource underpinning almost all artificial intelligence: Data. AI algorithms, particularly ML models, are fundamentally data-driven; their performance, reliability, fairness, and value are intrinsically linked to the quality, relevance, accessibility, and governance of the data they consume. Simply having large volumes is insufficient; organizations must treat data as a strategic asset and implement robust practices for its management, curation, and enablement specifically for AI purposes.

This chapter delves into data's critical role within the AI ecosystem (Compass South focus). We explore why data readiness is paramount, outline key pillars for data enablement (quality, accessibility, governance, architecture, curation), discuss the strategic role the AI Office often plays collaborating on enterprise data strategy (working with the Chief Data Officer - CDO), and highlight consequences of neglecting the data foundation. Ensuring data is fit-for-purpose for AI is essential for success and achieving higher EAMI maturity in Data Management and Data Quality dimensions.

Learning Objectives:

- Understand why high-quality, accessible, well-governed data is essential lifeblood for successful AI.
- Identify key pillars of data readiness for AI (Quality, Accessibility, Governance, Architecture, Curation).
- Recognize common challenges preparing/managing data for AI (silos, quality, governance gaps).
- Understand AI Office's strategic role influencing data strategy & collaborating with data governance (CDO).
- Appreciate critical linkage between robust data management and reliable, trustworthy, value-generating AI.

18.2 Data as a Strategic Asset for AI

In the age of AI, data transitions from operational byproduct to core strategic asset and potential competitive differentiator. Organizations effectively curating, managing, governing, leveraging unique data assets gain advantage over competitors. Realizing this requires a deliberate, strategic approach to data management tailored to AI's unique requirements:

- **Volume & Variety:** AI models (esp. deep learning) often need large, diverse datasets (structured, semi-structured, unstructured - text, image, audio, video, IoT). Data platforms must handle this scale/variety efficiently (Chapter 21).
- **Quality & Relevance:** Model performance is extremely sensitive to data quality. Inaccurate, incomplete, inconsistent, outdated, biased data leads to unreliable, biased models ("garbage in, garbage out"). Data must be relevant to the business problem. Ensuring quality (Chapter 23) is paramount. Affects EAMM Data Quality.
- **Accessibility & Timeliness:** AI teams need efficient, secure, timely access to right data (exploration, feature engineering, training, validation, inference). Delays (silos, complex access protocols, slow pipelines) hinder velocity. Real-time AI needs low-latency data. Affects EAMM Data Management.
- **Governance & Compliance:** Data used for AI (esp. personal/sensitive/proprietary) must comply with privacy regulations (GDPR, CCPA - Chapter 12) and ethical guidelines (Chapter 10). Robust data governance (lineage, metadata, usage policies, access controls, ethics) essential for responsible AI (Domain East, EAMM Governance, Compliance, Ethics).

> **(!)** Viewing data management solely as technical IT infrastructure concern is a critical mistake. Data strategy, quality improvement, governance must be core business priorities, sponsored by leadership, integrated with enterprise AI strategy (North) and governance (East) championed by the AI Office.

18.3 Pillars of Data Readiness for AI

Achieving data readiness for AI relies on maturing capabilities across interconnected pillars, requiring collaboration between AI Office, CDO function, IT data platforms, and business data owners/stewards:

Table 18.1: Pillars of Data Readiness for AI

Pillar	Focus Area	Key Activities & Considerations	Related Chapters	EAMM Dimension(s) Alignment
1. Data Quality	Ensuring data possesses necessary accuracy, completeness, consistency, timeliness, validity, uniqueness, representativeness for AI purpose.	Implementing automated profiling/validation rules; establishing cleansing processes; investing in Master Data Management (MDM); deploying quality monitoring/dashboards; addressing data drift proactively.	Chapter 23	Data Quality
2. Data Accessibility	Providing efficient, secure, well-documented, timely access to relevant, trustworthy data for AI teams/systems.	Breaking down silos; implementing modern platforms (lakes, warehouses, lakehouses); using data virtualization; establishing clear access policies & workflows; utilizing data catalogs for discovery; ensuring performant API access for inference.	Chapter 17, 21	Data Management, Technology
3. Data Governance	Establishing clear policies, standards, roles (ownership, stewardship), processes for managing data assets responsibly, securely, compliantly.	Defining ownership/ stewardship roles; implementing privacy controls (GDPR/CCPA); managing metadata/ lineage; establishing usage policies/consent; ensuring ethical sourcing/use protocols; integrating with AI governance framework (Domain East).	Chapter 12, 18	Data Management, Governance, Compliance, Ethics

4. Data Architecture	Designing scalable, flexible, cost-effective data infrastructure (storage, processing) and pipelines tailored for diverse AI workloads.	Choosing storage paradigms (lake/warehouse/lakehouse); designing efficient ETL/ELT pipelines (Chapter 22); implementing Feature Stores (Chapter 21); considering Data Mesh; ensuring scalability, performance, cost-efficiency (FinOps - Chapter 19).	Chapter 17, 21, 22	Data Management, Technology
5. Data Curation & Feature Engineering	Transforming raw data into meaningful, high-quality, optimized features suitable for effective ML model training/inference.	Conducting EDA; applying domain expertise for feature selection/extraction; handling missing values/outliers; data transformations; creating documented, reusable features (Feature Stores); ensuring alignment with fairness/ethics (Chapter 10).	Chapter 23, 21	Data Quality, Data Management

> **D**
> **DEFINITION**
>
> **Data Readiness for AI:** The organizational state characterized by possessing the necessary data quality standards, streamlined accessibility mechanisms, robust governance structures, appropriate architectural foundations, and effective curation processes required to efficiently, reliably, and responsibly fuel diverse enterprise AI initiatives. High data readiness is fundamental for progressing across critical EAMM dimensions (Data Management, Data Quality) and enables success in MLOps, Value Realization, Reliability.

18.4 Common Data Challenges in AI Implementations

Achieving data readiness remains a significant hurdle:

- **Pervasive Data Silos:** Data fragmented across disparate systems (CRM, ERP, marketing, legacy) with inconsistent definitions/formats hinders creation of unified, quality training datasets.
- **Endemic Data Quality Issues:** Operational data often suffers from inaccuracies, missing values, inconsistencies, duplicates, outdated info, requiring extensive manual cleaning.

- **Immature Data Governance:** Lack of clear ownership, consistent definitions, adequate metadata/lineage, enforced usage policies leads to confusion, distrust, compliance/ethical risks.
- **Data Accessibility Bottlenecks:** AI teams face delays gaining access due to complex/restrictive processes. Security concerns, if not addressed via RBAC/masking, can inappropriately limit access.
- **Inadequate Scalability:** Legacy infrastructure may lack scalability for massive volumes/diverse types (unstructured) needed for complex model training or real-time inference.
- **Feature Engineering Complexity & Redundancy:** Transforming raw data to ML features is time-consuming, requires domain expertise. Without reuse mechanisms (Feature Stores - Chapter 21), teams duplicate effort.
- **Ethical Data Sourcing & Usage:** Ensuring data (esp. third-party/personal) obtained ethically, with transparency/consent, used for permitted purposes is growing ethical/regulatory imperative (Domain East).

18.5 The AI Office's Role in Data Strategy and Enablement

While the CDO function typically owns overall enterprise data strategy, the AI Office plays a crucial collaborative, influencing, advocating role due to AI's unique demands (Compass South interaction with central data teams):

- **Advocating for AI-Specific Data Needs:** Articulating/championing specific data quality thresholds, accessibility/latency requirements, architectural features (lakehouse, feature stores), governance controls essential for prioritized AI initiatives to CDO/IT/platform teams.
- **Collaborating on Enterprise Data Strategy:** Providing input into data strategy formulation/evolution to ensure explicit support for current/future AI use cases (from AI roadmap - North). Highlighting where current strategy impedes AI.
- **Defining Data Quality Standards for AI:** Collaboratively defining specific, measurable data quality metrics/targets for critical AI applications with governance/domain experts, ensuring AI needs incorporated into enterprise DQ frameworks (Chapter 23).
- **Promoting Data & AI Literacy:** Partnering with HR/L&D/CDO on training to improve data/AI literacy, enabling users to understand data's role, quality impacts, interpret AI results responsibly (Chapter 26).
- **Facilitating Data Access for AI Teams:** Proactively working with owners/stewards/security to define/streamline secure, compliant access protocols for AI teams needing specific datasets, potentially using data catalogs.
- **Guiding Feature Engineering Practices & Reuse:** Promoting best practices, encouraging documented/reusable features, potentially advocating for Feature Stores (Chapter 21) with MLOps/Data Eng.
- **Ensuring Ethical Data Sourcing & Use:** Integrating ethical data checks (sourcing, consent, usage) into AI governance risk/review processes (Domain East), aligning with privacy policies.

> Cultivate strong, formalized partnership between AI Office and CDO function. Establish regular joint planning, shared objectives/KPIs for AI data readiness (linked to EAMI), clear RACIs (platform selection, DQ monitoring, access policy). Avoid separate "AI data silos" disconnected from enterprise data governance.

18.6 OmnioTech Case Study: Addressing Data Challenges for AI

Setting the Scene

As OmnioTech scales pilots (Chapter 25), data challenges become critical roadblocks. Lack of unified customer data limits HyperPers effectiveness. Inconsistent sensor data formats/quality impact PredMaint reliability. EAMI assessment confirmed Data Management and Data Quality as major weaknesses (Level 2). Priya Sharma (AI Office Head) initiates dedicated collaboration with CDO Maria Petrova.

Action & Application

- **Joint Problem Definition & Sponsorship:** Priya/Maria jointly present data challenges impacting AI initiatives (blocking EAMI Level 3 progress) to AI Review Team/sponsors. Secure executive sponsorship/budget for cross-functional "Data Enablement for AI" initiative (co-led AI Office/CDO Office).
- **Collaboration on Data Strategy:** Maria incorporates specific AI data requirements (from AI Office/roadmap) into V2.0 Enterprise Data Strategy. Key priorities added: phased Unified Customer Profile data mart (MDM principles), mandatory DQ standards/monitoring for critical IoT sensor data.
- **Targeted Data Quality Focus (Chapter 23):** Joint AI Office/CDO task force initiates formal DQ assessment (customer/sensor data) using profiling tools (Azure Purview). Define specific DQ rules/targets (>98% customer field completeness, <1% duplicates, valid sensor ranges). Implement checks in pipelines (ADF, Databricks). Power BI DQ dashboards created for Data Stewards (Marketing, Product Eng) with alerts.
- **Improved Accessibility & Governance:** CDO office accelerates Azure Purview rollout, prioritizing registration/metadata for key AI datasets (ownership/lineage documented). Streamlined data access request workflow piloted in Purview for approved AI projects.
- **Architecture Alignment for Features (Chapter 21):** Recognizing duplicated feature engineering for HyperPers/Chatbot, joint team pilots Databricks Feature Store. Cross-functional team defines/implements/serves initial core reusable customer features (purchase frequency, activity score) via store (online/offline APIs). Feature lineage tracked.

Outcome & Progress

Concerted, cross-functional effort (AI Office + CDO) systematically addresses foundational data challenges hindering AI progress. Joint initiative provides focus/resources/collaboration to improve DQ, accessibility, governance, curation for AI needs. Directly tackles pilot scaling roadblocks, lays groundwork for improving EAMI Data Management / Data Quality maturity (towards Level 3/4), enabling more reliable, impactful AI applications.

18.7 Conclusion: Fueling AI with Fit-for-Purpose Data

High-quality, well-curated data is the unshakeable foundation for reliable, ethical, effective AI. Neglecting data quality risks biased outcomes, inaccurate predictions, compliance failures, wasted resources, eroded trust, undermines AI success. Achieving data readiness requires sustained, disciplined, organization-wide commitment (Compass South), involving robust processes (quality assessment, cleansing, validation, monitoring), thoughtful curation (metadata, lineage, feature engineering), modern architectures, strong governance (East), and effective collaboration (AI Office, CDO, IT, Data Eng, DS, Business Stewards). Prioritizing data quality/curation, linking improvements to EAMI goals, creates the essential, trustworthy fuel needed for the AI maturity journey. Having considered talent, tech, data, Chapter 19 examines financial stewardship for sustainable AI operations.

Key Takeaways:

- Data is critical strategic asset/lifeblood of AI; quality/accessibility/governance impact success.
- **Key data readiness pillars:** Quality, Accessibility, Governance, Architecture, Curation.
- Common challenges (silos, quality, governance) must be proactively addressed.
- AI Office advocates for AI data needs, collaborates closely with CDO on strategy/governance.
- High EAMI maturity requires focus on Data Management / Data Quality.

Food for Thought / Application Exercise:

- Top 1-2 data challenges (silos, quality, accessibility, governance) hindering AI in your org?
- Collaboration strength between AI teams & central data/governance function? Improvement areas?
- How much effort on feature engineering? Could Feature Store improve efficiency/consistency?

CHAPTER 19

Financial Stewardship for Sustainable AI: Budgeting and Optimization

"Beware of little expenses. A small leak can sink a great ship."
— Benjamin Franklin

19.1 Introduction: The Economics of Enterprise AI

Previous chapters within Part 5 (Resource Management - South) addressed talent (Chapter 16), technology (Chapter 17), and data (Chapter 18). This chapter tackles the final, indispensable resource: Finance. While AI holds immense promise for value creation (Domain West), realizing that potential requires significant and often ongoing investment. Developing sophisticated models, building robust platforms, acquiring specialized talent, managing large datasets, ensuring comprehensive governance (Domain East), and driving continuous improvement (Domain Center) all carry substantial costs. Without diligent financial stewardship, AI initiatives can quickly become unsustainable budget drains, jeopardizing the entire program regardless of technical success or strategic alignment (Domain North).

Therefore, establishing sound financial management practices – encompassing appropriate funding models, realistic budgeting based on Total Cost of Ownership (TCO), disciplined cost tracking, proactive optimization techniques (like FinOps), and rigorous ROI validation – is a crucial aspect of the Resource Management (South) domain within the AI Office Compass™ framework. It ensures AI investments are made wisely, resources are used efficiently, costs are controlled, and financial value delivered (Domain West) demonstrably outweighs expenditures. This chapter explores practical approaches to funding, budgeting, cost optimization, and financial governance for AI, vital for long-term sustainability and achieving higher EAMI maturity in the critical Financial dimension.

Learning Objectives:

- Understand the imperative for dedicated financial stewardship for AI, given unique cost structures.
- Identify common financial challenges with AI initiatives (unpredictable cloud costs, hidden TCO).
- Explore different funding models for AI Offices/projects (centralized, chargeback, hybrid) and implications.
- Learn key considerations for accurately budgeting AI initiatives, emphasizing Total Cost of Ownership (TCO).
- Understand practical cost tracking/optimization techniques, including FinOps principles for AI/ML.
- Recognize importance of linking financial governance with value realization/ROI validation (Domain West).
- Appreciate AI Office's collaborative role promoting financial discipline and sustainability.

19.2 The Imperative for Financial Stewardship

AI initiatives often introduce different cost structures, making proactive financial management essential:

- **High Upfront Investment:** Significant initial CapEx/OpEx for infrastructure, hardware (GPUs/TPUs), platforms/tools, datasets, initial model development/training.
- **Variable and Potentially High Operational Costs:** Cloud compute costs (training large models, inference at scale) can be substantial and highly variable. Requires different management than fixed costs.
- **Specialized (Expensive) Talent Costs:** High salaries for scarce AI talent (MLE, DS, MLOps) are a major ongoing expense (Chapter 16).
- **Data Acquisition & Management Costs:** Acquiring external data, storing massive volumes, implementing robust data quality/processing pipelines incur significant ongoing infra/tooling costs (Chapters 18, 21-23).
- **Ongoing Maintenance, Monitoring & Retraining:** AI models often need continuous monitoring (drift) and periodic retraining, adding recurring MLOps operational expenses (Chapter 20).
- **Governance & Compliance Overhead:** Implementing/maintaining robust governance (Part 3) requires dedicated personnel time and tooling, adding to overall cost.

Without careful planning, tracking, optimization, costs can spiral. Proactive financial stewardship (Compass South) ensures:

- **Sustainability:** AI initiatives funded realistically, operate within budgets.
- **Efficiency:** Financial resources, cloud compute, storage, licenses used optimally.
- **Accountability:** Costs tracked, attributed (showback/chargeback), enabling clear ROI calculation (Domain West) and justifying investments (EAMM Financial).
- **Strategic Alignment:** Financial resources prioritized for highest strategic value/return initiatives (Domain North).
- **Transparency:** Clear cost visibility enables informed trade-off decisions.

19.3 Common Financial Challenges in AI

Organizations often face specific financial hurdles:

- **Unpredictable Cloud Costs:** Difficulty forecasting variable cloud compute costs (training experiments, inference, data processing). Challenges fixed budgeting.
- **Hidden TCO Elements:** Underestimating indirect costs: ongoing data storage/egress, data prep, MLOps

platform maintenance, monitoring/retraining, governance overhead, user training.

- **Talent Acquisition & Retention Costs:** High recruitment costs for specialized AI talent, plus ongoing expense of retention.
- **Platform & Tooling Licensing Costs:** Significant expenses for commercial AI platforms, MLOps tools, data software, labeling services, foundation model APIs.
- **"Pilot Purgatory" Financial Drain:** Wasted investment in pilots failing to scale due to lack of realistic financial planning for scaling/operational phases.
- **Lack of Cost Visibility & Allocation:** Difficulty tracking/attributing cloud/platform costs to projects/ teams/BUs. Hinders chargeback/showback and project ROI analysis.

19.4 Funding Models for AI Offices and Initiatives

How the AI Office and initiatives are funded impacts behavior and priorities. No single best model; depends on org structure, culture, EAMI level, objectives. Common models:

Table 19.1: Comparison of Funding Models for AI

Funding Model	Description	Pros	Cons	Best Suited For
Centralized Budget	AI Office (or central innovation/IT) receives dedicated, centrally allocated budget covering core staff, shared platforms (tech, data, MLOps), potentially seed/full funding for prioritized projects.	Strong central control over strategy/ spending; easier to fund foundational capabilities (platforms, governance) & cross-functional initiatives; simpler initial setup/ administration.	Can disconnect AI Office priorities from BU needs if governance isn't collaborative; BUs might perceive services as "free," leading to unconstrained demand; potential funding bottlenecks.	Early-stage AI programs (EAMI Level 2-3) building core capabilities; orgs favoring strong central control; funding foundational R&D.
Chargeback / Service Model	AI Office operates like internal service provider, charging BUs directly for resources consumed (DS time, compute hours, API calls) or project execution (SOWs).	Tightly links AI costs to BU demand/value; encourages efficient resource use by BUs; provides clear cost allocation for BU-level ROI.	Complex to implement/ manage accurately (needs usage tracking/metering/ billing); potentially discourages experimentation/ platform adoption by BUs; risk of AI Office becoming purely tactical/ reactive.	More mature AI programs (EAMI Level 4+) with defined services & reliable usage tracking; orgs with strong internal chargeback culture.

Hybrid Model	Blends centralized & chargeback. Central budget funds core AI Office (leadership, gov, strategy, base platform infra/ licenses). Specific projects funded jointly (AI Office + BU based on value split) or fully by sponsoring BU, potentially via chargeback for specific Hub services (compute, modeling support).	Balances central strategic investment (foundations) with BU accountability/ alignment; flexible funding based on project type/ importance; encourages partnership/shared ownership between AI Office & BUs.	Requires clear, transparent rules for cost allocation, co-funding agreements, prioritization; potentially more complex budgeting/ tracking/negotiation than purely centralized.	Many orgs, esp. with Hybrid/Federated AI model (Chapter 2) or transitioning to higher maturity (EAMI Level 3-4), seeking both strategic direction & BU accountability.
Project-Specific Funding	Individual AI initiatives funded via standard corporate CapEx/OpEx approval, competing against non-AI projects based on individual business cases/ROI. AI Office might only advise.	Ensures AI projects rigorously evaluated against standard investment criteria; forces strong business cases.	High risk of underfunding essential foundational AI capabilities (platforms, DQ, governance, MLOps, talent); potential for fragmented funding decisions lacking strategic coherence; may disadvantage exploratory/long-term AI R&D.	Orgs where AI is highly decentralized, viewed purely tactically project-by-project (often lower EAMI maturity).

★ Optimal funding model often evolves with EAMI level and operating model (Chapter 2). Centralized might suit initial capability building (Level 2-3). As maturity/demand grows (Level 4-5), Hybrid can better balance strategic investment with BU accountability. Clearly define, document, communicate the chosen model and cost allocation principles.

19.5 Budgeting for AI: Considering Total Cost of Ownership (TCO)

Realistic budgets require considering TCO over the lifecycle (typically 3-5 years), not just initial development cost. Failing to account for full TCO causes funding shortfalls.

Key TCO Cost Categories for AI (Recap & Expansion):

Table 19.2: AI Total Cost of Ownership (TCO) Categories

Cost Category	Examples	Considerations for Budgeting
1. Development Costs	Personnel (DS, MLE/DE, PMs, SMEs), Software Licenses (IDEs, libraries), Initial Data Acquisition/Preparation Effort, Cloud Compute for Training (large GPU/TPU costs), Experimentation Platform Costs.	Often significant upfront, but can be dwarfed by long-term operational costs. Training compute for large models can be very high.
2. Deployment Costs	Infrastructure Setup & Configuration (VPCs, Kubernetes), Integration development (APIs, databases), MLOps Pipeline Setup & Tooling Costs, Initial Deployment Labor & Testing.	Integration with legacy systems frequently underestimated. Robust, automated MLOps pipelines require expertise/tooling investment.
3. Operational Costs (Ongoing)	Cloud Inference Compute Costs (CPU/GPU), Cloud Storage Costs (models, logs, data), Monitoring Tool Licenses/Services, Platform Maintenance Personnel Time, Periodic Model Retraining Compute Costs, API Gateway Fees, Data Pipeline Running Costs.	Often largest TCO component, highly variable. Requires proactive monitoring/optimization (FinOps). Retraining frequency/cost must be factored in.
4. Governance & Compliance Costs	Personnel time (risk assessments, ethical reviews, audits, docs); Licenses for specialized Governance/Security/Privacy Tooling; Costs for external audits/certifications (ISO 42001).	Often "indirect" but essential for responsible AI. Requirements typically increase with AI application risk tier (Chapter 8).
5. Change Management & Training Costs	Costs for developing/delivering end-user training, internal communications, potentially process redesign consultants, managing workforce transition impacts (Chapter 26).	Critical for realizing value but often under-budgeted. Investment scales with degree of process change.

> Develop/mandate standardized TCO estimation templates for AI projects (prompting consideration of all categories over 3-5 years). Involve Finance and IT Infra/Cloud Ops early to validate assumptions, use standard rates, forecast cloud consumption. Use rolling forecasts for variable costs where feasible.

19.6 Cost Tracking and Optimization (FinOps for AI)

Effective financial stewardship requires tracking actual costs and optimizing, especially with variable cloud spend. This is FinOps applied to AI workloads (Compass South).

> **D** DEFINITION
>
> **FinOps (Financial Operations):** A cultural practice and operational framework, built on collaboration between Finance, Technology (including AI/ML and MLOps teams), and Business teams, designed to bring financial accountability and optimization to the variable expenditure model of cloud computing, enabling organizations to make data-driven trade-offs between deployment speed, operational quality, and cost.

Key FinOps Practices for AI Cost Optimization:
- **Cost Visibility & Allocation:** Implement robust, granular tagging for all AI-related cloud resources (compute, storage, services). Use cloud provider/third-party FinOps tools to visualize trends, analyze patterns, allocate costs (enabling showback/chargeback).
- **Resource Optimization:** Continuously identify/implement opportunities:
  - » **Right-sizing Compute Instances:** Select appropriate VM sizes (CPU/GPU/memory) for training/inference based on utilization data; avoid overprovisioning.
  - » **Leveraging Spot Instances:** Use cheaper (interruptible) spot instances for fault-tolerant, non-time-critical training where feasible (requires checkpointing).
  - » **Implementing Auto-scaling:** Configure inference endpoints to scale instances based on real-time volume, avoiding idle resources. Scale-to-zero optimizes further.
  - » **Model Efficiency Optimization:** Use techniques like quantization/pruning for smaller, computationally cheaper models (important for edge/high-volume APIs).
  - » **Optimizing Data Storage & Transfer:** Use appropriate storage tiers (standard, infrequent, archive). Implement lifecycle policies (delete/archive old data/models). Minimize costly data egress.
- **Budget Monitoring & Alerting:** Set budgets/forecasts for AI projects/teams/environments in cost tools. Configure automated alerts for exceeding thresholds, enabling timely intervention.
- **Fostering Cost Awareness Culture:** Make cost data visible/understandable to AI dev/MLOps teams (cost per prediction/training run dashboards). Incorporate cost into design reviews/decisions. Encourage engi-

neers to consider cost implications.

19.7 Financial Governance and ROI Validation

Integrate cost management with performance tracking and governance (Compass West/East links):

- **Budget Adherence Monitoring:** AI Office, with Finance/PMs, regularly reviews actual spending vs. budget, investigating variances.
- **ROI Validation Integration:** As part of performance/value reviews (Chapter 14), formally validate actual ROI vs. business case and calculated TCO. Confirm quantified benefits realized vs. full costs incurred.
- **Funding Gates & Decision Points:** Implement formal review gates (post-pilot, pre-scale) where continued funding depends on demonstrating adequate value progress and positive ROI potential based on real data.
- **Linking Financials to Prioritization:** Feed validated ROI/TCO insights back into opportunity prioritization (Chapter 5) and roadmap governance (Chapter 6). Inform future investment decisions (allocate more to high-ROI areas, de-fund underperformers).

19.8 The AI Office's Role in Financial Stewardship

The AI Office (Compass South) plays a vital coordinating/enabling role in financial discipline for AI:

- **Developing Funding Models & Budget Processes:** Collaborating with Finance on funding model; establishing standard processes/templates for AI budgeting/TCO.
- **Promoting Cost Transparency & FinOps:** Implementing/advocating for tools/practices (tagging) for cost visibility; championing FinOps adoption/culture.
- **Facilitating Optimization Efforts:** Working with tech teams to identify/prioritize cost optimizations (right-sizing, spot use, model efficiency).
- **Tracking & Reporting Financial Data:** Monitoring program budget adherence; contributing accurate TCO data to ROI/value tracking (West).
- **Informing Strategic Decisions:** Providing realistic cost/TCO insights and validated ROI data for strategic prioritization (North) and portfolio decisions.

19.9 OmnioTech Case Study: Establishing Financial Controls

Setting the Scene

With PredMaint/HyperPers scaled and cloud costs rising, CFO raises concerns about visibility/control. Priya Sharma (AI Office Head) collaborates with Finance/Cloud Ops (starting basic FinOps) to establish better financial stewardship (Compass South), crucial for maintaining EAMI Level 3/4 Financial maturity.

Action & Application

- **Funding Model Refinement:** Formally document/communicate Hybrid model: AI Office Hub core team/ base platforms centrally funded. Specific model scaling/operation needs joint funding proposal (sponsoring

BU + AI Office), reviewed by AI Review Team based on validated pilot ROI & projected TCO/benefits. Cloud resource "showback" (reporting allocated costs to BUs via tagging) implemented.

- **Mandatory TCO Estimation:** Standardized TCO template (AI Office + Finance) mandated for new AI project proposals beyond PoC. Requires estimating costs (Dev, Deploy, Ops, Gov, Change) over 3 years. Pilot teams retrospectively estimate TCO for baseline.
- **Enhanced Cost Tracking & FinOps:** Mandatory Azure resource tagging standards implemented/enforced (project, env, cost center, model). Cloud Ops configures Azure Cost Management reports for AI Office (monthly cost breakdowns per project/resource). Quarterly "AI FinOps Reviews" established (AI Office, Cloud Ops, lead MLEs) to identify/implement optimizations (instance types, auto-scaling).
- **ROI Validation Integrated into Reviews:** Existing quarterly "AI Value Review" meetings (Chapter 14) now formally include section where BU presents actual TCO vs. quantified value achieved. Leads to updated, evidence-based ROI. Continued funding contingent on sustained positive ROI trajectory vs. validated TCO.

Outcome & Progress

OmnioTech implements crucial financial controls/discipline. Hybrid model clarifies cost responsibilities. TCO estimation forces realistic planning. Enhanced tracking/FinOps provides visibility/enables optimization. Integrating formal ROI validation against TCO ensures accountability, links spending to value. Strengthens EAMI Financial maturity (solidifying Level 3/4), builds finance leadership confidence in AI investment sustainability.

19.10 Conclusion: Ensuring Sustainable AI Through Financial Discipline

Financial stewardship is a critical pillar supporting successful, sustainable enterprise AI (Compass South). Given unique cost structures, organizations need robust financial management: appropriate funding models, realistic TCO-based budgets, accurate cost tracking (FinOps), active optimization, rigorous ROI validation against value (West). The AI Office, partnering with Finance/IT Ops, enables this discipline, ensuring AI initiatives are innovative, impactful, AND economically viable long-term. Having addressed core resources, Chapter 20 dives into the operational discipline bridging AI dev/deployment: MLOps.

Key Takeaways:

- Financial stewardship (budgeting, optimization, ROI tracking) crucial for sustainable AI due to unique costs.
- **Common challenges:** unpredictable cloud costs, hidden TCO, talent expenses, lack of cost visibility.
- Choose funding models (Centralized, Chargeback, Hybrid) based on maturity/culture; Hybrid often balances.
- Budgeting requires estimating full Total Cost of Ownership (TCO).
- FinOps practices (visibility, optimization, governance) key for managing variable AI cloud costs.
- Link financial governance (budget adherence, funding gates) directly to validated ROI.
- AI Office collaborates with Finance/IT to drive financial discipline, essential for EAMI Financial maturity.

Food for Thought / Application Exercise:

- Current funding model for AI in your org? Pros/cons?
- Consider significant AI project. Potential TCO components (Dev, Deploy, Ops, Gov, Change)? Which likely most significant/hardest to estimate?
- Cloud cost visibility for AI workloads? FinOps practices applied? Biggest optimization opportunity?

~ CHAPTER 20 ~

Implementing Disciplined MLOps: Bridging Development and Deployment

"The gap between proving a concept and deploying a reliable service is often wider than anticipated, especially in AI. MLOps bridges that gap."

20.1 Introduction: The Need for Operational Discipline in AI

Previous chapters within Part 5 (Resource Management - South) covered talent (Chapter 16), technology platforms (Chapter 17), data enablement (Chapter 18), and financial stewardship (Chapter 19). This chapter addresses a critical process capability within the Resource Management (South) domain that underpins the reliable and scalable delivery of AI value: Machine Learning Operations (MLOps).

While developing a high-performing machine learning model in a lab environment is significant, successfully deploying, monitoring, managing, and retraining that model reliably in production presents distinct, often more complex challenges. Traditional DevOps practices provide a foundation, but MLOps extends these principles for ML systems – addressing their deep data dependency, experimental nature, and potential for performance degradation over time due to drift.

Establishing mature MLOps practices is crucial for moving beyond manual, error-prone deployments towards automated, repeatable, governed, and scalable processes. It enables organizations to accelerate AI value delivery (Domain West), improve model reliability, enhance collaboration (data science, ML engineering, data engineering, operations), and manage complex AI systems at scale. This chapter explores core MLOps principles and key practices, discusses tooling categories, highlights benefits (including links to DORA metrics), and outlines the AI Office's vital role in championing MLOps adoption – key for achieving higher EAMI maturity, specifically in the dedicated MLOps dimension, and supporting Technology, Reliability, and Governance dimensions.

Learning Objectives:

- Understand why specialized MLOps practices are necessary, extending DevOps for unique AI/ML challenges.
- Learn core MLOps principles (Automation, Reproducibility, Versioning Everything, Continuous Monitoring, Collaboration, Governance Integration).
- Identify key MLOps practices across the ML lifecycle (CI/CD/CT pipelines, version control, experiment tracking, testing, monitoring, registry).
- Recognize categories of tooling enabling MLOps practices.
- Appreciate MLOps benefits (faster deployment, reliability, governance, scalability).
- Understand the AI Office's role driving MLOps strategy, standardization, platform selection, adoption.

20.2 Why MLOps? Beyond Traditional DevOps

DevOps revolutionized software via automation, collaboration, rapid iteration (CI/CD). MLOps builds upon this but addresses unique ML challenges:

- **Data Dependency:** ML model performance is highly sensitive to data. MLOps must incorporate robust data versioning, validation pipelines, drift detection/response alongside code changes.
- **Model Lifecycle Complexity:** ML models have a distinct, dynamic lifecycle (experimentation, frequent retraining, complex validation - fairness/robustness, diverse deployment, continuous monitoring, retirement) more data-dependent than traditional software.
- **Experimentation & Reproducibility:** Data science is experimental. MLOps requires rigorous tracking (code/data versions, hyperparameters, environment, metrics) for reproducibility (debugging, audits, building on work).
- **Specialized Testing & Validation:** ML needs unique testing: data validation (schemas, distributions), model validation (performance metrics on holdouts), fairness/bias testing (Chapter 10), robustness testing (adversarial inputs - Chapter 11), compliance checks (Chapter 12).
- **Continuous Monitoring Needs:** Production monitoring must track operational metrics AND model-specifics: prediction accuracy degradation (model drift), input data changes (data drift), relationship changes (concept drift), potentially fairness/explainability over time.
- **Specialized Skills & Team Collaboration:** MLOps requires seamless collaboration between diverse roles: data scientists, ML engineers, data engineers, IT operations/SRE, governance/risk stakeholders.

Without dedicated MLOps, organizations face brittle manual deployments, inconsistent environments, lack of reproducibility, slow updates, undetected performance degradation ("silent failures"), difficulty managing models at scale – hindering AI value and keeping EAMI MLOps maturity low.

> **MLOps (Machine Learning Operations):** A set of practices, principles, and technologies focused on streamlining and automating the end-to-end machine learning lifecycle, from data preparation and model development to deployment, monitoring, and ongoing management in production. It aims to bridge the gap between data science, ML engineering, and IT operations, enabling organizations to deploy and maintain ML models reliably, efficiently, reproducibly, and at scale. Mature MLOps is key for higher EAMI MLOps maturity and enables progress in Technology, Reliability, Governance.

20.3 Core MLOps Principles

Effective MLOps extends DevOps philosophy for ML:

- **Automation Everywhere:** Automate as much of the ML lifecycle as feasible (data ingestion/validation, feature engineering, training/retraining, testing, deployment, infrastructure provisioning, monitoring/alerting). Improves speed, reduces errors, ensures consistency, enables scale.
- **Reproducibility is Non-Negotiable:** Ensure every component (experiments, training runs, deployed models) is fully reproducible via rigorous version control (code, data, models, hyperparameters, environment). Essential for debugging, audits, compliance, collaboration.
- **Version Everything:** Apply disciplined version control to source code (Git), datasets (DVC, Delta Lake time travel), and trained model artifacts (Model Registry).
- **Collaboration & Shared Responsibility:** Foster seamless collaboration, shared tooling, clear responsibilities among all teams (DS, MLE, DE, Ops, Business, Gov/Risk). Break silos.
- **Continuous Monitoring (CI/CD + CT + CM):** Extend CI/CD with Continuous Training (CT - automated retraining triggered by drift/new data) and Continuous Monitoring (CM - comprehensive automated production monitoring: operational health, prediction accuracy, drift, fairness, security).
- **Testing is Paramount:** Integrate rigorous, automated testing throughout the pipeline: data validation, model validation (performance, fairness, robustness), software unit/integration tests, infrastructure tests.
- **Governance Integration:** Embed governance controls (security scans, compliance checks, ethical/fairness tests, lineage tracking, docs) into automated MLOps workflows/platforms ("Governance-as-Code" - Chapter 9).
- **Scalability & Reliability by Design:** Architect MLOps pipelines, tooling, infrastructure (often cloud) to be scalable, resilient, highly available.

20.4 Key MLOps Practices Across the Lifecycle

Mature MLOps involves implementing specific practices and leveraging tooling at each lifecycle stage.

Table 20.1: Key MLOps Practices and Tooling Categories

MLOps Practice Category	Description	Example Tools/Techniques	EAMI Dimension Impact
1. Experiment Tracking & Management	Systematically logging all relevant info for every ML experiment (code/data versions, hyperparameters, env config, results, artifacts) for reproducibility, comparison, auditing.	Dedicated platforms (MLflow Tracking, Weights & Biases, Comet ML, Neptune.ai) or integrated cloud ML platform features (Azure ML Experiments, SageMaker Experiments, Vertex AI Experiments); Kubeflow Metadata.	MLOps, Collaboration, Reproducibility
2. Version Control (Code, Data, Model)	Applying rigorous version control to all components: source code (feature engineering, model training, pipelines), datasets (training/ eval), and trained model artifacts.	Git (GitHub, GitLab, Azure Repos) for code; DVC, Git LFS, data platform features (Delta Lake/Iceberg time travel) for data; dedicated Model Registry for models.	MLOps, Reproducibility, Governance
3. Feature Stores (See Ch 21)	Centralized platforms for defining, storing, managing, discovering, sharing, versioning, serving (online/ offline), and monitoring curated ML features consistently.	Open-source (Feast), Commercial (Tecton, Databricks Feature Store), Cloud-native (SageMaker Feature Store, Vertex AI Feature Store).	Data Management, MLOps, Reuse, Quality
4. CI/CD/CT Pipelines	Automating the end-to-end workflow for building, testing, validating, deploying, and potentially retraining ML models, triggered by code/ data changes or monitoring alerts.	Orchestration tools (Jenkins, GitLab CI, GitHub Actions, Azure Pipelines, Kubeflow Pipelines, Argo Workflows); Pipelines integrate data validation (Great Expectations), model testing, governance checks (OPA), deployment.	MLOps, Automation, Reliability, Speed
5. Model Registry	Central system cataloging models: storing, versioning, managing metadata (lineage, performance, parameters, governance status, deployment history), discovering, promoting approved models.	MLflow Model Registry, Cloud platform registries (Azure ML Model Registry, SageMaker Model Registry, Vertex AI Model Registry), potentially Artifactory configured for models.	MLOps, Governance, Deployment Management

6. Model Serving & Deployment	Deploying trained models from registry as scalable, reliable, performant prediction services (real-time APIs/microservices) or batch components.	Specialized serving frameworks (KFServing/KServe, Seldon Core, BentoML, TF Serving), Managed endpoints on cloud platforms (SageMaker/Azure ML/Vertex AI Endpoints), Kubernetes.	MLOps, Technology, Scalability, Reliability
7. Production Monitoring & Observability	Continuously tracking operational health (latency, errors) and predictive performance (accuracy, drift, fairness) of deployed models in the real world.	Standard monitoring (Prometheus/Grafana, Datadog), Cloud provider services (CloudWatch, Azure Monitor), specialized AI monitoring platforms (Arize, Fiddler, WhyLabs), Observability tools.	MLOps, Performance, Reliability, Governance
8. Infrastructure as Code (IaC)	Managing/provisioning underlying infrastructure (compute, storage, networks, Kubernetes) using declarative code for consistency/repeatability.	Terraform, AWS CloudFormation, Azure ARM/Bicep, Google Cloud Deployment Manager, Ansible, Pulumi; often combined with policy-as-code (Checkov).	Technology, Automation, Governance

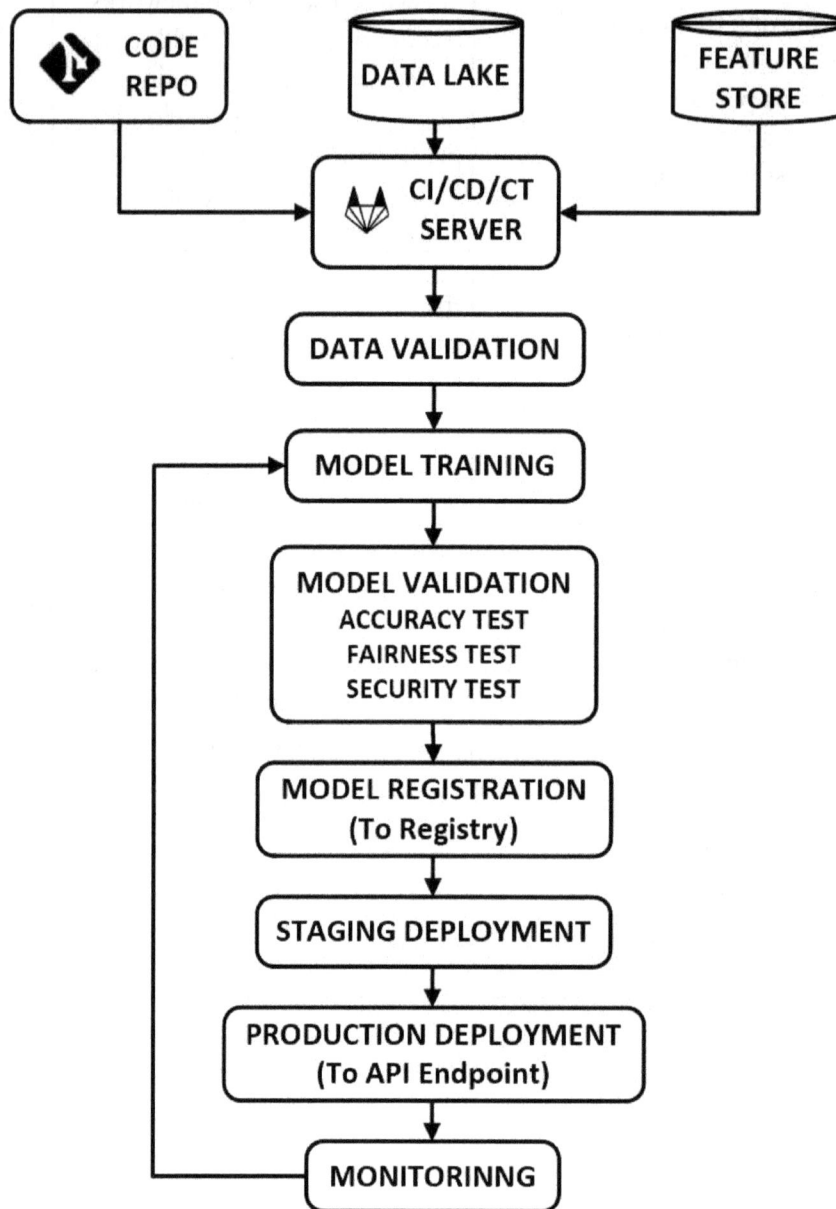

Figure 20.1: Detailed MLOps Pipeline Diagram

20.5 Benefits of Mature MLOps

Investing in MLOps delivers significant benefits (EAMM MLOps impact):

- **Faster Time-to-Market & Increased Deployment Frequency:** Automation reduces manual effort/time, enabling faster value delivery, more frequent updates (aligns with DORA metrics).
- **Improved Reliability & Model Quality:** Automated testing/validation, consistent environments reduce

errors, ensure deployed models meet quality/governance standards (impacts DORA Change Failure Rate).

- **Enhanced Scalability & Efficiency:** Automated pipelines/IaC allow efficient management of hundreds/thousands of models, enabling AI scale without linear overhead increase.
- **Increased Reproducibility & Auditability:** Rigorous versioning/tracking provides full traceability for debugging, audits (Governance East), reproducing results.
- **Better Collaboration & Productivity:** Shared platforms/workflows improve communication, reduce friction, foster collaboration (DS, MLE, DE, Ops), boost productivity.
- **Proactive Risk Management & Faster Recovery:** Integrated monitoring detects issues (drift, bias) early for proactive intervention. Automated pipelines enable faster recovery (DORA MTTR) from production issues. Embedded governance reduces risks.
- **Resource Optimization:** Automation reduces manual effort. Optimized deployment/monitoring helps control costs (FinOps - Chapter 19).
- **Higher EAMI Maturity:** Mature MLOps directly improves EAMI MLOps score, positively impacting Technology, Reliability, Governance, Value Realization dimensions.

> MLOps transforms ML development from artisanal craft into disciplined, repeatable, scalable, governable engineering. Provides essential operational backbone for delivering AI value reliably at enterprise scale.

20.6 Challenges and Prioritization

Implementing comprehensive MLOps presents challenges:

- **Complexity & Tooling Landscape:** Rapidly evolving, complex array of tools. Selecting, integrating, managing requires expertise/effort.
- **Skill Gaps:** Need specialized ML Engineering/MLOps skills, often scarce/hard to develop (Chapter 16).
- **Cultural Shift & Collaboration:** Requires breaking silos, fostering collaboration/shared responsibility across DS, SWE, Ops, DE teams.
- **Cost & Investment:** Requires upfront investment (technology, personnel). Justify based on future efficiency/risk reduction.
- **Integrating with Legacy Systems:** Embedding MLOps/integrating pipelines with legacy infra/apps can be technically difficult.

> ★ Don't attempt "big bang" MLOps implementation. Prioritize capabilities based on bottlenecks, roadmap needs, EAMI maturity. Start foundationally (version control, experiment tracking, basic registry/CI/CD). Gradually layer sophistication (automated testing, drift monitoring, CT, Feature Store) as program/skills mature. Solve biggest delivery pain points first.

20.7 The AI Office's Role in MLOps Adoption

The AI Office (Platform/MLOps Eng roles - Compass South), collaborating with IT/DevOps/Platform Eng, drives MLOps adoption/maturation:

- **Defining MLOps Strategy & Standards:** Setting vision/strategy for MLOps adoption aligned with AI maturity goals. Defining enterprise standards, best practices, reference architectures.
- **Selecting & Providing Core MLOps Tooling:** Leading/influencing evaluation, selection, implementation, management of core enterprise MLOps platform/toolchain (key to Tech ecosystem - Chapter 17).
- **Developing Reusable Templates & Accelerators:** Creating standardized, reusable pipeline templates (YAML, workflow definitions), code libraries, container images, best practice guides to accelerate adoption/consistency.
- **Training & Enablement:** Providing targeted training, workshops, coaching on MLOps principles, standard tools, best practices (Talent enablement - South).
- **Facilitating Collaboration:** Promoting communication/collaboration between technical teams (DS, MLE, DE, Ops) via shared platforms/CoPs.
- **Integrating Governance:** Working with Governance Lead (East)/Security to embed controls (security scans, compliance checks, fairness tests, docs) into standard MLOps pipelines.
- **Measuring & Improving MLOps Maturity:** Tracking metrics (adoption rates, pipeline efficiency, reliability) and EAMI MLOps scores. Using insights to drive continuous refinement (Compass Center).

20.8 OmnioTech Case Study: Implementing Foundational MLOps

Setting the Scene

As OmnioTech scales pilots (Chapter 25), the Platform/MLOps Lead (AI Office Hub) implements foundational MLOps using standardized Azure ML platform (Compass South activity). Goal: Move beyond manual deployments, establish reliable processes for EAMI MLOps Level 3.

Action & Application

- **Standardization on Azure ML:** Mandate use of Azure ML workspaces for end-to-end lifecycle (experimentation, training, tracking, registry, deployment, basic monitoring) for scaled pilots/new projects.
- **Version Control Enforcement:** Mandate Azure Repos (Git) for all code, pipeline definitions, IaC. Establish

branching strategies. Link commits to experiments. Mandate Azure ML Model Registry use.

- **Experiment Tracking:** Require data scientists rigorously use Azure ML Experiments to log parameters, metrics, code versions, data sources, artifacts for reproducibility.
- **Basic CI/CD Pipeline Templates:** Develop initial reusable Azure Pipelines templates (YAML):
  - » **CI:** Triggered on commit. Executes linting, basic unit tests (Pytest), potentially SAST.
  - » **CD:** Triggered manually post-CI/validation. Retrieves registered model, packages container, basic staging integration tests, awaits mandatory manual approvals (incl. governance checks), deploys container to AKS via Azure ML Managed Endpoint (canary strategy).
- **Initial Monitoring & Alerting:** Configure Azure Monitor for basic operational metrics (latency, error rate, utilization) for deployed endpoints. Set basic alerts (via Action Groups) notifying MLOps lead/owners on threshold breach. Plan documented for Azure ML data drift monitoring (Phase 2).
- **Training & Rollout:** MLOps Lead conducts workshops for pilot teams on Azure Repos, Azure ML Experiments/Registry, V1.0 CI/CD templates. Documentation/guides shared on internal SharePoint.

Outcome & Progress

OmnioTech establishes foundational MLOps capabilities on Azure ML/DevOps. While lacking advanced features (full CT, sophisticated testing, Feature Store), V1.0 provides version control, basic reproducibility, significant deployment automation, initial monitoring. Improves reliability/speed for scaled pilots, reduces manual errors, establishes core processes/tooling for EAMI MLOps Level 3, setting stage for further advancement.

20.9 Conclusion: Engineering Reliability and Scale into AI

Machine Learning Operations (MLOps) provides critical engineering discipline bridging AI development and reliable production operations at scale (Compass South). Embracing core principles (automation, reproducibility, versioning, monitoring, collaboration) and implementing key practices/tooling enables faster value delivery, improves reliability/quality, manages risks, scales efficiently. The AI Office drives MLOps strategy, standards, platforms, adoption – essential for higher EAMI maturity and sustainable enterprise AI success. Following chapters delve deeper into data/tech foundations managed via MLOps.

Key Takeaways:

- MLOps extends DevOps for unique ML challenges (data dependency, lifecycle, experiments, testing, monitoring).
- **Core principles:** Automation, Reproducibility, Versioning Everything, Continuous Monitoring (CI/CD/ CT/CM), Collaboration, Testing, Governance Integration.
- **Key practices span lifecycle:** Experiment Tracking, Version Control, Feature Stores, CI/CD/CT Pipelines, Model Registry, Serving, Monitoring, IaC.
- **Mature MLOps benefits:** faster time-to-market, reliability, scalability, reproducibility, collaboration, risk management, resource efficiency.
- Implementing MLOps requires addressing complexity, skills, culture, cost via phased approach prioritized by AI Office.
- AI Office drives MLOps strategy, standards, tooling, training, governance integration, crucial for EAMI MLOps maturity.

Food for Thought / Application Exercise:

- Evaluate current ML model deployment process. How much automated vs. manual? Biggest bottlenecks/ errors?
- Which MLOps practices (Table 20.1) implemented (even partially)? Biggest gaps/opportunities?
- Collaboration strength between DS, MLE, DE, IT Ops? How improve?

~ CHAPTER 21 ~

Data Stores and Data Lakes: Architecting the Data Foundation of AI

"The ability to store, manage, and access data effectively is the bedrock upon which successful AI systems are built."

21.1 Introduction: The Architectural Backbone for Data

Previous chapters within Part 5 (Resource Management - South) addressed the 'what' of data as a strategic asset (Chapter 18) and the 'how' of ensuring its quality (Chapter 23, upcoming). This chapter focuses on the 'where' and 'how' of storing and structuring this vital resource: the Data Architecture, specifically examining data stores like lakes, warehouses, and the increasingly influential lakehouse paradigm. The architectural choices made here profoundly impact data accessibility, scalability, flexibility, cost-efficiency, and governance capabilities – all critical factors for enabling sophisticated AI development and achieving higher EAMI maturity levels, particularly in the Data Management and Technology Selection dimensions.

Designing the right data storage and processing architecture is a foundational element managed within the Resource Management (South) domain. It involves moving beyond traditional database approaches, often optimized solely for structured transactional data, to embrace solutions capable of handling the sheer volume, high velocity, and diverse variety (structured, semi-structured, unstructured text, images, audio, video, sensor data) typically required for training complex machine learning models. We will explore different architectural patterns, discuss the rise of the lakehouse (a hybrid approach combining lake flexibility with warehouse management features), delve into the crucial role of Feature Stores in streamlining ML workflows, and touch upon emerging concepts like Data Mesh and Data Observability. Making informed architectural decisions ensures the data infrastructure can effectively support the organization's AI ambitions, from basic analytics to advanced deep learning applications, and contributes to overall EAMI Technology maturity.

Learning Objectives:

- Understand the specific architectural requirements for data storage and processing needed to support enterprise AI workloads effectively.
- Compare and contrast different data storage paradigms, the traditional Data Lakes, Data Warehouses, and the modern Lakehouse architecture, understanding their implications for AI.
- Recognize the growing importance and core functions of Feature Stores in standardizing, accelerating, and improving the reliability of ML feature development and serving.
- Understand the basic concepts of emerging architectural patterns like Data Mesh and Data Observability and assess their potential relevance to AI strategies in mature organizations.
- Appreciate the collaborative role of the AI Office in influencing enterprise data architecture decisions to ensure long-term AI readiness and alignment with EAMI goals.

21.2 Data Architecture Requirements for AI

Traditional data warehousing, primarily designed for structured data and predefined business intelligence (BI) reporting, often falls short of meeting the diverse and demanding needs of modern AI development and deployment. An effective data architecture specifically tailored for AI needs to provide:

- **Support for Diverse Data Types:** Seamless capability to ingest, store, manage, and process structured data, semi-structured data (JSON, XML), and large volumes of unstructured data (text, images, audio, video, IoT logs).
- **Massive Scalability:** Ability to cost-effectively store and efficiently process potentially massive datasets (TBs/PBs/EBs). Cloud object storage (S3, ADLS, GCS) often provides the foundation.
- **Flexibility for Exploration & Feature Engineering:** Provide easy, performant access for data scientists/ MLEs to explore raw/minimally processed data for pattern discovery and feature engineering, unlike rigid warehouse schemas.
- **Efficient & Varied Data Processing:** Support diverse workloads:
  - » Large-scale batch processing (Spark, cloud services like Glue, ADF, Databricks) for training data prep or batch inference.
  - » Potentially low-latency real-time/streaming processing (Kafka, Spark Streaming, Flink) for applications needing immediate insights (real-time fraud, personalization).
- **Integrated Data Governance & Security:** Robust capabilities for metadata management, lineage tracking, data cataloging, fine-grained access control (RBAC), compliance (Domain East), data quality monitoring (Chapter 23). Governance is essential.
- **Seamless Integration with AI/ML Tools:** Efficient connectivity and optimized access for AI/ML platforms (SageMaker, Azure ML, Vertex AI) and libraries (TensorFlow, PyTorch, Spark MLlib).
- **Cost-Effectiveness & Optimization:** Offer cost-efficient storage (tiering) and processing, aligning with FinOps principles (Chapter 19) to manage significant infrastructure costs.

21.3 Data Storage Paradigms: Lake vs. Warehouse vs. Lakehouse

Three primary architectural patterns represent the evolution of enterprise data storage, each with implications for AI workloads:

Table 21.1: Comparison of Data Storage Paradigms

Paradigm	Description	Data Types Handled Primarily	Schema Handling Approach	Primary Historical Use Cases	Pros for AI/ML	Cons for AI/ML
Data Warehouse	Central repository storing highly structured, cleansed, transformed data, optimized for SQL-based BI reporting.	Structured	Schema-on-Write (Structure defined before loading)	Enterprise BI reporting, historical analysis, operational dashboards.	Good for structured feature sources; mature SQL interface.	Poor handling of unstructured/semi-structured data; inflexible schema; data often overly processed/aggregated; costly storage for massive datasets.
Data Lake	Central repository storing vast amounts of raw data (structured, semi-structured, un-structured) in original format, typically on low-cost object storage.	All Types	Schema-on-Read (Structure applied during query/processing)	Data science exploration, ML training on raw/diverse data, big data batch processing, archiving.	Highly flexible (stores any type); cost-effective storage; scalable; excellent for exploration & training on raw data.	Can become "data swamp" if poorly governed; requires significant processing for BI; lacks transactional integrity & robust management features inherently.
Data Lakehouse	Modern approach combining data lake flexibility/cost with warehouse management/reliability/performance features directly on lake storage.	All Types	Schema-on-Read (with schema enforcement/evolution)	Unified platform for BI, data science, ML on all data types; single source of truth.	Best of both worlds: flexible storage, direct SQL/ML access on same data, ACID transactions, schema enforcement, versioning.	Newer paradigm (tooling maturing); potential complexity; relies on open table formats (Delta Lake, Iceberg, Hudi).

D
DEFINITION

Data Lakehouse: A modern data management architecture implementing traditional data warehouse structures and capabilities (ACID transactions, schema enforcement, indexing, caching, time travel/versioning) directly on low-cost, scalable data lake object storage (S3, ADLS, GCS). Uses open table formats (Delta Lake, Iceberg, Hudi) enabling SQL-based BI and advanced AI/ML on the same unified data repository. Platforms like Databricks and Snowflake are leading proponents.

The Lakehouse architecture is increasingly favored for organizations aiming for higher EAMI maturity (Level 4/5) in Data Management and Technology. It provides a unified, flexible, cost-effective foundation avoiding data duplication between separate lake/warehouse systems, streamlining pipelines, improving freshness, reducing storage costs, and enabling analytics/AI teams to work more effectively.

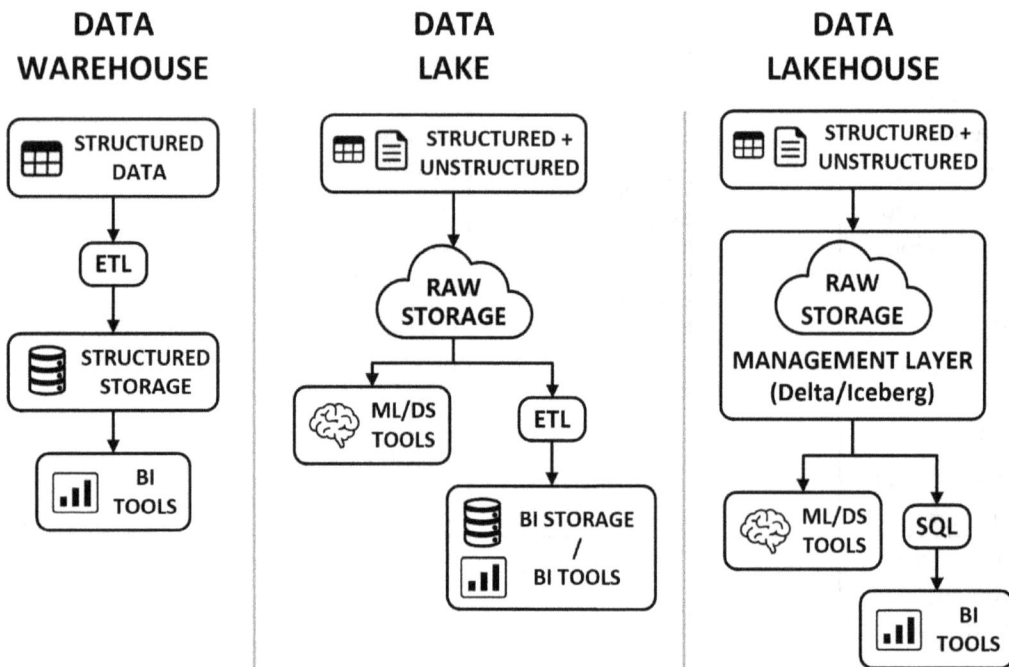

Figure 21.1: Lake vs. Warehouse vs. Lakehouse Diagram

21.4 Feature Stores: Standardizing Features for ML

While lakes/lakehouses provide foundational storage, Feature Stores are a specialized architectural component (Compass South) addressing critical ML workflow challenges: managing features used for training and serving models consistently and efficiently.

> **D** DEFINITION
>
> **Feature Store:** A dedicated data system within the MLOps infrastructure serving as a central repository for documenting, storing, discovering, sharing, versioning, and serving curated features (variables derived from raw data, engineered for ML) consistently across offline model training and online, low-latency inference services.

Why Feature Stores are Increasingly Critical for Mature AI (EAMI Level 4/5):

- **Preventing Training-Serving Skew:** Ensures exact feature calculation logic/data snapshot used during training is reliably retrieved for real-time inference, preventing a common cause of performance degradation.
- **Accelerating Feature Development & Promoting Reuse:** Provides central catalog to discover/understand/reuse curated features, avoiding massive duplication of effort (often 60-80% of DS time), speeding development, ensuring consistency.
- **Enhancing Data Governance for ML Features:** Dedicated place to manage metadata, track lineage, enforce quality, control access, manage versions specifically for ML features, improving governance (Domain East).
- **Enabling Efficient Online Serving:** Includes high-performance "online" component (key-value store like Redis) serving pre-computed feature vectors quickly (ms latency) for real-time inference, decoupling complex calculations.
- **Ensuring Point-in-Time Correctness:** Facilitates retrieving consistent historical feature values for accurate training dataset generation without data leakage.

Key Capabilities of a Feature Store:

- **Feature Registry & Discovery:** Searchable catalog (define, document, version, discover features).
- **Feature Transformation & Ingestion:** Define/execute/schedule logic transforming raw data to features, ingest into store.
- **Dual Storage (Offline & Online):** Optimized storage for historical features (Offline Store - lake/lakehouse) for training, and low-latency access to latest values (Online Store) for inference.
- **Feature Serving API:** Efficient APIs retrieving feature vectors (by entity ID) for batch training and online prediction.
- **Monitoring & Validation:** Monitor feature quality, detect drift, track usage/importance.

Implementing a Feature Store signifies higher MLOps and Data Management maturity (EAMI Level 4/5). Options include open-source (Feast), commercial (Tecton), cloud services (SageMaker/Vertex AI/Databricks Feature Store).

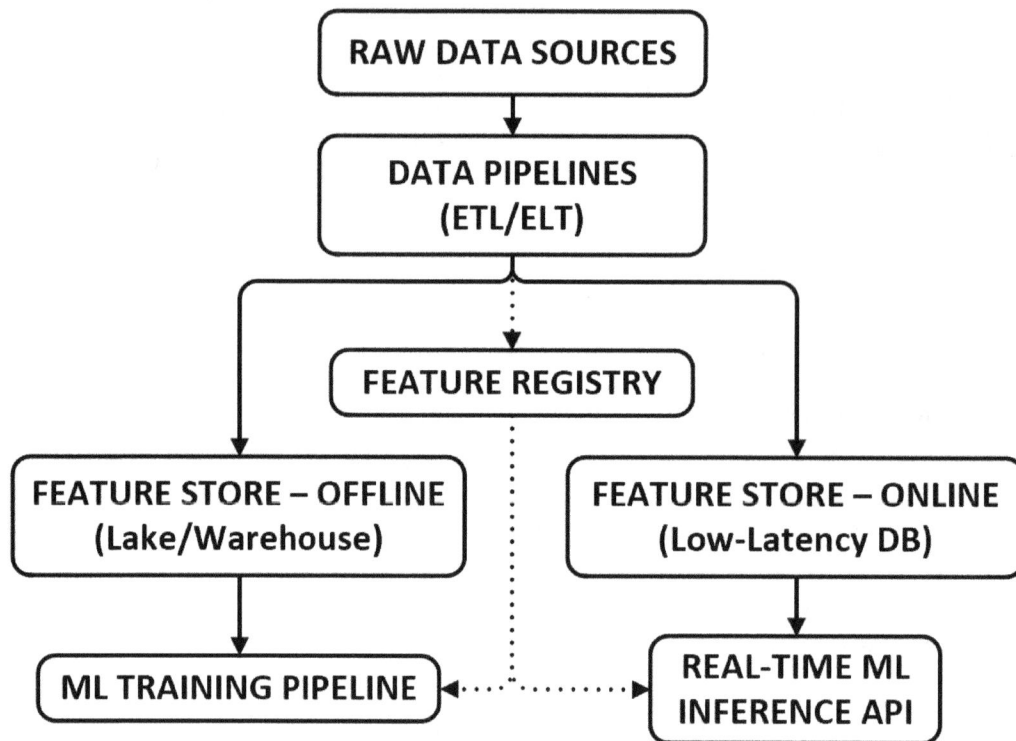

Figure 21.2: Feature Store Architecture Diagram

21.5 Emerging Concepts: Data Mesh and Data Observability

Two related concepts gain traction, especially in large organizations aiming for higher data maturity:

- **Data Mesh:** Socio-technical approach challenging monolithic central data platforms. Advocates shifting ownership of analytical data to business domains generating it (e.g., Marketing owns 'Customer Engagement Data Product'). Domains treat data as product (discoverable, understandable, trustworthy, secure, consumable via standard APIs) served on shared self-serve infra. Federated governance sets global standards (interoperability, quality, security). Aims to improve scalability, agility, accountability, quality via decentralization. Aligns with federated AI models (Chapter 2) and high EAMI maturity (Level 5).

- **Data Observability:** Applies software observability principles (metrics, logs, traces) to data pipelines/assets. Focuses on automated, end-to-end visibility into data system health/quality/usage. Monitors freshness, distribution, volume, schema, lineage. Platforms (e.g., Monte Carlo, Databand) use automation/ML to detect issues (quality, pipeline breaks, schema changes) proactively before impacting consumers (AI models, BI). Shifts data quality from reactive debugging to proactive reliability management, crucial for AI trust (EAMI Data Quality / Reliability).

> Fully implementing Data Mesh is significant undertaking for large orgs (Level 5). Principles (domain ownership, data-as-product, self-serve infra, federated gov) can inform strategy even earlier. Adopting Data Observability, even basic automated monitoring, significantly improves reliability over manual validation, enhancing AI trustworthiness (EAMI Data Quality).

21.6 The AI Office's Role in Data Architecture

While CDO/Enterprise Architecture typically own data architecture, the AI Office (Compass South) collaborates closely:

- **Articulating AI Architectural Requirements:** Communicating specific AI workload needs (unstructured data support, scalability, latency, feature consistency) to data architects/platform owners.
- **Platform Evaluation & Selection:** Participating in evaluation/PoC/selection for core data platforms (lakes, lakehouses), integration tools, catalogs, Feature Stores, ensuring they meet AI roadmap technical needs.
- **Promoting Data Catalog Adoption & Standards:** Championing catalog use by AI teams (discovery, lineage, trust). Collaborating on AI-relevant metadata standards (e.g., bias assessment tags).
- **Advocating for Feature Stores & Reuse:** Making business case for Feature Stores where appropriate (consistency, efficiency, governance). Promoting feature reuse culture.
- **Staying Informed & Advising on Trends:** Monitoring data architecture trends (Lakehouse, Mesh, Observability, Vector DBs), advising on applicability/adoption path.
- **Ensuring Governance Integration:** Collaborating with data governance/security to ensure architecture supports necessary controls, lineage, security for AI data (Domain East linkage).

21.7 OmnioTech Case Study: Evolving the Data Architecture

Setting the Scene:

OmnioTech's initial V1.0 stack (ADLS + ADF ETL) created silos and inflexibility, becoming roadblocks for scaling pilots (Chapter 25). Lack of unified customer data hurt HyperPers; sensor data issues hit PredMaint. EAMI assessment confirmed Data Management maturity stuck at Level 2. Priya Sharma (AI Office) collaborates with CDO Maria Petrova.

Action & Application:

- **Strategic Shift to Lakehouse:** AI Office/CDO/IT Arch make successful case for Lakehouse architecture. Choose Databricks on Azure (Delta Lake on ADLS). Rationale: Unified platform (BI/ML on same data), improved reliability (ACID, schema enforcement), better Spark ML performance, integrated Feature Store. Planned as major Y2 initiative targeting EAMI Data Management Level 4.
- **Feature Store Pilot Implementation:** Recognizing duplicated feature engineering for HyperPers/Chatbot

(needing similar real-time features), Priya champions pilot of Databricks Feature Store. Cross-functional team defines/implements/serves initial 10-15 core reusable customer features (purchase frequency, activity score) via store (online/offline APIs) for both projects. Feature lineage tracked.

- **Data Catalog Enhancement:** CDO office, with AI Office support, intensifies Azure Purview population (metadata, lineage, ownership) for key curated AI datasets ('gold' Delta layer). Improves discovery/trust/ governance.
- **Exploring Data Observability:** MLOps/Data Eng team begins PoC evaluating third-party Data Observability tools (e.g., Monte Carlo) integrated with Databricks pipelines. Goal: Test feasibility of automated monitoring (freshness, volume, schema, distribution) for proactive quality issue detection.

Outcome & Progress:

OmnioTech makes strategic decisions to evolve data architecture. Move to Databricks Lakehouse provides unified, governed, performant platform. Feature Store pilot addresses ML workflow inefficiencies/risks. Enhanced catalog improves discovery/trust. Exploring Data Observability is proactive step for reliability. Requires investment (Compass South - Finance), but crucial for improving EAMI maturity (Data Management, Technology Selection towards Level 4/5), reducing tech debt, enabling more sophisticated, reliable AI.

21.8 Conclusion: Architecting the Foundation for AI Fuel

The underlying data architecture (storage platforms like lakes/lakehouses, pipelines, governance, Feature Stores) forms the essential technical foundation for reliable, scalable, trustworthy enterprise AI (Compass South). Strategic architectural choices prioritizing scalability, flexibility, governance, accessibility, integration are paramount. Lakehouse offers unified approach; Feature Stores target ML workflow bottlenecks; Mesh/Observability promise further maturity. AI Office collaborates with Data/IT Arch to ensure infrastructure effectively supports evolving AI strategy/maturity. Having addressed data storage/structure, Chapter 22 examines efficient data processing (ETL/ELT).

Key Takeaways:

- Modern AI needs data architectures handling diverse data, scaling massively, enabling exploration, supporting varied processing, integrating governance, connecting to ML tools, being cost-effective.
- Lakes offer flexibility; Warehouses structure for BI; Lakehouses aim to unify both on the lake.
- Feature Stores crucial for mature MLOps (feature consistency, reuse, governance, low-latency serving).
- Data Mesh (domain ownership) & Data Observability (proactive monitoring) are emerging concepts.
- AI Office collaborates with CDO/Architecture to advocate AI needs, evaluate platforms, promote catalogs/ Feature Stores.

Food for Thought / Application Exercise:

- Evaluate current data storage architecture (Warehouse, Lake, Lakehouse?). How well supports AI's diverse data/exploration needs?
- **Feature Store concept:** Do teams re-engineer similar features? Could store improve efficiency/consistency?
- Data discoverability? Central catalog? Lineage/quality understanding? How might Mesh/Observability help?

~ CHAPTER 22 ~

ETL vs. ELT: Optimizing Data Processing Workflows for AI

"Efficiency is doing things right. Effectiveness is doing the right things. Optimize both in your data pipelines."
— Inspired by Peter Drucker

22.1 Introduction: Processing Data for AI Consumption

Chapter 21 explored the architectural foundations for storing AI data – data lakes, warehouses, and the unifying lakehouse paradigm. However, storing data is only part of the equation; data rarely arrives in a format perfectly suited for machine learning model training or inference. It needs to be moved, cleaned, transformed, aggregated, and structured appropriately. This chapter, continuing within the Resource Management (South) domain of the AI Office Compass™ framework, focuses on the critical data processing workflows that prepare data for AI consumption, specifically comparing two dominant paradigms: ETL (Extract, Transform, Load) and ELT (Extract, Load, Transform).

The choice between ETL and ELT significantly impacts pipeline architecture, flexibility, scalability, cost, and the suitability of the data infrastructure for supporting diverse AI workloads. Understanding the trade-offs and selecting the right approach (or combination) based on the data architecture (Chapter 21) and specific AI use case requirements is essential for building efficient, reliable, and cost-effective data pipelines that fuel AI initiatives. These pipelines are the arteries carrying the data lifeblood (Chapter 18) to fuel AI systems. This chapter explores the characteristics of each approach, discusses considerations like real-time vs. batch processing (including Change Data Capture - CDC), touches upon the role of DataOps practices in managing these workflows, and considers the implications for AI development efficiency and overall EAMI maturity in the Data Management and Technology dimensions.

Learning Objectives:

- Understand the fundamental difference between ETL (Extract, Transform, Load) and ELT (Extract, Load, Transform) data processing paradigms.
- Compare the pros, cons, and typical use cases for ETL versus ELT, particularly in the context of AI and modern data platforms (like data lakes/lakehouses).
- Recognize the importance of real-time data processing capabilities (e.g., using Change Data Capture - CDC) for certain AI applications.
- Appreciate the role of DataOps principles in managing data pipelines effectively and reliably.
- Understand how the choice of data processing workflow impacts AI development efficiency and data readiness, influencing EAMI Data Management and Technology maturity.

22.2 Understanding ETL (Extract, Transform, Load)

ETL has been the traditional standard for data warehousing for decades, primarily focused on preparing data for structured business intelligence (BI) reporting. The process involves sequential steps:

- **Extract:** Data is pulled from various source systems (e.g., operational databases like Oracle or SQL Server, applications like Salesforce CRM, flat files).
- **Transform:** The extracted data undergoes significant cleaning, validation, standardization, aggregation, enrichment, and structuring before being loaded into the target system. This complex transformation logic typically occurs in a dedicated ETL tool (like Informatica PowerCenter, Talend, IBM DataStage) or a separate staging database environment.
- **Load:** The final, cleaned, transformed, and structured data is loaded into the target data warehouse (like Teradata, Oracle Exadata, or cloud warehouses like Redshift in its traditional configuration), ready for analysis and reporting using SQL.

Characteristics of ETL:

- **Schema-on-Write:** Data structure and complex transformations are defined and enforced before loading.
- **Structured Target:** Primarily designed for loading highly curated, structured data into relational data warehouses.
- **Mature Tooling Ecosystem:** Benefits from established commercial and open-source graphical ETL tools.
- **Data Latency:** Often implemented as batch processes (e.g., nightly), introducing latency.

ETL in the AI Context: While effective for traditional BI, ETL can present challenges for modern AI. AI often benefits from raw, less processed data. The rigid, upfront transformation might discard subtle signals or introduce biases. Handling large volumes and diverse formats (unstructured data) common in AI can be cumbersome with traditional ETL tools compared to cloud data platforms.

22.3 Understanding ELT (Extract, Load, Transform)

With scalable cloud data lakes and lakehouses (Chapter 21), ELT has gained prominence, especially for analytics and AI. ELT reverses the order:

- **Extract:** Data is pulled from source systems (often using lightweight tools or CDC).
- **Load:** Raw or minimally processed data is loaded directly and quickly into the target data lake or lakehouse (e.g., S3, ADLS).
- **Transform:** Significant data transformation, cleaning, structuring, aggregation, and feature engineering logic is applied after loading, leveraging the target platform's compute power (e.g., Spark, Databricks SQL, Snowflake, BigQuery).

Characteristics of ELT:

- **Schema-on-Read:** Raw data loaded first; structure/transformations applied later (though modern lakehouse formats allow schema enforcement).
- **Flexible Target:** Suited for diverse data types into flexible lake/lakehouse storage.
- **Leverages Target Platform Power:** Uses scalable cloud compute (Spark, MPP SQL) for efficient transformations post-load. Tools like dbt manage the 'T' stage using SQL, bringing engineering practices.
- **Faster Raw Data Availability:** Raw data available quickly for exploration/ML use without waiting for upfront ETL.
- **Flexibility for AI:** Data scientists apply custom transformations/feature engineering on raw/semi-processed lakehouse data, offering more flexibility than pre-transformed warehouse data.

ELT in the AI Context: ELT is generally preferred for modern data architectures supporting AI. It aligns well with handling diverse data at scale, provides rapid access to raw data, leverages cloud processing power, and fits naturally with the lakehouse paradigm (Chapter 21). This supports the iterative nature of AI development and enables higher EAMI maturity in Data Management and Technology.

The table below summarizes the key differences between ETL and ELT:

Table 22.1: Comparison of ETL vs. ELT

Feature	ETL (Extract, Transform, Load)	ELT (Extract, Load, Transform)
Transformation Stage	Before Loading (in staging/ETL tool)	After Loading (in target data lake/lakehouse)
Data Loaded to Target	Transformed, Structured	Raw or Minimally Processed
Typical Target System	Data Warehouse (often relational)	Data Lake or Lakehouse (object storage + compute)
Schema Handling	Schema-on-Write (Structure defined before load)	Schema-on-Read (Structure applied after load)
Flexibility for AI/DS	Lower (Predefined structure, raw data lost)	Higher (Access to raw data, transform as needed)
Raw Data Availability	Limited / None in target system	Readily available in target system
Processing Engine	Often dedicated ETL server/tool	Target platform's engine (Spark, Cloud SQL engines, etc.)
Popular Tools	Informatica, Talend, DataStage	Spark, Databricks, Snowflake, BigQuery, dbt, Airflow
Maturity Context	Traditional BI standard	Modern standard for cloud analytics & AI

> **D** **DEFINITION**
>
> **dbt (data build tool):** An increasingly popular open-source command-line tool focusing on the 'T' (Transform) in ELT workflows on data already in a cloud warehouse/lakehouse/database. Allows analysts/engineers to build, test, document, deploy complex data transformation pipelines using primarily SQL, incorporating software engineering best practices (version control, modularity, automated testing, dependency management). Works well with platforms like Snowflake, BigQuery, Redshift, Databricks.

22.4 Real-Time vs. Batch Processing & Change Data Capture (CDC)

ETL/ELT workflows can use different processing cadences based on latency needs:

- **Batch Processing:** Data processed/loaded in large chunks at intervals (nightly, hourly). Suitable for traditional reporting and periodic ML model training. Simpler to implement/manage than real-time.
- **Real-Time / Streaming Processing:** Data processed continuously (event-by-event or micro-batches) as it arrives, low latency (seconds/milliseconds). Essential for AI needing immediate insights/actions (real-time fraud, dynamic pricing, streaming personalization). Requires specialized frameworks (Kafka, Flink, Spark Streaming, Kinesis, Dataflow, Stream Analytics), more complex to operate reliably.
- **Change Data Capture (CDC):** Techniques capturing only changes (inserts, updates, deletes) in source databases, avoiding full extracts. CDC tools (Debezium, Fivetran, Qlik Replicate, cloud services like AWS DMS CDC, ADF CDC) monitor logs/triggers to stream changes. Reduces source system load, minimizes data transfer, enables lower latency replication, supporting near real-time analytics/AI even without full streaming downstream.

> ★
>
> Carefully evaluate the business need for real-time data before investing in complex streaming architectures. Many cases adequately served by efficient batch (perhaps more frequent) or micro-batching via CDC, often less complex/costly. Choose cadence based on target AI application's latency needs/value. Leverage CDC for efficiency with frequently changing sources.

22.5 DataOps: Bringing Discipline to Data Pipelines

Building/maintaining reliable data pipelines (ETL/ELT, batch/streaming) needs operational discipline beyond code. DataOps applies DevOps/Agile principles to data analytics/engineering workflows, improving pipeline quality, speed, reliability, collaboration.

> **D**
> **DEFINITION**
>
> **DataOps:** A collaborative data management practice focused on improving communication, integration, quality, security, automation of data flows between data engineers, scientists, analysts, operations, consumers. Aims to shorten analytics development cycle, deliver pipelines/insights faster, more reliably, with higher verifiable quality, leveraging principles like automation (CI/CD for pipelines), testing (quality, logic), monitoring (health, freshness), version control, collaboration.

Key DataOps practices relevant for managing robust ETL/ELT workflows for AI:

- **Automation:** Automating pipeline execution (orchestrators like Airflow, Prefect, ADF), deployment (CI/CD), monitoring/alerting for failures/quality issues.
- **Version Control (Code & Config):** Using Git rigorously for all pipeline code (SQL, Python/Spark, dbt models), definitions (DAGs, ADF JSON), infra configs (IaC), docs.
- **Automated Testing:** Implementing automated tests:
  - » **Data Quality Tests:** Check source/transformed data against rules (completeness, accuracy) using tools like Great Expectations, dbt tests.
  - » **Transformation Logic Tests:** Unit/integration tests for custom code (Python/Spark) or SQL logic (dbt tests).
- **Pipeline Integrity Tests:** End-to-end tests verifying pipeline runs correctly.
- **Monitoring & Observability:** Continuously monitoring execution status, run times, data freshness/latency, volume, quality metrics, resource use. Automated alerting for failures/delays/quality breaches (potentially using Data Observability platforms - Chapter 21).
- **Collaboration & Communication:** Fostering close collaboration between data engineers, scientists/analysts, platform ops using shared tools (Git, Jira, Slack/Teams) and agile methods.
- **Environment Management:** Maintaining consistent dev/test/prod environments using containerization (Docker) and IaC.

Adopting mature DataOps practices significantly improves reliability, maintainability, security, efficiency of data pipelines fueling AI, contributing to higher EAMI maturity in Data Management, MLOps (as reliable data pipelines are foundational), and Technology Selection dimensions.

22.6 AI Office Role and Considerations

While Data Engineering (under CDO/IT) typically implements ETL/ELT pipelines, the AI Office (Compass South) collaborates:

- **Defining Processing Requirements:** Articulating specific data transformation, logic, quality, latency needs of prioritized AI initiatives to Data Engineering.
- **Advocating for Appropriate Patterns:** Championing modern ELT patterns/tools (dbt, Spark on lakehouse) where advantageous for AI flexibility, scalability, accessibility vs. legacy ETL.

- **Collaboration on Pipeline Design:** Participating in design reviews for critical AI data pipelines, ensuring quality checks, transformations, latency needs met. Providing feedback on ML-optimized data structures.
- **Promoting DataOps Adoption:** Advocating for robust DataOps practices (automation, testing, monitoring) by Data Engineering to improve reliability/quality of data for AI systems.
- **Ensuring Tooling Alignment:** Collaborating with Data Eng/IT Arch to ensure chosen processing tools integrate effectively with data storage architecture (Chapter 21) and AI/ML platforms (Chapter 17).

22.7 OmnioTech Case Study: Optimizing Data Pipelines for AI

Setting the Scene:

OmnioTech's initial V1.0 ADF pipelines (ETL pattern to Azure SQL) were inflexible, created silos. Adopting Databricks Lakehouse (Chapter 21) requires redesigning core processing workflows, led by Data Engineering (under CDO) collaborating with AI Office Platform Lead.

Action & Application:

- **Strategic Shift to ELT:** Jointly decide to standardize on ELT for most new ingestion into Databricks Lakehouse. Raw data (CRM, IoT, Web Logs) loaded with minimal processing into bronze Delta tables (ADLS) using ADF/Databricks Auto Loader.
- **Leveraging dbt for Transformations:** Data Eng adopts dbt on Databricks SQL warehouses for all SQL-based transformations creating silver/gold Delta tables. Enables version control (Git), automated tests, documentation, dependency visualization. Data scientists access both raw bronze and curated gold data.
- **Implementing CDC for Near Real-Time Needs:** For AI Support Chatbot needing fresh customer data, implement CDC for Dynamics 365 CRM (using ADF CDC connector) streaming changes to bronze Delta table with low latency.
- **Adopting Key DataOps Practices:** Team implements:
  - » **Version Control:** All dbt models, ADF definitions, Spark scripts managed in Azure Repos.
  - » **Automation:** Databricks Workflows / ADF pipelines for orchestration/scheduling.
  - » **Testing:** dbt tests implemented for key quality checks/referential integrity on core gold tables.
  - » **Monitoring:** Basic Databricks job monitoring / Azure Monitor alerts for pipeline failures/delays. Plans to evaluate Data Observability tools later.
- **AI Office Collaboration:** AI Office provided requirements (raw/curated access, latency needs driving CDC), collaborated on gold table structures (Unified Customer Profile) ensuring necessary features for models.

Outcome & Progress

OmnioTech transitions to modern, flexible, scalable ELT architecture (Databricks lakehouse), improving data accessibility for AI. Adopting dbt brings discipline (version control, testing). CDC addresses near real-time needs. Initial DataOps enhances reliability. This Resource Management (South) shift significantly improves data processing, contributing to higher EAMI Data Management / Technology Selection maturity, providing robust foundation for diverse AI use cases.

22.8 Conclusion: Efficiently Preparing Data for AI

Optimizing data processing (ETL vs. ELT), leveraging appropriate cadences (batch/real-time/CDC), adopting disciplined DataOps is crucial for reliably fueling AI (Compass South). While ETL retains relevance, modern AI needing diverse raw data, flexibility, scalable cloud compute generally benefits from ELT on lakehouses (using Spark, dbt). Ensuring critical data pipelines are robust, efficient, tested, monitored delivers fit-for-purpose data essential for downstream success (MLOps - Chapter 20) and reliable AI value (West). With data stored (Chapter 21) and processed (Chapter 22), Chapter 23 focuses on ensuring ongoing quality/curation.

Key Takeaways:

- ETL (Transform then Load) traditional for warehouses; ELT (Load then Transform) modern for lakes/lakehouses, often better for AI needs.
- ELT leverages scalable cloud compute for transformations post-load, enabling faster raw data access.
- Processing can be Batch or Real-Time/Streaming; CDC enables efficient near real-time replication.
- DataOps principles (automation, version control, testing, monitoring, collaboration) crucial for reliable pipelines.
- Choice of ETL/ELT and DataOps adoption impacts data readiness and EAMI maturity.

Food for Thought / Application Exercise:

- Examine key data pipeline. ETL or ELT? Pros/cons in your context?
- AI use cases needing lower latency? Could CDC help efficiently?
- DataOps maturity (automation, testing, monitoring, version control)? Biggest improvement opportunity for AI reliability?

~ CHAPTER 23 ~

Data Quality and Curation: Ensuring the Reliable Foundation for AI

"Quality is never an accident.
It is always the result of high intention, sincere effort, intelligent direction and skillful execution."
— William A. Foster

23.1 Introduction: The Uncompromising Need for Quality Data

Previous chapters within Part 5 (Resource Management - South) discussed data storage architectures (Chapter 21) and processing workflows (Chapter 22). This chapter addresses perhaps the most persistent and critical challenge in leveraging data for AI: ensuring Data Quality and effective Data Curation. The old adage "garbage in, garbage out" is amplified significantly in the context of machine learning. AI models are highly sensitive to the quality of the data they are trained on; inaccuracies, inconsistencies, missing values, or biases in the input data will inevitably lead to unreliable, inaccurate, and potentially harmful or unfair model outputs, regardless of algorithmic sophistication.

Achieving and maintaining high data quality is not a one-time cleanup task but an ongoing, rigorous discipline requiring dedicated processes, specialized tools, strong governance (Domain East), and a pervasive culture of data stewardship across the organization. Furthermore, raw data, even if technically accurate, often needs careful curation – including selection, cleaning, transformation, annotation, enrichment, and feature engineering (creating informative input variables) – to transform it into a state optimal for effective ML model training and inference.

This chapter delves into the key dimensions of data quality critical for AI, outlines practical techniques for assessing, improving, and monitoring quality, discusses essential data curation practices including metadata management and data lineage tracking, and highlights emerging approaches like Data Observability. Mastering data quality and curation, managed primarily within the Resource Management (South) domain but fundamentally impacting Governance (East - fairness, compliance) and Performance (West - reliable value), is foundational for building trustworthy AI systems and achieving higher EAMI maturity levels in the critical Data Quality dimension.

Learning Objectives:

- Understand why high data quality is a non-negotiable prerequisite for reliable, fair, effective AI.
- Identify and define key dimensions of data quality relevant to AI (accuracy, completeness, consistency, timeliness, validity, uniqueness, representativeness).
- Learn practical techniques for systematically assessing, cleaning, validating, monitoring data quality within AI data pipelines.
- **Understand importance of data curation practices:** metadata management, data lineage, feature engineering support (linking to Feature Stores - Chapter 21).
- Recognize value proposition of emerging concepts like Data Observability platforms for proactive data reliability.
- Appreciate essential collaboration needed across AI Office, central data governance (CDO), data engineering, business domain experts for high data quality/curation standards.

23.2 Why Data Quality is Foundational for AI

Neglecting data quality is a common reason AI projects fail or cause harm. Poor data quality undermines AI initiatives:

- **Inaccurate and Unreliable Models:** Models trained on flawed data learn incorrect patterns, leading to unreliable predictions and poor real-world performance (impacts Domain West).
- **Biased and Unfair Outcomes:** Biases in training data (historical, sampling, measurement, labeling) are learned/amplified, leading to discriminatory outcomes, creating ethical/legal risks (Chapter 10, Domain East).
- **Poor Generalization to Production Data:** Training data not accurately representing real-world conditions (poor representativeness) leads to models failing to generalize effectively when deployed.
- **Wasted Development Effort and Resources:** Data scientists spend excessive time (60-80% est.) identifying, cleaning, preparing poor-quality data before modeling, slowing cycles and wasting talent resources (Domain South).
- **Failed Deployments and Operational Issues:** Models developed on poor data may fail validation. Data quality issues surfacing in production can break pipelines, crash models, produce nonsensical predictions, causing disruptions and losing user trust.
- **Eroded Trust and Adoption Barriers:** Users lose trust in systems producing incorrect, inconsistent, biased results often traced to data quality. Lack of trust hinders adoption/value realization (Domain West, Center).
- **Compliance Violations and Legal Risks:** Using inaccurate, incomplete, improperly governed data (esp. personal) can violate regulations (GDPR mandates accuracy), resulting in fines/legal liabilities (Domain East).

> Proactively investing in continuous, robust data quality management tailored for AI is not optional; it's fundamental for reliable models, mitigating risks, efficient development, realizing value. Demonstrating high maturity in the EAMM Data Quality dimension strongly correlates with overall AI success.

23.3 Key Dimensions of Data Quality for AI

Evaluating data quality requires assessing multiple distinct dimensions, particularly critical for AI/ML:

Table 23.1: Key Dimensions of Data Quality for AI

Data Quality Dimension	Description	Relevance & Impact on AI
1. Accuracy	The degree to which data values correctly represent the true, real-world facts, events, or characteristics they describe.	Critical. Inaccurate features or labels directly lead to models learning incorrect patterns and making flawed predictions.
2. Completeness	The degree to which all required data records and attribute values are present and not missing (i.e., absence of nulls where expected).	Crucial. Missing data can bias models, force discarding records, or require complex imputation which may introduce errors or distort distributions.
3. Consistency	The degree to which data is free from contradictions and represented uniformly across different data sources, records, or time periods.	Essential for reliable processing. Inconsistent formats (dates), units, labels ('USA' vs 'U.S.A.'), or lack of referential integrity can break joins, confuse models, require extensive cleaning.
4. Timeliness / Freshness	The degree to which data is sufficiently up-to-date for its intended use and available within the required operational timeframe.	Very important. Models trained on stale data may not reflect current realities (concept drift). Real-time AI applications (fraud, dynamic pricing) require very low data latency.
5. Validity / Conformity	The degree to which data values conform to defined business rules, constraints, acceptable value ranges, data types, or formats.	Important for pipeline reliability/model stability. Invalid data (text in numeric field, negative age, out-of-range sensor readings) can cause pipeline failures, skew calculations, lead models to learn spurious patterns. Ensures data fits expected schemas.
6. Uniqueness	The degree to which each record or entity within a dataset represents a distinct real-world entity without duplication.	Important for avoiding skewed results. Duplicate records can artificially inflate data point importance during training, leading to biased models or inaccurate stats. Requires MDM or effective deduplication.

	The degree to which the dataset accurately reflects the diversity and statistical properties of the real-world population it represents, without systemic under- or over-representation or biases (Chapter 10).	Absolutely critical for fairness and generalization. Non-representative or biased datasets are primary cause of unfair outcomes and models performing poorly on different populations. Requires specific focus beyond standard checks.
7. Representa-tiveness / Lack of Bias		

> **D**
> **DEFINITION**
>
> **Data Quality for AI:** Data Quality in the context of AI refers to the state where data possesses the necessary accuracy, completeness, consistency, timeliness, validity, unique-ness, and representativeness required to reliably train fair, accurate, robust machine learning models and support trustworthy AI-driven decision-making for a specific intended use case. It goes beyond traditional database integrity to encompass fitness for complex analytical and predictive purposes. Achieving high quality is central to the EAMM Data Quality dimension.

23.4 Practical Techniques for Improving Data Quality

Achieving/maintaining high data quality is ongoing, requiring techniques integrated throughout the data lifecycle, often via DataOps (Chapter 22):

- **Data Profiling:** Systematically analyzing datasets (using automated tools) to discover structure, content, distributions, nulls, outliers, identifying potential quality issues early.
- **Data Validation Rules & Testing:** Defining specific, measurable, automatable rules (business logic, quality dimensions). Implementing as automated checks in pipelines (Great Expectations, dbt tests, Deequ, custom scripts) to detect/flag/reject non-conforming data proactively.
- **Data Cleaning & Remediation:** Applying targeted techniques:
  - » **Handling Missing Values:** Simple imputation (mean/median - caution!), model-based imputation, flagging missingness, deleting records/features (last resort).
  - » **Correcting Inaccuracies/Inconsistencies:** Using reference data (address validation), rule-based transformations (standardize codes), fuzzy matching, human-in-the-loop correction workflows.
  - » **Standardizing Formats:** Applying consistent rules for dates, currencies, addresses, labels, units.
  - » **Deduplication:** Implementing algorithms (record linkage, fuzzy matching, MDM IDs) to identify/merge/remove duplicates.
- **Master Data Management (MDM):** Establishing/governing authoritative "golden records" for core entities (Customers, Products). Referencing master data ensures consistency/uniqueness.
- **Data Quality Monitoring & Reporting:** Ongoing monitoring of key DQ metrics over time (% complete-ness, % passing validation, freshness). Visualizing on dashboards, automated alerts for drops below thresh-olds or anomalies, notifying owners/stewards. Enables proactive issue handling.

23.5 Data Curation: Preparing Data for ML Consumption

Beyond quality, Data Curation involves selecting, enriching, transforming, managing data specifically for effective ML model consumption. Key activities:

- **Relevant Data Selection:** Carefully selecting appropriate datasets, time periods, populations relevant to the AI problem.
- **Data Annotation / Labeling (for Supervised Learning):** Accurately assigning target labels (image tags, text classifications). Label quality/consistency paramount. Often involves human annotators, labeling platforms (Labelbox, Scale AI, SageMaker Ground Truth), QC processes (consensus, audits). Programmatic labeling (heuristics, Snorkel) can augment.
- **Feature Engineering:** Critical process transforming raw data into informative features better representing patterns for models. Needs domain expertise + DS skills. Techniques:
  - » **Creating Derived Variables:** Ratios, differences, interactions, polynomials.
  - » **Binning/Discretizing Continuous Variables:** Grouping numeric values.
  - » **Encoding Categorical Variables:** Converting text labels (One-Hot, Target Encoding, Embeddings).
  - » **Extracting Features from Unstructured Data:** TF-IDF, word embeddings, transformers (BERT), CNN embeddings for text/images.
  - » **Generating Time-Based Features:** Lags, rolling window aggregates, seasonality indicators.
- **Metadata Management:** Capturing comprehensive metadata about curated datasets/features (definitions, types, logic, ownership, usage, quality, frequency, relationships). Crucial for understanding, trust, reuse. Data catalogs essential.
- **Data Lineage Tracking:** Automatically tracking/visualizing end-to-end data flow (source → pipelines → curated datasets → features → models). Indispensable for debugging, impact analysis, root cause analysis, audits/compliance (Domain East). MLOps/pipeline tools increasingly offer lineage.

> Treat data curation/feature engineering as highly collaborative. Foster close work between data scientists, data engineers, business domain experts. Document curated datasets/features thoroughly in shared repository (catalog, Feature Store - Chapter 21) for understanding, reuse, consistency.

23.6 Emerging Concepts: Data Observability

Building on traditional DQ monitoring, Data Observability aims for deeper, holistic, proactive insights into data system health/behavior, inspired by software observability (metrics, logs, traces).

> **D** **DEFINITION**
>
> **Data Observability:** An organization's ability to fully understand the health and state of data in its systems. Platforms typically automate monitoring across pillars: Freshness, Distribution, Volume, Schema, Lineage. Uses statistical analysis/ML to detect quality issues, pipeline breaks, drift proactively, before downstream impact. Shifts DQ management from reactive to proactive reliability. Dedicated platforms (Monte Carlo, Databand) emerging, enabling higher EAMI maturity in Data Quality / Data Management (Level 4/5).

23.7 Collaboration for Data Quality and Curation

Ensuring high-quality, curated data is fundamentally cross-functional:

- **AI Office / Governance Lead:** Sets organizational standards/expectations for DQ needed for AI tiers; integrates DQ/ethical sourcing checks into governance reviews (Domain East).
- **CDO Office / Data Governance Team:** Defines enterprise DQ policies/frameworks/metrics; manages core governance infra (catalog, MDM); oversees stewardship programs.
- **Data Engineers:** Build/maintain reliable pipelines incorporating automated DQ checks/transformations; ensure accessibility/performance; implement pipeline monitoring.
- **Data Scientists / ML Engineers:** Define specific data requirements (features, quality, volume); perform EDA; conduct feature engineering; assess curated feature quality; provide feedback on data issues.
- **Business Domain Experts / Data Stewards:** Act as owners/custodians; provide context on data meaning/rules/characteristics; help define DQ rules; validate accuracy; resolve domain-specific quality issues.
- **Platform / IT Operations:** Ensure underlying data storage/processing infra is reliable, performant, secure.

> **★**
>
> Establish formal Data Steward roles in key business domains. Empower stewards (with domain expertise) with responsibility, authority, tools/training to monitor/champion data quality in their domain, working with central governance/engineering to resolve issues. Federated approach often crucial for scaling DQ efforts.

23.8 OmnioTech Case Study: Tackling Data Quality Issues

Setting the Scene

As OmnioTech scales pilots (Chapter 25), data quality issues become critical roadblocks. Inconsistent sensor formats/nulls degrade PredMaint. Duplicate/incomplete customer records limit HyperPers. EAMI assessment confirmed Data Quality as major weakness (Level 2). Priya Sharma (AI Office) launches joint initiative with CDO Maria Petrova.

Action & Application

- **Joint Initiative & Governance:** Formal, jointly sponsored "AI Data Quality Improvement" initiative launched. DQ expectations added to AI Governance pre-deployment checklist.
- **Define Standards & Metrics:** Specific DQ targets set for critical datasets: >99% validity/<1% nulls (sensors); >98% completeness (customer profile); <1% duplicates post-MDM. Linked to EAMI Data Quality KPIs.
- **Implement Profiling & Monitoring:** Data Eng implements automated profiling (Databricks) on curated 'silver' tables. Power BI DQ dashboards created for Data Stewards (Marketing, Product Eng) tracking metrics vs. targets. Azure Monitor alerts notify stewards of deviations.
- **Cleaning, MDM & Curation:**
  - » **Customer Data:** Project implements lightweight MDM (ADF + custom logic) identifying/merging duplicates into Unified Customer Profile 'gold' table. Marketing Data Stewards review exceptions.
  - » **Sensor Data:** Stricter validation rules in ADF ingestion pipeline flag/filter implausible readings (rules defined with Product Eng). Work with Product Eng on improving future sensor protocols.
- **Feature Store Enhancement:** Feature Store pilot team (Chapter 21) now sources features exclusively from validated 'gold' data tables. Implement feature-level quality validation checks within store. Feature lineage tracked by Databricks.

Outcome & Progress

Concerted, cross-functional effort (AI Office + CDO) systematically addresses foundational data quality challenges. Joint initiative provides focus/resources/collaboration. Proactive monitoring, standards, MDM, curation improvements directly enhance AI fuel reliability. Directly tackles pilot scaling roadblocks, lays groundwork for improving EAMI Data Quality maturity (towards Level 3/4), enabling more trustworthy, fair, impactful AI.

23.9 Conclusion: The Unshakeable Foundation

High-quality, well-curated data is the absolute, unshakeable foundation for reliable, ethical, effective, value-generating enterprise AI. Neglecting data quality introduces risks (bias, inaccuracy, compliance failures), wastes resources, erodes trust, undermines success. Achieving data readiness requires sustained, disciplined, organization-wide commitment (Compass South), involving robust processes (assessment, cleansing, validation, monitoring), thoughtful curation (metadata, lineage, features), modern architectures, strong governance (East), and effective collaboration (AI Office, CDO, IT, DE, DS, Business Stewards). Strategically prioritizing data quality/curation, linking improvements to EAMI goals, creates the essential, trustworthy fuel for the AI maturity journey. Having covered resources, Part 6 returns to the central orchestrating function: Maturity Management (Center).

Key Takeaways:

- High data quality is fundamental for reliable, fair, valuable AI; not optional.
- **Key quality dimensions:** Accuracy, Completeness, Consistency, Timeliness, Validity, Uniqueness, Represen-

tativeness/Lack of Bias.
- Improve quality via Profiling, Validation Rules, Cleaning, MDM, Monitoring.
- Data Curation (Selection, Annotation, Feature Engineering, Metadata, Lineage) prepares quality data for ML.
- Data Observability offers proactive monitoring; Feature Stores streamline features.
- Data quality is cross-functional (AI Office, CDO, DE, DS, Stewards), vital for EAMI Data Quality maturity.

Food for Thought / Application Exercise:

- Consider key dataset for important AI model. Biggest potential weaknesses across 7 quality dimensions (Table 23.1)?
- How is DQ monitored? Reactive or proactive (automated)? How could Data Observability help?
- Who holds DQ/curation responsibility? Stewards defined/empowered? Collaboration effectiveness?

PART 6 - AI MATURITY

MANAGEMENT

~ CHAPTER 24 ~

Assessing Enterprise AI Maturity: Using EAM Framework (EAMM and EAMI)

"You can't improve what you don't measure.
Objective assessment is the first step towards meaningful progress."

24.1 Introduction: Understanding Your Starting Point

Welcome to Part 6, focusing on the central orchestrating function of the AI Office Compass™ Framework: Maturity Management (Center). Previous parts detailed how the AI Office drives Strategic Alignment (North), ensures robust Governance (East), demonstrates Value (West), and manages essential Resources (South). The Center domain integrates these efforts by systematically assessing the organization's current AI capabilities and driving targeted improvements to achieve higher levels of effectiveness, responsibility, and impact across all other domains.

A critical enabler for this crucial Center domain function is the ability to objectively measure the organization's current state of AI maturity. Without a clear, data-driven understanding of existing strengths, weaknesses, and specific capability gaps across all dimensions of enterprise AI (Strategy, Governance, Data, Tech, Talent, Value, etc.), improvement efforts risk being unfocused, inefficient, reactive rather than proactive, and difficult to track or justify.

This chapter introduces the core assessment tools used throughout this book, designed specifically for this purpose: the Enterprise AI Maturity Matrix (EAMM) and the Enterprise AI Maturity Index (EAMI). We will delve into the structure of the EAMM, outlining its key capability dimensions (which map closely to the Compass domains and underlying resources/processes) and the specific characteristics defining each of the five distinct maturity levels. We will then explain how the EAMI provides a quantitative scoring mechanism, utilizing specific Key Performance Indicators (KPIs) tied to each EAMM dimension to generate an objective index score. This EAMI score reflects the organization's overall AI maturity and allows progress resulting from uplift initiatives to be tracked transparently over time. This chapter details the EAMM/EAMI framework itself, the practical process for conducting assessments, guidance on interpreting the results to inform strategy, and clarifies the AI Office's central role in administering these vital assessment tools. Mastering EAMM/EAMI assessments provides the data-driven foundation for the targeted uplift strategies discussed in Chapter 25.

Learning Objectives:

- Understand the critical need for objectively assessing enterprise AI maturity as a core function of the Maturity Management (Center) domain.
- Learn the structure and components of the Enterprise AI Maturity Matrix (EAMM), including its 14 key capability dimensions and five defined maturity levels, based on the provided framework.
- Understand how the Enterprise AI Maturity Index (EAMI) provides a quantitative score based on specific, measurable KPIs linked to EAMM dimensions, enabling objective progress tracking.
- Explore practical approaches for conducting EAMI assessments within an organization, leveraging data collection methods across Compass domains.
- Learn how to interpret EAMI assessment results (overall score, dimension scores, heatmaps, spider charts) to identify specific strengths, critical weaknesses, and prioritize improvement efforts.
- Recognize the AI Office's central role in managing the EAMI assessment process and utilizing its outputs to drive the continuous improvement cycle.

24.2 The Enterprise AI Maturity Matrix (EAMM)/EAM Framework

The EAMM provides a structured, qualitative framework for evaluating an organization's capabilities across the key areas required for successful, scalable, and responsible enterprise AI adoption. It defines specific capability dimensions and describes the typical characteristics, processes, and behaviors observed at five distinct levels of maturity, from nascent beginnings to transformative integration.

> **D**
> **DEFINITION**
>
> **EAMM (Enterprise AI Maturity Matrix)/EAM Framework:** A comprehensive framework, based on the provided source materials, that defines 14 key capability dimensions essential for enterprise AI success (covering Strategy, Governance, Data, Technology, Talent, Value, etc.) and describes the observable characteristics associated with five distinct maturity levels (Level 1: Initial/Exploratory, Level 2: Emerging/Developing, Level 3: Defined/Established, Level 4: Optimized/Integrated, Level 5: Transformative/Adaptive). It provides the qualitative structure for assessing AI maturity.

EAMM Dimensions: The EAMM framework outlined in the source materials utilizes the following 14 dimensions, which comprehensively cover the aspects discussed across the AI Office Compass™ domains:

1. **Strategic Alignment:** Clarity, communication, C-level sponsorship, integration of AI strategy with core business objectives. (Compass: North)
2. **Governance:** Existence, formalization, adoption, effectiveness of policies, processes, roles, risk manage-

ment for responsible AI. (Compass: East)

3. **Data Management:** Maturity of data platforms, data accessibility, integration capabilities, architectural support for AI (lakehouse, feature stores). (Compass: South)

4. **Data Quality:** Processes/maturity ensuring data accuracy, completeness, consistency, timeliness, validity, representativeness for AI. (Compass: South)

5. **Technology Selection:** Approach to selecting, implementing, managing, integrating AI/ML platforms, libraries, infrastructure. (Compass: South)

6. **MLOps (Machine Learning Operations):** Maturity/automation level of practices managing end-to-end ML lifecycle (versioning, testing, deployment, monitoring, retraining). (Compass: South)

7. **Project Management:** Discipline, methodologies (Agile), effectiveness in managing AI projects and portfolio. (Compass: North/Center)

8. **Performance Metrics:** Definition, tracking, reporting of relevant technical/business KPIs for AI initiatives. (Compass: West)

9. **Value Realization:** Processes/rigor for quantifying, tracking, validating (ROI/TCO), communicating tangible business value delivered by AI. (Compass: West)

10. **Talent Acquisition:** (Mapped to Team Competency in source tables) Effectiveness in attracting, recruiting, hiring diverse AI skills. (Compass: South)

11. **Skill Development:** (Mapped to Team Competency) Maturity of programs for upskilling/reskilling internal talent, fostering AI literacy, supporting continuous learning. (Compass: South/Center)

12. **Culture & Change Management:** (Mapped to Cultural Adoption) Organizational readiness, leadership behaviors, data-driven culture, collaboration effectiveness, managing change associated with AI adoption. (Compass: Center)

13. **Innovation Ecosystem:** Processes/resources for exploring emerging trends, fostering responsible experimentation, managing R&D, building partnerships. (Compass: Center)

14. **Ethics & Compliance:** Specific focus on embedding ethical principles (fairness, transparency), bias mitigation, security protocols, ensuring adherence to legal/regulatory requirements. (Compass: East)

EAMM Maturity Levels: The EAMM describes typical organizational characteristics at each of the five levels:

- **Level 1:** Initial / Exploratory: Fragmented, opportunistic efforts; no formal strategy/governance/AI Office; manual/ad-hoc processes; basic/inconsistent tools; siloed/poor data; unmanaged risks; isolated successes; unclear value; high variability.

- **Level 2:** Emerging / Developing: Increased awareness; basic strategy draft (limited alignment); informal/reactive governance; foundational tech used but manual/inconsistent processes (deployment, monitoring); data quality/access challenges persist; initial AI roles emerge; pilots struggle to scale reliably; rudimentary value measurement (technical metrics).

- **Level 3:** Defined / Established: Formal AI strategy/roadmap defined/communicated (partial alignment - North); dedicated AI Office established (Chapter 2); formal governance framework (policies, risk assessment) established/starting adoption (East); standardized platforms/tools implemented (cloud ML, basic MLOps - South); focused data quality/access initiatives underway (South); consistent PM practices; value tracking frameworks defined/piloted (West); foundational AI literacy programs launched (Center); some successful scaling of pilots demonstrates repeatable value. Represents core foundations for scalable, governed AI.

- **Level 4:** Optimized / Integrated: AI strategy fully integrated with/informs business strategy; robust governance embedded/automated in workflows ("Governance-by-Design"); data consistently treated as strategic asset (high quality, broad access via mature platforms like lakehouse/Feature Stores, strong lineage); advanced MLOps enables reliable scaling/continuous improvement (automated retraining, sophisticated monitoring); value systematically tracked/validated (clear ROI/TCO); AI capabilities integrated into core processes/decisions; strong cross-functional collaboration; AI-aware culture developing; EAMI score consistently high (e.g., >60-61%).

- **Level 5:** Transformative / Adaptive: AI drives continuous business model innovation/operational excellence; AI strategy shapes business strategy; adaptive/proactive governance; highly optimized data ecosystem (real-time curation, federated); cutting-edge AI tools, fully automated MLOps; talent continuously developed/deployed dynamically; pervasive/optimized value creation; adaptive learning system culture; potential external ecosystem leadership; EAMI score consistently very high (e.g., >80-81%).

A summary table outlining these dimensions and levels can also be found in Appendix C. The table below provides a high-level overview:

Table 24.1: EAMM Overview Table Summary

EAMM Area	Level 1: Initial	Level 2: Emerging	Level 3: Defined	Level 4: Optimized	Level 5: Transformative
Strategic Alignment	Ad-hoc, Reactive	Basic Strategy, Limited Align	Defined Strategy, Partial Align	Integrated & Aligned Strategy	Strategy Drives, Business Innovation
Governance	None, High Risk	Informal, Basic Policies	Formal Policies, Initial Adoption	Robust, Embedded, Automated Checks	Adaptive, Predictive Ethics
Data Management & Quality	Siloed, Poor Quality	Fragmented, Basic DQ Awareness	Standardized Stores, DQ Initiatives	Integrated Lakehouse/FS, High DQ	Optimized Ecosystem, Proactive DQ
Technology & MLOps	Manual, Inadequate Tools	Basic Tools, Limited CI/CD	Scalable Platforms, Core MLOps	Advanced Tools, Mature MLOps	Cutting-edge, Fully Automated
Talent & Culture	Limited Skills, Resistance	Few Experts, Basic Awareness	Defined Roles, Initial Training	Strong Teams, Developing AI Culture	Pervasive Literacy, Adaptive Culture
Value Realization	Unclear Impact	Basic Metrics, Anecdotal Value	Defined KPIs, Initial Tracking	Systematic Tracking, Clear ROI	Optimized & Pervasive Value Delivery
Innovation Ecosystem	No Formal Innovation	Ad-hoc Experimentation	Defined Process, Some Exploration	Systematic Innovation, Partnerships	Continuous & Transformative

24.3 The Enterprise AI Maturity Index (EAMI)

While EAMM is qualitative, EAMI translates assessment into a quantitative score for objective measurement, benchmarking, and progress tracking.

> **D**
> **DEFINITION**
>
> **EAMI (Enterprise AI Maturity Index):** A quantitative score, typically 0-100% or 1.0-5.0 (aligning with EAMM levels), calculated based on assessing performance against specific, measurable Key Performance Indicators (KPIs). Each KPI links to an EAMM dimension, providing objective measure of overall AI maturity and proficiency per dimension.

EAMI KPIs: Each of the 14 EAMM dimensions has one or more measurable Primary KPIs and potentially Sub-KPIs, quantifying capabilities described in EAMM levels. The detailed KPI list, measurement methods, value sources, and significance are crucial for assessment and are provided in Appendix B. Examples previously discussed:

- **Strategic Alignment KPI:** % AI initiatives linked to strategic goals.
- **Governance KPI:** % Projects passing governance reviews pre-deployment.
- **Data Quality KPI:** % Critical data elements meeting quality thresholds.
- **MLOps KPI:** Average lead time for ML model changes.
- **Value Realization KPI:** % AI initiatives with positive ROI post-scaling.
- **Talent KPI:** % Critical AI roles filled.
- **Culture KPI:** Employee survey score on AI readiness.

(Refer to Appendix B for the comprehensive KPI list and details).

EAMI Scoring Methodology: The process typically involves:
1. **Assess Primary KPIs:** Collect quantitative data (surveys, logs, financials, PM tools, audits, HR systems per 'Value Source' in KPI table) for each Primary KPI across all 14 dimensions.
2. **Score Dimensions:** Convert KPI measurement for each dimension to a percentage score (0-100%). Map measured KPI value to EAMM level description and percentage range (e.g., Level 1: 0-20%, L2: 21-40%, L3: 41-60%, L4: 61-80%, L5: 81-100%). Use qualitative evidence/Sub-KPIs to validate/adjust score, ensuring holistic reflection.
3. **Calculate Overall EAMI Score (Simple Average):** Average the 14 validated dimension scores:
   » EAMI Score (%) = (Sum of all 14 Dimension Scores %) / 14.
4. **Calculate Overall EAMI Score (Weighted Average - Optional):** Assign weights (summing to 100%) reflecting strategic priorities:
   » Weighted EAMI Score (%) = Sum of (Dimension Score % * Dimension Weight %).
5. **Map to EAMI Level:** Map final score (simple or weighted avg) back to EAMM Level (1-5) using defined bands (e.g., L1: 0-20%, L2: 21-40%, L3: 41-60%, L4: 61-80%, L5: 81-100%).

EAMI Calculation Example Snippet: Assume 14 dimension scores sum to 840. Simple Average EAMI = 840 / 14 = 60%. Maps to EAMI Level 3 (41-60% band). If 'Governance' (scored 75%) weighted 20%, others 5.91%, weighted score might differ slightly but likely remain Level 3. Validation confirms Level 3.

24.4 Conducting an EAMI Assessment

The AI Office (Maturity Management - Center) typically orchestrates the EAMI assessment:

1. **Define Scope & Frequency:** Determine scope (enterprise-wide, BU pilot) and cadence (annually for strategy, semi-annually/quarterly for uplift tracking).
2. **Gather Data Systematically:** Collect quantitative/qualitative KPI data using diverse methods across Compass domains:
   - » **Surveys:** Assess culture, training effectiveness, perceptions.
   - » **System Data:** Extract metrics from MLOps, PM tools, monitoring, HR, finance systems.
   - » **Documentation Review:** Analyze strategic plans, governance policies/audits, tech docs, DQ reports, value reports.
   - » **Interviews & Workshops:** Conduct structured interviews/workshops with stakeholders (AI Office leads, BU leaders, tech leads, stewards, gov staff, HR, Finance) for qualitative insights, KPI validation, context, collaborative scoring.
3. **Score & Analyze Results:** Facilitate scoring workshops reviewing evidence against EAMM/KPIs for consensus dimension scores (0-100%). Analyze scores to identify strengths (high scores) and critical weaknesses (low scores, gaps vs. target).
4. **Visualize & Communicate Findings:** Present findings clearly:
   - » **Overall EAMI Score & Level:** Headline status.
   - » **Dimension Score Heatmap:** Visualizes maturity levels (colors) across dimensions, highlighting strengths/weaknesses. Compare current vs. past/target.
   - » **Spider/Radar Charts:** Plots dimension scores on axes, showing profile vs. target/past assessments to visualize progress/gaps.
5. **Develop Actionable Recommendations:** Translate findings into formal report with scores, visualizations, AND specific, prioritized, actionable recommendations for improvement initiatives targeting identified gaps (feeding Chapter 25 uplift strategy).

Treat EAMI assessment as collaborative diagnostic/alignment tool, not just scoring. Involve cross-section of stakeholders in data gathering, scoring validation, interpretation. Builds shared understanding, ownership, buy-in for improvement actions. Communicate results/uplift plan transparently via AI Office leadership.

24.5 Interpreting EAMI Results

EAMI score and dimension details provide strategic insights:

- **Objective Baseline:** Initial assessment provides data-driven starting point, replacing anecdotes.
- **Targeted Gap Analysis:** Dimension scores pinpoint specific strengths (leverage) and weaknesses (focus improvement). Prioritize resources effectively (e.g., fix low DQ before advanced modeling).
- **Tracking Progress & ROI of Uplift:** Comparing successive EAMI scores/KPIs gives quantitative evidence of progress from uplift initiatives (Chapter 25). Helps demo ROI of foundational investments (governance, platforms, talent, DQ).

- **Internal Benchmarking:** Comparing scores across BUs/functions identifies pockets of excellence (share best practices) and lagging areas (need support).
- **External Benchmarking (Use with Caution):** Third-party benchmarks offer some external perspective but use cautiously due to methodology/context differences. Internal progress tracking often more actionable.
- **Informing Strategic Planning & Uplift Strategy:** EAMI gaps/strengths are primary input for developing data-driven, targeted maturity uplift strategies (Chapter 25), ensuring focus where most needed/impactful.

24.6 The AI Office's Role in Maturity Assessment

The AI Office (Maturity Management - Center) is typically central owner/orchestrator of EAMI assessment:

- **Owning & Maintaining the Framework:** Custodian of tailored EAMM dimensions, EAMI KPIs, scoring method, templates. Ensures relevance, updates framework.
- **Orchestrating the Assessment Process:** Planning, scheduling, coordinating, managing regular EAMI assessments. Coordinates data collection, facilitates scoring workshops, ensures consistent methodology.
- **Analyzing & Synthesizing Results:** Analyzing data, calculating scores, identifying key findings (strengths, weaknesses, trends), synthesizing into clear reports.
- **Communicating Findings & Recommendations:** Presenting EAMI results/actionable recommendations clearly to leadership, governance bodies, stakeholders.
- **Facilitating Uplift Planning:** Using EAMI results as primary input for cross-functional workshops developing targeted improvement plans (Chapter 25).
- **Tracking Maturity Progress Over Time:** Maintaining historical EAMI data, monitoring progress vs. uplift plans, reporting changes in scores/levels to demo impact.

24.7 OmnioTech Case Study: Conducting the Baseline EAMI Assessment

Setting the Scene:

As an early key deliverable, OmnioTech's new AI Office Hub (led by Priya Sharma) undertakes formal baseline EAMI assessment (Compass Center activity). Goal: move beyond anecdotal Level 1.5 estimate, get objective starting point to guide 2-year strategy targeting EAMI Level 3.

Action & Application:

- **Process Execution:** AI Office team uses 14 EAMM dimensions/KPIs from adopted framework. Data collection (4 weeks): structured interviews (VPs), doc review (limited formal strategy/policies found), online survey (~50 managers/staff on awareness/challenges/culture), analysis of PM tool/IT log data.
- **Scoring Workshop & Results:** Workshop with AI Review Team reviews evidence vs. EAMM levels. Consensus scores reached:
  - » **Strengths (Low L2):** Basic Technology Selection (Azure ML), Data Management (initial ADLS).
  - » **Major Weaknesses (L1 / Low L2):** Strategic Alignment (none formal), Governance (absent), MLOps (manual), Data Quality (known issues, no process), Value Realization (untracked), Ethics/Compliance (informal), Innovation (ad-hoc).
  - » **Other Areas (Low L2):** Talent/Skills (few experts), Culture/Change (mixed), Project Management (inconsistent).

- **Overall EAMI Calculation:** Dimension scores averaged (simple average adopted). Sum across 14 dimensions (e.g., 310 points / 1400 max).
  - » EAMI Score = 310 / 14 ≈ 22.1%. This score falls in 21-40% band, formally confirming baseline as EAMI Level 2 (Emerging).
- **Visualization & Reporting:** Priya's team creates spider chart comparing baseline scores vs. target EAMI Level 3 profile (scores 41-60%). Summary report for Execs/AI Review Board highlights overall Level 2 score, critical weaknesses (Gov, MLOps, DQ, Value), proposes recommendations aligning with AI Office mandate/roadmap aiming for Level 3.
- **Communication & Action:** Results presented transparently to leadership. Objective data validates need for AI Office and its prioritized focus (V1.0 governance, platform/MLOps, value metrics, DQ program). Baseline EAMI score becomes key benchmark for measuring AI Office success over first 1-2 years.

Outcome & Progress

Formal EAMI assessment gives OmnioTech invaluable, objective baseline (Level 2). Moves beyond anecdotes, identifies specific gaps, validates priorities for AI Office, provides quantifiable benchmark (22.1%) for measuring progress towards EAMI Level 3. Data-driven approach strengthens change case, focuses improvement efforts.

24.8 Conclusion: Measuring the Journey to AI Maturity

Objectively assessing AI maturity via EAMM/EAMI (Maturity Management - Center) is essential starting point for AI transformation. EAMM provides qualitative roadmap; EAMI translates to measurable score using KPIs, enabling progress tracking, gap identification, benchmarking. Regular assessment (orchestrated by AI Office, involving stakeholders) provides critical insights to diagnose current state, prioritize improvements, demo uplift impact, navigate journey to higher maturity/value. With current state understood via EAMI, Chapter 25 focuses on developing targeted uplift strategies.

Key Takeaways:

- Objective AI maturity assessment (EAMM/EAMI) crucial for understanding state, guiding improvement (Maturity Management - Center).
- EAMM defines 14 dimensions, 5 levels (Initial to Transformative) - qualitative framework.
- EAMI provides quantitative score (0-100%) based on measurable KPIs linked to EAMM dimensions - tracks progress.
- Assessments involve collecting data (surveys, systems, interviews), scoring dimensions, calculating EAMI score, visualizing (heatmaps, spider charts).
- EAMI results provide baseline, identify gaps, track progress, inform uplift strategies.
- AI Office typically owns/orchestrates EAMI assessment process.

Food for Thought / Application Exercise:

- Review 14 EAMM dimensions. Which 2-3 feel intuitively strongest in your org? Weakest? Why?
- Consider 5 EAMI levels. Which best describes your org's current overall state? Evidence?
- Biggest challenges conducting EAMI assessment in your org (collecting reliable KPI data)? How overcome?

~ CHAPTER 25 ~

Driving Continuous Improvement and Scaling AI: Uplift Strategies

"Maturity is not a final destination,
but a continuous journey of learning, adapting, and improving."

25.1 Introduction: From Assessment to Action

Chapter 24 detailed the critical process of assessing enterprise AI maturity using the EAMM framework and the quantitative EAMI score. This assessment provides an invaluable snapshot of the organization's current capabilities, highlighting both strengths to leverage and weaknesses to address across various dimensions like strategy, governance, data, technology, talent, and value realization. However, the assessment itself is only the starting point; its true value lies in enabling targeted action. This chapter, firmly rooted in the Maturity Management (Center) domain of the AI Office Compass™ framework, focuses on translating assessment insights into concrete uplift strategies designed to systematically improve AI maturity and effectively scale successful AI initiatives across the enterprise.

We will explore how to develop tailored improvement plans based on specific EAMI assessment results, discuss common challenges and proven strategies for scaling AI solutions beyond initial pilots (moving from "pilot purgatory" to enterprise impact), address the complexities of managing technical debt and maintaining governance integrity as AI scales, consider approaches for enabling federated AI development safely in mature organizations, and emphasize the importance of measuring the success of uplift efforts. The AI Office plays a crucial role in orchestrating this continuous improvement cycle, ensuring that the organization not only adopts AI but matures its capabilities over time to maximize sustainable value and maintain a competitive edge. This chapter directly addresses the EAMM Uplift Strategy dimension.

225

Learning Objectives:

- Understand how to translate EAMI assessment results into actionable maturity uplift strategies prioritized based on gaps and strategic goals.
- Learn tailored approaches and specific initiatives suitable for improving AI capabilities based on the organization's current assessed EAMI level (1-5).
- Identify common challenges (technical, data, MLOps, governance, change management) encountered when scaling successful AI pilot projects to enterprise-wide deployment and learn effective mitigation strategies.
- Recognize the importance of proactively managing ML technical debt and maintaining governance standards consistently during scaling phases.
- Explore different approaches for enabling federated AI development responsibly in mature organizations (linking to hybrid operating models - Chapter 2).
- Understand how to measure the success and impact of AI maturity uplift initiatives, linking back to EAMI score improvements and business value (Domain West).

25.2 Translating EAMI Assessments into Uplift Strategies

The EAMI assessment (Chapter 24) provides a data-driven diagnosis of the organization's AI health. The next step, driven by the Maturity Management (Center) function, is prescribing the treatment – developing a targeted maturity uplift strategy. This typically involves:

1. **Analyze Gaps & Strengths:** Deeply analyze EAMI results (overall score, dimension scores, KPIs, qualitative findings). Pinpoint 1-3 most critical capability gaps hindering progress towards target EAMI level (e.g., moving L2→L3). Identify existing strengths to leverage.
2. **Prioritize Dimensions for Uplift:** Based on gap analysis and strategic priorities (Domain North), identify EAMM dimensions where focused improvement yields most strategic benefit, addresses pressing risks, or unlocks critical dependencies. E.g., scaling pilots might prioritize improving MLOps (South) and Data Quality (South) over Innovation Ecosystem (Center) initially.
3. **Define Specific Uplift Initiatives:** Brainstorm and define concrete, actionable, SMART initiatives addressing prioritized gaps. Examples:
   - » **Governance Gap:** "Implement V1.0 AI Risk Assessment Process & Policy Training"
   - » **MLOps Gap:** "Pilot Standardized CI/CD Pipeline Template using Azure ML Pipelines"
   - » **Data Quality Gap:** "Launch Data Quality Monitoring Dashboard for Critical Customer Data"
   - » **Talent Gap:** "Develop Role-Based AI Literacy Training Modules"
4. **Create an Uplift Roadmap / Integrate with AI Roadmap:** Integrate uplift initiatives into the overall Enterprise AI Roadmap (Chapter 6) as distinct projects/work streams. Assign ownership, define timelines, estimate resources (budget, personnel), establish measurable success KPIs (often linked to improving specific EAMI dimension KPIs).
5. **Secure Buy-in & Resources:** Clearly communicate EAMI findings, rationale for prioritized initiatives, expected benefits (incl. projected EAMI score improvement) to leadership/stakeholders to secure sponsorship, budget, cross-functional collaboration.

> **D** **DEFINITION**
>
> **AI Maturity Uplift Strategy:** A planned and prioritized portfolio of specific, measurable initiatives, derived directly from objective EAMI assessment findings and aligned with strategic business goals. This strategy aims to systematically improve capabilities across targeted EAMM dimensions, increase the overall EAMI score, and advance the organization to a higher AI maturity level over a defined timeframe. It is a core output of the Maturity Management (Center) domain.

25.3 Tailored Uplift Strategies by EAMI Level

Uplift strategies adapt based on current EAMI level, building capabilities progressively.

Table 25.1: Priority Uplift Focus Areas by EAMI Level

EAMI Level	Typical Characteristics (Recap)	Priority Uplift Focus Areas for Moving to Next Level	Example Uplift Initiatives
Level 1: Initial	Ad-hoc, siloed experiments, high risk.	Focus: Awareness & Foundation. Securing executive sponsorship, basic AI literacy for leaders, defining initial AI vision/potential, identifying 1-2 high-potential pilot use cases, starting basic data discovery, forming initial cross-functional AI working group.	Conduct executive AI strategy workshops; Draft initial AI vision & principles doc; Form AI Steering Committee/Task Force; Conduct high-level use case brainstorming; Initiate inventory of key potential data sources.
Level 2: Emerging	Basic strategy/ governance drafts, inconsistent processes, pilot struggles.	Focus: Formalization & Enablement. Formalizing AI strategy/ roadmap, establishing foundational AI Office/CoE structure & roles, implementing V1.0 Governance framework (policies, risk assessment), selecting/implementing initial standard platform/tools (cloud ML, basic MLOps), launching first controlled pilots with clear value KPIs, delivering basic AI literacy training.	Finalize & approve initial AI roadmap (Ch 6); Formally charter AI Office & hire core leads (Ch 2); Implement core Responsible AI policy, risk assessment process & AI Review Team (Ch 8/9); Select & implement standard cloud ML platform (Ch 17); Run 2-3 pilots with defined value tracking (Ch 13/14); Deliver intro AI awareness training.

Level 3: Defined	Formal strategy/governance/platforms exist, initial scaling successes, basic MLOps.	Focus: Operationalization & Scaling. Rolling out/ensuring adoption of governance framework, significantly maturing MLOps (automation, CI/CD/CT, monitoring), systematically improving data quality/accessibility (catalog, pipelines), implementing rigorous value tracking/reporting, scaling proven pilots, broadening role-based skills development.	Implement enterprise rollout/training for governance; Implement standard MLOps CI/CD pipelines & monitoring dashboards (Ch 20); Launch enterprise DQ program & populate data catalog (Ch 23/18); Implement standardized ROI reporting & quarterly value reviews (Ch 14); Scale 2-3 successful pilots; Launch role-based technical AI training (Ch 16).
Level 4: Optimized	Integrated strategy, robust embedded governance, mature MLOps/data, clear ROI demonstrated.	Focus: Optimization & Federation. Optimizing end-to-end AI lifecycle processes, enhancing platforms (Feature Stores, advanced testing/monitoring), enabling federated development safely (empowering 'Spokes' with guardrails), fostering deeper AI integration into core processes/decisions, strengthening innovation pipeline, driving pervasive AI culture.	Implement Feature Store (Ch 21); Adopt advanced MLOps (automated A/B tests, integrated fairness/security tests, automated retraining); Establish formal 'Spoke' enablement program (Ch 2/16); Launch enterprise AI fluency program & Champion network (Ch 26); Formalize AI R&D funding process & partnerships (Ch 27); Integrate AI insights into core workflow tools.
Level 5: Transformative	AI drives business strategy/innovation, adaptive governance, fully automated lifecycle, pervasive AI culture.	Focus: Adaptation & Leadership. Maintaining leadership via continuous adaptation; exploring cutting-edge AI (quantum, neuro-symbolic); developing adaptive governance (anticipating future ethics/regulations); optimizing human-AI collaboration; contributing to external ecosystem/standards/ethics.	Establish strategic external AI research partnerships; Pilot emerging AI tech; Develop AI-driven adaptive governance monitoring; Implement advanced human-AI teaming interfaces; Contribute to open-source/standards; Publish on responsible AI leadership.

> Moving between EAMI levels requires focused investment and cross-functional commitment, orchestrated by the AI Office (Maturity Management - Center). Uplift strategy should be realistic, phased, aligned with strategy/capacity. It's a marathon, often 12-24+ months per level, especially L3+.

25.4 Scaling AI: From Pilot Success to Enterprise Impact

A critical challenge (often EAMI Level 2/3) is moving successful pilots from lab/limited deployment to full-scale, reliable production – escaping "pilot purgatory". Successfully scaling requires addressing challenges beyond initial model development:

- **Technical Scalability & Reliability:** Ensuring infrastructure (compute, storage) and serving architecture handle production volumes/SLAs reliably/cost-effectively. Requires robust platform engineering (Chapter 17), mature MLOps (Chapter 20).
- **Data Pipelines & Quality at Scale:** Building/maintaining reliable, scalable data pipelines (Chapter 22) feeding production models with high-quality, fresh data. Addressing DQ issues (Chapter 23) critical at scale.
- **MLOps Maturity for Production:** Implementing mature MLOps is non-negotiable: robust CI/CD, comprehensive monitoring (performance, drift, bias), alerting, registries, versioning, potentially automated retraining.
- **Integration with Core Business Processes & Systems:** Seamlessly embedding AI outputs/insights/actions into operational workflows, UIs (CRM, ERP), decision processes for actual use/impact. Often requires process redesign, deep integration.
- **Governance Application at Scale:** Consistently/efficiently applying governance policies (ethics, bias, security, compliance) as model count grows. Manual reviews bottleneck; needs automation (MLOps pipelines) & potentially federated governance.
- **Change Management & User Adoption:** Proactively managing impact on users/processes. Requires clear communication, comprehensive user training, addressing concerns, demonstrating value for widespread adoption (Chapter 26).
- **Cost Management & Optimization (FinOps):** Actively managing significant/variable operational costs (cloud compute, storage, monitoring) using FinOps principles (Chapter 19).

Strategies for Successful Scaling:

- **Design for Scale from the Start:** Encourage considering scalability (tech, data, MLOps, gov) even during pilot design.
- **Pilot with Production Intent:** Use pilot to test key production architecture, data pipelines, MLOps processes, monitoring on smaller scale.
- **Implement "Scaling ROI Gates":** Establish formal checkpoints post-pilot. Require evidence of validated pilot value, realistic scaling plan/budget (TCO), confirmed operational readiness (MLOps, data), clear business case before approving enterprise deployment.
- **Adopt Modular & Reusable Architecture:** Design using modular components (microservices), promote reusable assets (Feature Stores, MLOps templates, libraries) for easier maintenance/updates/scaling.
- **Prioritize Foundational Platforms & MLOps:** Invest in robust, scalable, reusable data platforms (lakehouse, feature store) and mature MLOps infrastructure (CI/CD, monitoring, registry) as prerequisite for scaling multiple applications. Include in roadmap.
- **Consider Phased Rollout Strategies:** Instead of "big bang", roll out incrementally (BU, geography, user group) to manage risk, gather feedback, refine processes.
- **Establish Clear Ownership for Scaled Systems:** Define clear ownership/accountability for ongoing operation, maintenance, monitoring, performance of scaled AI systems (collaboration between dev team, central MLOps/Platform, sponsoring BU).

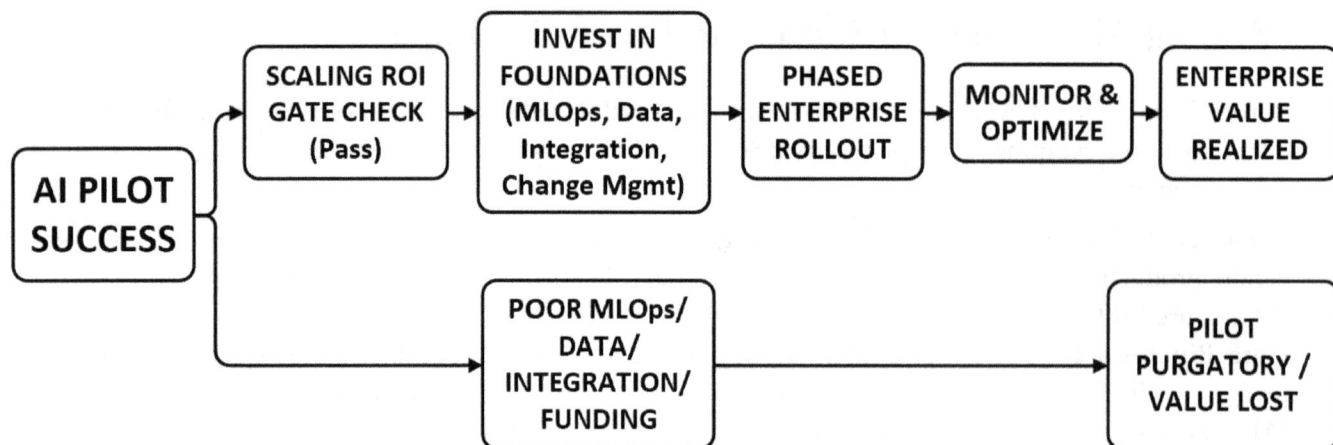

Figure 25.1: Pilot Purgatory vs. Scaling Pathway Diagram

25.5 Managing Technical Debt and Governance at Scale

As AI deployment grows, proactively manage two key challenges:

- **ML Technical Debt:** Implicit cost of rework from easy short-term solutions vs. better long-term ones. Manifests as poor docs, lack of tests, monolithic code, tight dependencies, manual deployments, inadequate monitoring, neglecting retraining. Cripples agility over time.
  - » **Mitigation:** Embed good engineering (MLOps), prioritize code quality/modularity, mandate automated tests, enforce docs standards, invest in refactoring, manage model decay/decommissioning.
- **Maintaining Governance Integrity at Scale:** Consistently/effectively applying standards (ethics, bias, security, compliance) as development becomes distributed (federated models). Central manual reviews impossible.
  - » **Mitigation:** Heavy reliance on automating controls in standard MLOps pipelines (Chapter 9); clear accountability (roles, trained local Spokes/Champions); regular audits; continuous training; centralized tooling (registry, gov platforms) for visibility/enforcement. Strong leadership support essential (Compass Center).

25.6 Enabling Federated AI Development Responsibly

As orgs reach higher EAMI maturity (Level 4/5) and adopt Hybrid/Federated models (Chapter 2), enabling 'Spoke' teams in BUs to develop/deploy AI more autonomously is key for agility/scale. Requires strong central Hub (AI Office) enablement/governance:

- **Provide Standardized Platforms & Tools:** Hub provides well-documented, easy-to-use, centrally managed platforms (cloud ML, MLOps pipelines, Feature Stores) Spokes expected/required to use.
- **Establish Clear Guardrails & Policies:** Hub defines clear enterprise policies, standards, guidelines (technical, ethical, security, risk assessment) all teams (incl. Spokes) must follow. Risk-tiering critical (Chapter 8).
- **Offer Reusable Components & Accelerators:** Hub actively develops/shares reusable assets (features, MLOps templates, models, libraries, best practices via CoPs/repos) to accelerate Spoke development.

- **Deliver Robust Training & Certification:** Provide comprehensive, role-based training/certification for BU 'Spoke' roles (BU DS, AI Translator) ensuring skills/understanding of standards/governance.
- **Implement a Federated Governance Model:** Define model where Spokes might self-assess/approve low-risk apps using standard pipelines/tools. Higher-risk apps or non-standard approaches still need formal Hub review/approval (AI Review Board). Requires trust, strong auditing, clear escalation paths.

25.7 Measuring Uplift Success

Effectiveness of AI maturity uplift strategies should be measured quantitatively/qualitatively, linking back to EAMI (Compass Center tracking):

- **EAMI Score Progression:** Primary measure is demonstrable improvement in overall EAMI score and targeted dimension scores, tracked via regular reassessments (Chapter 24).
- **Improvement in Key EAMI KPIs:** Monitor specific operational KPIs associated with targeted dimensions (e.g., reduced model deployment lead time for MLOps; increased % projects with positive ROI for Value Realization; improved DQ metrics). (Refer to Appendix B).
- **Achievement of Uplift Initiative Goals:** Track completion/success metrics for specific uplift projects in roadmap (e.g., % adoption of new governance process, % completing MLOps training, Feature Store pilot launch).
- **Business Outcome Linkage:** Ultimately, demonstrate how maturity improvements correlate with/contribute to tangible business metric improvements (faster innovation cycles, reduced fines, increased revenue). Closes loop to value (Domain West).

25.8 OmnioTech Case Study: Driving Maturity Uplift Towards Level 3

Setting the Scene

Following baseline EAMI assessment (Level 2, 22.1%) highlighting critical gaps (Chapter 24), OmnioTech's AI Office Hub implements its prioritized 12-month uplift strategy (embedded in roadmap - Chapter 6), explicitly targeting EAMI Level 3 across key dimensions (Compass Center activity).

Action & Application - 12 Month Progress Review

- **Uplift Initiative Execution:**
  - » **Governance (Target L3):** V1.0 framework (policies, risk assessment, AI Review Team) rolled out/applied to all new projects (KPI: 100% project compliance). Training delivered.
  - » **MLOps (Target L3):** Azure ML platform operational. Standardized CI/CD templates used for scaled pilots/new pilots. Basic registry/monitoring established. (KPI: 80% prod models deployed via std pipeline; Avg deploy time weeks → days).
  - » **Data Quality (Target L3):** Joint DQ program launched. Monitoring dashboards implemented. MDM pilot reduced customer duplicates ~60%. (KPI: Key DQ metrics improved avg. 40%; DQ dashboards operational).
  - » **Value Realization (Target L3):** Standardized value tracking implemented. Quarterly Value Reviews institutionalized. Initial ROI trends positive. (KPI: 100% scaled projects have quarterly value reviews; Initial

pilot ROI positive).
- » **Talent/Culture (Target L2/3):** Core AI Office roles filled. Foundational AI literacy module completed by 75% target staff. (KPIs met).
- **Scaling Progress:** PredMaint scaled to one product line, showing measurable service call reduction. Hyper-Pers scaling proceeding but data integration challenges slow rollout. Two new medium-risk pilots initiated using framework/platform.
- **EAMI Re-Assessment (12 Months):** Follow-up assessment conducted:
  - » **Governance:** ~55% (Level 3)
  - » **MLOps:** ~50% (Level 3)
  - » **Data Quality:** ~45% (Level 3)
  - » **Value Realization:** ~45% (Level 3)
  - » **Talent/Skills:** ~40% (Level 2/3)
  - » **Culture/Change:** ~35% (Level 2) Overall EAMI Score: Average now ~48% (firmly EAMI Level 3 - Defined).
- **Reporting & Next Steps:** Priya presents positive EAMI progress/pilot value to executives. Results justify continued investment, validate structured approach. AI Review Team uses new EAMI results to prioritize next uplift phase (further MLOps maturity, broader training, data accessibility).

Outcome & Progress

Executing targeted uplift strategy informed by baseline EAMI assessment, focusing on foundational capabilities (Governance, MLOps, DQ, Value Tracking), OmnioTech successfully, demonstrably advances AI maturity (Level 2 → Level 3) within timeframe. Systematic approach (Maturity Management - Center) builds momentum, sets stage for tackling EAMI Level 4 capabilities.

25.9 Conclusion: The Continuous Improvement Journey

Achieving enterprise AI maturity is a continuous journey of assessment, targeted improvement, responsible scaling, orchestrated by Maturity Management (Center). Leveraging EAMM/EAMI frameworks provides structure to diagnose gaps, develop data-driven uplift strategies. Scaling pilots successfully requires addressing technical scalability, data, MLOps, governance, cost, change management. Embracing disciplined cycle (assess → plan → act → measure → adapt), guided by AI Office, enables progressive capability enhancement, maximizing value, managing risks, transforming into mature, adaptive AI-driven enterprise. Final element: fostering right culture/future readiness via innovation (concluding chapters).

Key Takeaways:

- AI maturity requires continuous improvement driven by EAMI assessment and targeted uplift strategies (Maturity Management - Center).
- Uplift priorities tailored to current EAMI level, building foundations first.
- Scaling AI pilots needs addressing tech scalability, data pipelines, MLOps maturity, governance, change management, cost.
- Implement "Scaling ROI Gates," manage ML technical debt proactively during scaling.
- Mature orgs enable federated AI responsibly (central platforms, guardrails, training, federated gov).
- Measure uplift success via EAMI score/KPI improvements, link maturity gains to business outcomes.

Food for Thought / Application Exercise:

- Based on org's estimated EAMI level (Ch 24), top 2-3 priority dimensions for uplift strategy (Table 25.1)?
- Think of AI pilot successful technically but failed to scale. Which scaling challenges (Section 25.4) likely primary contributors?
- How could org better enable responsible AI dev by teams outside central AI office? Needed platforms, guardrails, training?

~ CHAPTER 26 ~

Cultivating an AI-Driven Culture: Change Management and Literacy

"The secret of change is to focus all of your energy not on fighting the old, but on building the new."
— Socrates (adapted)

26.1 Introduction: The Human Side of AI Transformation

Previous chapters, particularly within Part 5 (Resource Management - South) and Part 6 (Maturity Management - Center), focused heavily on the strategic, technical, data, and process elements required for enterprise AI success. However, arguably the most significant determinant of whether AI truly transforms an organization lies not just in the technology itself, but in its people and culture. Implementing sophisticated AI systems without concurrently addressing the human element – fostering broad AI literacy, proactively managing inevitable resistance to change, cultivating organizational trust, and adapting the prevailing corporate culture – is a well-documented recipe for stalled adoption, underutilized capabilities, and ultimately, unrealized business value.

This chapter dives deep into these critical aspects of Culture & Change Management, core components of the Maturity Management (Center) domain within the AI Office Compass™ framework. We will explore the fundamental imperative for cultural transformation to truly support and sustain AI adoption at scale. We will identify common human and cultural barriers (such as fear of displacement, skill gaps, lack of trust, or simple inertia) that often derail AI initiatives. We will discuss practical frameworks and effective strategies for building appropriate levels of AI literacy across the entire workforce, from executives to front-line employees. We will also examine key principles and proven techniques for managing the significant organizational change associated with implementing AI systems that often alter workflows and roles. Finally, we highlight the indispensable role of visible and committed leadership in championing and modeling an AI-driven, data-informed mindset. Successfully cultivating the right culture is absolutely essential for unlocking the full potential of AI investments and achieving the higher, more integrated and adaptive levels of maturity described in the EAMM framework (Levels 4 and 5), particularly in the critical Culture and Change Management dimensions.

Learning Objectives:

- Understand why cultural transformation and effective, proactive change management are critical, non-negotiable enablers for successful enterprise AI adoption and achieving higher EAMI maturity.
- Identify common cultural and human barriers that frequently hinder AI adoption within organizations (e.g., fear of job loss, lack of trust in AI systems, significant skill gaps, general resistance to workflow changes).
- Explore practical frameworks and concrete strategies for building appropriate levels of AI literacy and data-driven decision-making skills across different roles and levels within the organization.
- Learn key principles and established techniques (e.g., Kotter's model, ADKAR) for effectively managing the organizational change associated with implementing potentially disruptive AI systems.
- Recognize the vital, visible role that leadership – from the C-suite down to team managers – must play in championing, modeling, and reinforcing an AI-supportive culture.
- Appreciate the AI Office's crucial collaborative role in partnering with Human Resources (HR), Learning & Development (L&D), and Corporate Communications to design and drive essential cultural and change initiatives related to AI.

26.2 The Cultural Imperative for AI Success

Implementing AI effectively often requires more than just installing new software; it frequently necessitates fundamental shifts in how people work, collaborate, make decisions, and learn. An organization's existing culture – its shared values, beliefs, norms, behaviors – can either accelerate or impede AI adoption and maturity (EAMM Culture):

- **Decision-Making Culture:** Moving from intuition/experience-based decisions towards embracing data-driven insights and AI recommendations requires a cultural shift. Fostering data literacy, critical evaluation of sources/outputs, building appropriate trust while retaining human judgment is key.
- **Collaboration Norms:** AI projects demand unprecedented cross-functional collaboration (business SME, DS, MLE, DE, Ops, Gov, users). Strongly siloed cultures hinder this synergy. (EAMM Collaboration)
- **Attitude towards Experimentation & Learning:** AI development involves experimentation, iteration, accepting uncertainty, learning from failures. Risk-averse cultures stifling experimentation hinder AI innovation/adaptation. Psychological safety is crucial. (EAMM Innovation Ecosystem)
- **Adaptability & Agility:** AI evolves rapidly. Orgs/individuals need adaptability, continuous learning, ability to adjust strategies quickly. Rigid cultures struggle.
- **Trust & Transparency:** Widespread adoption hinges on trust. Requires reliable systems AND culture valuing transparency (how AI works, data usage, governance), open dialogue about capabilities/limits.
- **Human-AI Teaming Mindset:** Maximizing value often involves workflows where humans and AI collaborate. Requires cultural acceptance of AI as augmentative tool, not just threat/replacement.

> Attempting large-scale AI without proactively addressing cultural readiness is common/costly. AI initiatives succeed more in orgs consciously cultivating culture that is data-curious, collaborative, adaptable, ethically aware, embraces continuous learning, provides psychological safety. Building this supportive cultural foundation (EAMI Levels 4/5) is as critical as technical infrastructure.

26.3 Common Cultural and Human Barriers to AI Adoption

Organizations often encounter significant human-centric barriers:

- **Fear of Job Displacement / Automation Anxiety:** Common anxiety AI will eliminate jobs/devalue skills. Can manifest as resistance, non-cooperation, reluctance to share expertise.
- **Lack of Trust in AI Systems:** Skepticism about accuracy, reliability, fairness, security, ethics. Stems from lack of understanding ("black box"), negative past experiences, media hype, genuine concerns.
- **Skill Gaps & Insufficient AI Literacy:** Workforce may lack skills or foundational understanding to use AI tools, interpret insights, collaborate effectively, understand data quality/governance importance. Leads to frustration, underutilization. (EAMM Talent/Skills)
- **General Resistance to Change:** Natural human/organizational inertia resists changes to routines, workflows, power structures. AI often requires significant process redesign, triggering resistance.
- **Misaligned Incentives or Performance Metrics:** Existing systems may discourage needed behaviors (cross-functional collaboration, data sharing, experimentation) or adoption of new AI processes.
- **Lack of Visible Leadership Commitment & Communication:** Insufficient/inconsistent/superficial support from leadership regarding AI importance, vision, responsible implementation commitment. Lack of clear "why" breeds uncertainty.
- **Valid Ethical Concerns:** Genuine worries (employees, customers, public) about AI implications (surveillance, fairness, environmental impact, lack of oversight). Must be addressed openly via governance (Domain East).
- **"Not Invented Here" Syndrome / Silo Mentality:** Resistance from departments/teams to adopting centrally developed platforms, tools, policies, preferring local approaches.

Navigating AI transformation requires proactively identifying and mitigating these common barriers.

26.4 Frameworks for Building AI Literacy and Culture

Cultivating an AI-ready culture requires deliberate, sustained effort, orchestrated by AI Office (Maturity Management - Center) collaborating with HR, L&D, Comms, leadership. Key initiatives:

Table 26.1: Initiatives for Building AI Literacy and Culture

Initiative Type	Description	Examples	EAMM Dimension Impact
1. AI Literacy & Training Programs	Providing targeted, role-appropriate education on foundational AI concepts, specific internal tools, data literacy, ethical considerations, org AI strategy/governance framework.	Executive AI strategy briefings;"AI for Managers" training; Foundational "AI 101" online modules; Specific tool training; Role-based technical upskilling (Ch 16); Mandatory Responsible AI & ethics training (Ch 7).	Talent/Skills, Culture, Ethics
2. Strategic Communication & Awareness	Consistently/transparently communicating org AI vision, strategy, progress (successes/challenges), value delivered, commitment to responsible AI principles.	Regular internal newsletter/blog features ("AI Spotlights"); CEO/Leadership messages in Town Halls; Dedicated AI intranet portal (strategy, FAQs, training); Sharing compelling "AI Value Stories" (Ch 15); Open communication about workforce impact & support.	Culture, Change Management, Leadership
3. Leadership Engagement & Role-Modeling	Ensuring senior execs/managers visibly champion AI, actively use AI insights, allocate resources, consistently model desired data-driven/collaborative behaviors.	C-suite discusses AI roadmap in reviews; Leaders ask data-driven questions; Managers encourage team AI training/experimentation; Public celebration of AI successes/responsible practices by leaders.	Leadership, Culture, Change Management
4. Fostering Collaboration & Knowledge Sharing	Creating formal/informal mechanisms encouraging cross-functional teamwork, sharing AI knowledge, best practices, code, lessons learned across silos.	Establishing/nurturing AI CoPs (DS, MLOps, Ethics); Running internal AI challenges/datathons/hackathons; Mandating cross-functional project teams; Utilizing shared collaboration platforms (Slack/Teams channels, Git repos).	Collaboration, Culture, Innovation

5. Aligning Incentives & Recognition	Modifying performance management/rewards to explicitly encourage/ reward desired behaviors (collaboration, data sharing, responsible experimentation, new tool adoption).	Including AI-related objectives in performance reviews; Bonuses/ awards for successful AI implementation delivering value; Publicly celebrating internal "AI Champions" promoting adoption/literacy.	Culture, Performance Metrics
6. Promoting Ethical Dialogue & Psychological Safety	Creating formal/informal safe spaces for employees to discuss ethical concerns, ask questions, challenge assumptions, report issues without fear. Fostering learning from mistakes.	Establishing accessible AI Ethics Councils/Office Hours (Ch 8); Regular ethics training/reporting channels; Blameless post-mortems for failures; Leaders encouraging questions/dissent; Protecting whistleblowers.	Ethics, Governance, Culture
7. Dedicated Change Management Support	Providing resources, structured methodologies, tailored support to help navigate significant changes from AI implementation (workflows, roles, skills).	Assigning experienced change managers to major AI initiatives; Developing clear communication plans (impacts/timelines); Providing targeted retraining/ transition support; Establishing feedback loops during rollout.	Change Management, Culture, Talent

Implement a holistic, integrated AI Cultural Transformation Program coordinating these initiatives under unified strategy (often co-led AI Office + HR/L&D). Ensure consistent messaging, shared goals (linked to EAMM Culture/Change), reinforcement. Measure impact regularly (surveys, adoption metrics, training completion, feedback), refine program based on insights.

26.5 Managing Organizational Change for AI

Implementing significant AI often requires fundamental changes to processes, structures, roles, work (EAMM Change Management). Managing this effectively is critical. Structured change methodologies increase success. Kotter's 8-Step Process provides a useful framework:

- **Create Urgency:** Articulate compelling reasons for AI adoption/change now. Link to strategy (North), competition, pain points, risks. Explain the "why".
- **Build a Guiding Coalition:** Assemble powerful group (leadership authority, credibility, expertise - tech/ domain/change, influence) to lead change.
- **Form a Strategic Vision & Initiatives:** Develop clear, compelling vision for future state post-AI change. Outline key strategic initiatives to achieve it.

- **Enlist a Volunteer Army (Communicate for Buy-in):** Communicate vision/strategy frequently, broadly, using multiple channels (Chapter 15), tailored to stakeholders. Address concerns, explain benefits ("WIIFM"), empower employee contribution.
- **Enable Action by Removing Barriers:** Identify/remove obstacles (conflicting structures, skill gaps, misaligned incentives, resistance). Empower employees with tools/permissions.
- **Generate Short-Term Wins:** Plan for/create visible, unambiguous early successes. Celebrate publicly to build credibility, reward efforts, undermine cynicism.
- **Sustain Acceleration (Consolidate Gains & Produce More Change):** Use momentum to tackle larger obstacles, implement further changes, refine processes, deepen AI integration. Don't declare victory too soon.
- **Institute Change (Anchor New Approaches in the Culture):** Reinforce new ways of working (communication, recognition, training integration, performance management adjustments) until they become the established norm ("the way we do things").

> Change management for AI is sensitive due to job displacement anxiety. Communication must be transparent, empathetic, proactive about workforce impacts. Frame AI as augmentation where possible. Invest heavily/visibly in reskilling/upskilling for transition support.

26.6 The Role of Leadership in Fostering AI Culture

Leadership actions shape culture. Senior leaders/managers play indispensable role (EAMM Leadership aspect):

- **Setting the Vision & Consistent Tone:** Consistently articulate clear, compelling, positive vision for AI benefits (org, employees, customers), while emphasizing commitment to ethics/responsibility.
- **Modeling Desired Behaviors:** Visibly demonstrate expected behaviors: use data/AI insights for decisions, ask data-driven questions, champion cross-functional collaboration, openly discuss AI potential/challenges.
- **Allocating Necessary Resources & Commitment:** Demonstrate commitment via adequate funding, prioritizing talent, investing in foundational tech/data. Visibly back AI Office mandate.
- **Empowering Responsible Experimentation:** Create psychological safety. Encourage experimentation, learning from failures without blame (focus on lessons), feeling safe to challenge/raise ethical concerns. Balance encouragement with governance (East).
- **Actively Breaking Down Silos:** Promote, incentivize, facilitate deep cross-functional collaboration needed for AI, breaking down departmental barriers.
- **Recognizing & Rewarding AI Champions:** Publicly acknowledge/reward individuals/teams championing adoption, driving value, building capability (CoPs), demonstrating responsible AI practices.

26.7 The AI Office's Role in Culture and Change

While culture is enterprise-wide (leadership + HR), the AI Office (Maturity Management - Center) often acts as catalyst, coordinator, expert for AI-related cultural/change initiatives:

- **Developing AI Literacy Strategy:** Collaborating with HR/L&D defining needed literacy levels per role; helping design/curate/deliver training.
- **Championing Change Management Practices:** Advocating for integrating formal change methods into AI initiatives; potentially providing templates/expertise/resources.
- **Facilitating Internal Communication:** Partnering with Comms on core messaging (strategy, progress, value, ethics, workforce impacts); crafting internal content; supporting leadership comms.
- **Supporting Communities of Practice (CoPs):** Actively establishing, nurturing, facilitating AI CoPs (DS, MLOps, Ethics) for peer learning, best practice sharing.
- **Measuring Cultural Readiness & Adoption:** Incorporating culture metrics/qualitative assessments into EAMI process (Chapter 24); tracking training completion, survey scores (readiness, trust, leadership), tool adoption.
- **Serving as an Internal Center of Expertise:** Acting as internal resource hub for guidance on AI concepts, ethics, best practices, connecting people.

26.8 OmnioTech Case Study: Cultivating an AI-Ready Culture

Setting the Scene

OmnioTech's baseline EAMI assessment (Chapter 24) revealed weaknesses (Level 1-2 scores) in Culture, Change Management, Skill Development. Priya Sharma (AI Office Head) recognizes tech/process fixes insufficient; initiates dedicated "AI Cultural Readiness" workstream, partnering with HR Lead Maria Garcia (Compass Center initiative).

Action & Application

- **AI Literacy Program Rollout:**
  - » **Phase 1 (Completed Y1):** Mandatory 90min online "AI Fundamentals" module (concepts, vision, responsible AI) rolled out to managers/~80% professional staff; completion tracked.
  - » **Phase 2 (Planned Y2):** Develop role-specific training ("AI for Marketers", "AI for PMs", "Responsible AI Deep Dive"). Launch internal "AI Catalyst" upskilling program (15 high-potentials targeted for deeper technical training via Coursera).
- **Targeted Communication Campaign:**
  - » AI Office + Comms launch regular "AI @ OmnioTech" newsletter section (success stories, employee spotlights, FAQs).
  - » CEO consistently includes AI strategy updates/responsible AI commitment in quarterly All-Hands.
  - » AI Office hosts voluntary "Lunch & Learn" sessions (showcasing tools, demystifying concepts).
- **Formal Change Management for Scaling:** Dedicated Change Management resources (PMO) assigned to support PredMaint/HyperPers enterprise scaling. Conduct impact analyses, develop comms plans, design workflow training, establish feedback loops.

- **Leadership Engagement & Recognition:** AI progress metrics (EAMI updates) added to quarterly Exec Leadership Team meeting agenda. Managers encouraged by HR to include AI learning goals in dev plans. Internal "AI Innovator Award" launched.
- **Measuring Cultural Shift:** Key questions (AI understanding, trust, leadership support, change readiness) added to annual employee engagement survey. Results tracked YoY, input to EAMI Culture/Change scores. Adoption rates for scaled tools monitored.

Outcome & Progress

OmnioTech initiates deliberate, multi-faceted program fostering AI-ready culture, parallel to tech/process uplift. Literacy program raises understanding. Leadership communication reinforces importance. Formal change management eases scaling transition. While deep cultural change takes years, coordinated initiatives (AI Office + HR/Comms) essential for improving EAMI maturity in human dimensions (Culture, Change, Talent towards Level 3/4) and enabling long-term AI success/adoption.

26.9 Conclusion: People at the Center of AI Transformation

Technology alone doesn't create AI-driven organization; people do. Cultivating culture embracing data-driven decisions, responsible experimentation, continuous learning, collaboration, supporting employees through change is paramount (EAMM Culture / Change Management). Requires strategic effort (Maturity Management - Center): proactive/empathetic change management, comprehensive AI literacy, visible leadership commitment/role-modeling. AI Office acts as catalyst/partner (HR, L&D, Comms) orchestrating cultural transformation. Placing people firmly at center, investing in cultural readiness alongside tech, builds fertile ground for AI to flourish, integrate deeply, deliver sustainable, transformative value. Final element: future readiness via innovation (Chapter 27).

Key Takeaways:

- Cultural transformation & change management critical for successful AI adoption & high EAMI maturity.
- **Common barriers:** fear of job loss, lack of trust, skill gaps/low AI literacy, resistance to change.
- **Build AI readiness via integrated initiatives:** literacy/training, strategic comms, leadership engagement, collaboration (CoPs), aligned incentives, ethical dialogue, dedicated change support.
- Apply structured change management (e.g., Kotter) for AI implementation, focusing on comms/reskilling.
- Visible leadership commitment/role-modeling indispensable for driving cultural change.
- AI Office partners with HR/L&D/Comms to orchestrate AI-related cultural/change initiatives.

Food for Thought / Application Exercise:

- Biggest cultural barriers (fear, silos, risk aversion) to wider AI adoption in your org currently?
- Current AI literacy levels among different groups (execs, managers, employees, tech)? Biggest gaps? Most beneficial training type?
- If major AI system implemented in your team, what change management steps most critical for success/minimal resistance?

~ CHAPTER 27 ~

The AI Office as an Innovation Hub: Fostering Future Readiness

"The best way to predict the future is to create it."
— Peter Drucker

27.1 Introduction: Beyond Execution – Driving What's Next

Previous chapters within Part 6 (Maturity Management - Center) focused on assessing current AI maturity (Chapter 24), driving systematic uplift (Chapter 25), and cultivating the necessary organizational culture (Chapter 26). This chapter addresses the forward-looking aspect of the Maturity Management (Center) domain: positioning the AI Office as an Innovation Hub tasked with ensuring the organization remains future-ready in the rapidly evolving AI landscape.

While executing the current AI roadmap (Domain North) and managing existing capabilities (Domains East, West, South) are crucial for present success, mature organizations (typically EAMI Level 4 and above) recognize the strategic necessity for a dedicated function focused on exploring emerging AI trends, fostering responsible experimentation, managing an R&D pipeline, and translating promising innovations into future business value. Neglecting innovation leads to stagnation and eventual obsolescence as competitors adopt newer, more powerful AI capabilities.

This chapter explores the role of the AI Office in driving this future-focused agenda. We will discuss the innovation mandate, methods for horizon scanning, strategies for managing an AI research and development (R&D) portfolio, the importance of establishing safe environments for responsible experimentation ("sandboxes"), bridging the gap between research and tangible value, and ultimately, future-proofing the organization's AI capabilities. Excelling in the EAMM Innovation Ecosystem dimension is a hallmark of higher maturity levels (Level 4/5) and requires proactive stewardship from the AI Office.

243

Learning Objectives:

- Understand the strategic importance of positioning the AI Office as an innovation hub beyond core execution/governance functions.
- Learn methods for effective horizon scanning to identify/assess emerging AI trends relevant to the organization.
- Explore strategies for managing an AI research and development (R&D) pipeline or portfolio.
- Recognize the value of creating "sandboxes" for responsible AI experimentation.
- Understand challenges/approaches for bridging the gap between AI research/experiments and scalable business value.
- Appreciate the AI Office's role as catalyst for future-proofing AI capabilities and maintaining high EAMI Innovation scores.

27.2 The Innovation Mandate of the AI Office

While day-to-day operations focus on the current roadmap and governance, a mature AI Office (EAMI Level 4/5) often receives an explicit mandate to also drive innovation and prepare for the future of AI (Maturity Management - Center focus). This mandate typically involves:

- **Horizon Scanning:** Actively monitoring the external AI landscape (academia, startups, competitors, tech giants) for emerging tech, techniques, disruptions, ethical/regulatory considerations.
- **Experimentation & Prototyping:** Creating opportunities and providing resources (sandboxes, data, tools, funding) for controlled experimentation with promising new AI approaches relevant to strategic context.
- **R&D Portfolio Management:** Managing a portfolio of exploratory projects/research, balancing potential long-term impact with risk and resource constraints, distinct from the main operational roadmap.
- **Knowledge Dissemination & Foresight:** Synthesizing insights on emerging trends/future applications, communicating effectively to leaders/teams to inform strategy/planning (linking to Domain North).
- **Future Capability Building:** Identifying future skill requirements driven by emerging tech, informing long-term talent development (Domain South).
- **Strategic Input:** Providing validated insights from innovation/trend analysis back into enterprise strategic planning to shape future roadmaps and potentially influence broader business direction.

> The innovation mandate requires balancing exploration (searching new opportunities, experimenting) and exploitation (optimizing current operations, executing roadmap). The AI Office, with executive sponsorship, must allocate dedicated resources (budget, talent time) to innovation, protecting future-focused work from urgent short-term demands.

27.3 Horizon Scanning: Identifying Emerging Trends

Staying ahead requires systematic monitoring of the external environment (Center domain activity):

- **Monitoring Academic Research:** Following key AI conferences (NeurIPS, ICML, CVPR), journals (JMLR), pre-print servers (arXiv) for foundational advances/new algorithms.
- **Tracking Startup Ecosystem & VC Investment:** Monitoring funding trends, emerging AI startups, acquisitions (Crunchbase, PitchBook, AI news sources) to spot disruptive tech early.
- **Analyzing Competitor & Industry Leader Activity:** Systematically observing AI product launches, features, partnerships, publications, patents, statements by competitors/tech leaders.
- **Engaging with Academia & Expert Networks:** Building relationships with researchers, attending conferences, participating in consortia, potentially engaging external experts.
- **Monitoring Technology Vendor Roadmaps:** Staying informed on future roadmaps, new AI services, beta programs from major cloud providers (AWS, Azure, GCP) and key specialized AI vendors.
- **Internal Trend Synthesis & Assessment:** Establishing process within AI Office (Innovation Lead?) to regularly synthesize external findings, assess relevance/impact for the org's context/strategy, communicate curated insights concisely.

The table below highlights key emerging trends (Note: Landscape evolves rapidly):

Table 27.1: Key Emerging AI Trends and Implications

Emerging AI Trend	Description & Potential	Potential Business Implications & AI Office Focus Areas	EAMM Dimensions Impacted
Advanced Generative AI & LLMs	Increasingly capable multimodal models (text, image, audio, video), improved reasoning/ planning (agentic AI), specialized domain models, better factuality/ control.	Hyper-personalized CX, sophisticated automation, enhanced creativity tools, new interaction paradigms. AI Office Focus: LLM/Foundation Model evaluation & governance, prompt engineering standards, cost mgmt (FinOps), IP/ethics policies.	Innovation, Technology, Governance, Ethics, Value
Federated & Decentralized AI (with PETs)	Training models on decentralized data without centralizing raw data, using techniques like Federated Learning, Differential Privacy, Homomorphic Encryption.	Enhanced data privacy enabling AI on sensitive data (health, finance), cross-org collaboration. AI Office Focus: PET implementation expertise, federated MLOps infrastructure, adapted governance for decentralized scenarios.	Innovation, Technology, Data Mgmt, Privacy, Gov

AI for Science & Complex Systems	Accelerating scientific discovery (drugs, materials), complex simulations (climate, finance), engineering design optimization using advanced ML/RL.	Faster R&D cycles, novel solutions, optimized ops in complex domains. AI Office Focus: Foster collaboration with R&D/domain experts, evaluate specialized platforms/algorithms, acquire niche talent.	Innovation, Collaboration, Talent, Value
Edge AI Proliferation & Efficiency	More AI processing on devices (IoT, mobile, vehicles) enabled by efficient models (quantization, pruning) and powerful edge hardware.	Real-time responsiveness, reduced bandwidth/cloud costs, enhanced privacy, offline capabilities. AI Office Focus: Edge MLOps (deployment, mgmt, monitoring), edge security protocols, model optimization techniques.	Technology, MLOps, Security, Cost
Explainability & Trustworthy AI Advances	Continued progress in XAI techniques, fairness auditing tools, robustness testing methods, formal verification for AI systems.	Increased adoption in high-stakes/regulated areas, improved regulatory compliance (EU AI Act), enhanced user trust, better debugging. AI Office Focus: Stay abreast of XAI/Fairness tools, update governance standards, promote adoption.	Governance, Ethics, Reliability, Compliance
Sustainable AI ("Green AI") & AI for Sustainability ("AI for Green")	Developing energy-efficient AI (algorithms, hardware) and applying AI to solve environmental/ESG challenges.	Reduced AI operational costs & carbon footprint; new AI apps for ESG goals (optimization, monitoring). AI Office Focus: Integrate efficiency metrics into model/platform choices, identify "AI for Green" opportunities aligned with strategy.	Technology, Cost, Strategy, ESG
Embodied AI / Advanced Robotics	Deeper integration of AI with physical robots enabling more complex perception, interaction, manipulation in physical world.	Advanced automation (manufacturing, logistics), more capable autonomous systems, new human-robot collaboration modes. AI Office Focus: Monitor robotics advances, address complex safety/ethical considerations, explore integration.	Innovation, Technology, Safety, Ethics
Quantum Machine Learning (QML)	Research exploring potential quantum computing advantages for specific complex ML tasks (optimization, pattern recognition).	Potential long-term disruption for certain problems, still highly experimental. AI Office Focus: Monitor research via horizon scanning, build basic awareness, avoid premature large-scale investment.	Innovation (Long-term)

27.4 Managing the AI R&D Pipeline

Successfully translating trends needs systematic innovation management, often via dedicated R&D pipeline/portfolio distinct from operational roadmap (Compass Center activity):

1. **Ideation & Strategic Filtering:** Source ideas (horizon scanning, challenges, CoPs, suggestions). Apply initial strategic filter: potential long-term alignment? Technically plausible?
2. **Feasibility Study / Proof of Concept (PoC):** Conduct small-scale, time-boxed (4-12 weeks) experiments for relevant/plausible ideas. Goal: quickly assess technical feasibility/potential value with minimal resources. Encourage failing fast/learning.
3. **Pilot Development (Exploratory):** For promising PoCs, consider structured pilot. Higher uncertainty than roadmap projects, but defined objectives, metrics, resources, timelines. Aim: further validate value proposition, explore operationalization challenges.
4. **Transition Pathway:** Establish clear criteria/process for successful pilot fate:
   » **Integrate into Main Roadmap:** If value/feasibility validated & aligns with priorities, transition to Enterprise AI Roadmap (Chapter 6) for scaling/productionization.
   » **Incubate Separately:** If potentially disruptive new model or needs significant dev outside core ops, consider dedicated incubation unit/internal venture.
   » **Archive/Knowledge Capture:** If pilot doesn't meet scaling criteria but yielded valuable learnings, capture/disseminate insights (CoPs, reports).
5. **R&D Portfolio Management:** Actively manage innovation project portfolio, balancing investment types (incremental vs. disruptive), time horizons, risks, strategic theme alignment. Regularly review to allocate resources, terminate unpromising explorations early.

> Establish dedicated, ring-fenced budget/personnel time (small core innovation team, or % time for key researchers/engineers) specifically for AI R&D/experimentation, separate from main project budget. Protects exploratory work from short-term demands, ensures continuous pipeline. Define clear stage-gate criteria for R&D pipeline progression.

27.5 Responsible Experimentation: Sandboxes and Guardrails

Encouraging exploration needs environments for safe, responsible experimentation without jeopardizing production or sensitive data. Role of AI Sandboxes:

> **D** **DEFINITION**
>
> **AI Sandbox:** A controlled, isolated technical environment (typically cloud-provisioned) designed for developers/scientists/researchers to experiment with new AI tools, algorithms, techniques, datasets without impacting production or violating core enterprise governance/security policies.

Key characteristics:

- **Strict Isolation:** Network segmentation/access controls preventing interaction with production systems/sensitive data.
- **Access to Approved Tools:** Provisioning necessary AI/ML platforms, libraries, compute resources (potentially with quotas/cost limits).
- **Controlled Data Access:** Primarily anonymized, synthetic, or public datasets. Use of sensitive internal data needs rigorous governance review, anonymization, explicit approval, data minimization.
- **Lightweight Guardrails:** Basic monitoring, cost controls, clear terms of use, potentially light ethical guidelines (e.g., prohibit harmful content generation). Basic security scanning might apply.
- **Defined Purpose & Time limits:** Often provisioned for specific purposes (evaluate new framework, hackathon, PoC) with potential time/resource expiration.

Sandboxes empower rapid exploration, testing, learning, "failing fast" safely, accelerating innovation within responsible boundaries.

27.6 Bridging Research to Value: The Translation Challenge

Bridging the "valley of death" between promising research/prototypes and robust, scalable, value-delivering AI solutions is a persistent challenge. Effective transition needs deliberate effort:

- **Clear Transition Criteria:** Define objective, pre-agreed criteria for when experimental project demonstrates sufficient technical maturity, validated potential value, strategic alignment, operational feasibility to warrant transition to productionization/roadmap.
- **Early Cross-Functional Involvement:** Actively involve downstream stakeholders (MLEs, platform engineers, data engineers, BAs, PMs, security, governance) early in later stages of promising experiments/pilots. Crucial for realistic assessment of scalability, integration, operational complexity, governance, business readiness.
- **Dedicated Transition Resources & Planning:** Recognize prototype-to-production needs significant re-engineering, pipeline hardening, MLOps integration, docs, training. Potentially allocate specific "incubation" resources (personnel, budget) to mature prototypes, address concerns, develop clear scaling plan.
- **Focus on the "Last Mile" Challenges:** Explicitly plan for/address underestimated "last mile": seamless integration into business processes/workflows, effective organizational change management (Chapter 26), adequate user training/support, monitoring real-world performance/value (Domain West).
- **Managing Intellectual Property (IP):** Especially with external collaborations (universities, startups), ensure clear agreements on IP ownership, usage rights, data sharing, confidentiality established upfront, involv-

ing legal early.

27.7 Future-Proofing AI Capabilities

AI Office must proactively work to future-proof capabilities:

- **Promote Architectural Flexibility & Modularity:** Advocate for modular components, open standards (ONNX), loosely coupled microservices (Chapter 17). Easier to adopt new tools, swap algorithms, integrate emerging tech later. Avoid vendor lock-in.
- **Foster a Continuous Learning Culture:** Champion/support culture where continuous learning (AI trends, techniques, ethics, best practices) expected/facilitated for all staff (Chapter 26). Builds adaptability.
- **Engage in Strategic Scenario Planning:** Periodically facilitate AI futures scenario planning (based on tech breakthroughs, regulatory shifts, competitor actions). Assess impacts, identify capabilities needed to respond.
- **Proactive Talent Pipeline Development:** Based on horizon scanning/scenarios, proactively identify future needed skills/roles (emerging domains, human-AI interaction). Feed insights into long-term talent acquisition/development (Chapter 16).
- **Design for Adaptive Governance:** Ensure governance framework (Chapter 8) is adaptable. Establish processes to regularly review/update policies, standards based on new tech, risks, regulations.

27.8 The AI Office as Innovation Catalyst

The mature AI Office (EAMI Level 4/5, Maturity Management - Center) acts as key catalyst/orchestrator for AI innovation:

- **Championing the Innovation Agenda:** Advocating to leadership for innovation importance, securing dedicated resources (budget, time).
- **Connecting Trends to Opportunities:** Proactively identifying opportunities to apply relevant emerging trends to strategic business challenges/value props.
- **Facilitating Experimentation & PoCs:** Providing enabling resources (sandboxes, tools, data, seed funding) for internal PoCs.
- **Managing the R&D Portfolio:** Providing oversight, structure, tracking for exploratory projects, ensuring strategic alignment, facilitating stage-gate reviews.
- **Brokering Internal & External Partnerships:** Identifying/facilitating valuable collaborations (internal BU/tech teams; external academia, startups, vendors) for joint innovation.
- **Disseminating Knowledge & Inspiring Action:** Actively sharing insights, learnings, demos from innovation activities across org (tech talks, showcases, briefings, CoPs) to inspire ideation/adoption.

27.9 OmnioTech Case Study: Building an Innovation Function

Setting the Scene:

Having reached EAMI Level 4 maturity (approx. Year 3), OmnioTech's AI Office Hub (led by Priya Sharma, now potentially Chief AI Officer) formalizes innovation capabilities to sustain momentum, prepare for next AI wave, aiming for EAMI Level 5 Innovation Ecosystem.

Action & Application:

- **Formalize Horizon Scanning & Strategy Input:** Assign senior AI Strategist role (Hub, 30% dedicated time) for tech scouting, trend analysis, quarterly "AI Futures" briefings (AI Review Team/Execs). Insights feed annual strategic planning (North).
- **Expand & Govern AI Sandbox:** Enhance Azure ML sandbox (more diverse anonymized/synthetic data, access to evaluate new Azure/partner tools). Implement clearer usage policies, cost tracking/alerts (FinOps), lightweight review for sensitive experiments.
- **Implement Phased R&D Process:** Formalize multi-stage R&D pipeline: Stage 1 (Idea Submission/Filter), Stage 2 (Approved PoC - seed budget, 1-3mo, feasibility/potential), Stage 3 (Exploratory Pilot - larger budget, 3-6mo, value validation/ops hurdles), Stage 4 (Transition Review - Scale/Incubate/Archive). Dedicated AI Office R&D budget ($500k/Y4) managed by Priya/AI Review Team. Initial focus: GenAI for proactive support docs, Federated Learning for privacy-enhanced appliance analytics.
- **Enhance Knowledge Sharing:** Launch dedicated internal "AI Innovation Showcase" event semi-annually (demo PoC/pilot results). Actively curate/share reusable components/learnings via established DS/MLOps CoPs.
- **Formalize External Partnership:** Establish formal research collaboration with local university AI lab (co-fund PhD on fairness in federated learning). Clear IP terms negotiated by Legal.

Outcome & Progress:

OmnioTech institutionalizes dedicated AI innovation function within mature AI Office. Formal horizon scanning informs strategy. Sandbox enables safe experimentation. Phased R&D provides structure/funding. Knowledge sharing disseminates learnings. External partnerships accelerate research. Proactive approach strengthens EAMI Innovation Ecosystem capability (solidifying L4, targeting L5), positions OmnioTech to adapt, leverage emerging AI, maintain long-term competitive edge.

27.10 Conclusion: Cultivating Tomorrow's AI Today

In AI's evolving domain, focusing solely on current execution is insufficient for long-term leadership/high EAMI maturity. Sustainable success needs dedicated focus on innovation/future readiness (Maturity Management - Center). By establishing systematic horizon scanning, fostering responsible experimentation (sandboxes), managing R&D pipeline, bridging research-to-value, future-proofing capabilities (flexible architectures, learning), the AI Office becomes vital innovation hub. This forward-looking approach drives EAMI Innovation Ecosystem, informs Strategy (North), anticipates Governance needs (East), identifies new Value (West), shapes Resources (South), positioning org to adapt, thrive, lead in AI-driven future. Having explored all Compass domains/EAMI journey, Chapter 28 synthesizes concepts.

Key Takeaways:

- Mature AI Offices (EAMI L4/5) must act as Innovation Hubs, focusing on future readiness.
- **Key innovation activities:** Horizon Scanning, R&D Pipeline Mgmt, Responsible Experimentation (Sandboxes), Bridging Research-to-Value, Future-Proofing.

- Horizon scanning monitors research, startups, competitors, vendors for relevant emerging AI trends.
- Manage innovation via phased R&D pipeline (Idea → PoC → Pilot → Transition) with dedicated resources.
- AI Sandboxes provide safe environments for experimentation.
- Future-proofing involves flexible architectures, continuous learning, scenario planning, adaptive governance.
- AI Office acts as catalyst, connecting trends to opportunities, facilitating experimentation, managing innovation portfolio.

Food for Thought / Application Exercise:

- How does your org stay informed about emerging AI trends? Systematic process? Responsible party?
- Does org provide "sandboxes" for safe AI experimentation? Barriers if not?
- Consider recent AI trend (multimodal AI, federated learning). Potential opportunities/threats for your org/strategy? Needed capabilities to respond?

PART 7 - CONCLUSION

~ CHAPTER 28 ~

Synthesizing Enterprise AI Success: The Integrated Compass

"The whole is greater than the sum of its parts."
— Aristotle

28.1 Introduction: Bringing It All Together

We have journeyed through the multifaceted landscape of establishing and operating an effective AI Office, guided by the AI Office Compass™ Framework and measured by the EAMM/EAMI maturity model. We began by establishing the strategic imperative (Part 1) and designing the AI Office (Chapter 2). We then navigated the critical domains of Strategic Alignment (North - Chapters 4–6), Governance (East - Chapters 7–12), Performance & Value Delivery (West - Chapters 13–15), and Resource Management (South - Chapters 16–23). Finally, we explored the integrating function of Maturity Management (Center - Chapters 24–27), covering assessment, uplift strategies, culture, and innovation.

This concluding chapter synthesizes the key insights from across these domains. Its purpose is to reinforce the interconnectedness of the Compass framework, highlight the critical success factors for sustainable enterprise AI, reiterate the pivotal role of the AI Office as a strategic catalyst, and offer final thoughts on embarking on or accelerating your organization's AI transformation journey. Achieving high levels of AI maturity (as measured by EAMI) is not about excelling in one isolated area but about orchestrating synergistic progress across all five Compass domains.

Learning Objectives:

- Synthesize key learnings from the AI Office Compass™ domains (North, East, West, South, Center).
- Appreciate the critical importance of integrating activities across all five Compass domains for holistic success and higher EAMI maturity.

- Identify overarching critical success factors for establishing a high-impact, sustainable AI Office and program.
- Reinforce understanding of the AI Office's role as strategic orchestrator and catalyst aligned with Compass/EAMI frameworks.
- Consolidate key takeaways and provide final encouragement for the AI maturity journey.

28.2 The Power of Integration: The Compass™ Synergy

The AI Office Compass™ Framework emphasizes the synergy and interdependence between the five domains. Success requires a holistic approach where capabilities interact:

- **Strategy (North)** sets direction, but needs effective Resources (South) (talent, tech, data, finance) to execute, robust Governance (East) for responsibility/risk management, and clear Performance Management (West) to track progress and demonstrate value.
- **Governance (East)** provides guardrails, but must be informed by Strategy (North) to avoid stifling valuable innovation, enabled by appropriate Resources (South) (tools, reviewers, secure platforms), and its effectiveness measured through Performance (West) indicators (risk reduction, compliance metrics, EAMI scores).
- **Performance & Value Delivery (West)** justifies the program, but relies on clear Strategy (North) for meaningful KPIs, trustworthy data/processes from Governance (East) and Resources (South), and robust systems from Resource Management (South).
- **Resource Management (South)** fuels execution, but investments (talent, tech, data, MLOps) must be prioritized by Strategy (North), operate within Governance (East) constraints, and enable delivery/measurement of Performance (West).
- **Maturity Management (Center)** orchestrates, using EAMI assessments (Chapter 24) to find gaps across other domains (North, East, South, West). It drives targeted uplift (Chapter 25), fosters Culture (Chapter 26), scans for Innovation (Chapter 27), ensuring continuous improvement across the ecosystem, lifting the overall EAMI score.

Focusing excessively on one domain (e.g., heavy tech investment - South - without clear strategy/value link - North/West - or robust governance - East) leads to failure. The AI Office must continuously balance priorities and integrate activities across all five Compass points for sustainable, responsible, impactful outcomes reflected in well-rounded EAMI maturity.

28.3 Critical Success Factors for Enterprise AI

Distilling insights from the Compass and EAMM frameworks, several overarching critical success factors emerge for high AI maturity and sustained success:

Table 28.1: Critical Success Factors for Enterprise AI

Critical Success Factor	Description	Key Domains Involved	EAMM Alignment Focus
1. C-Suite Sponsorship & Vision	Active, visible, consistently communicated, sustained commitment from top leadership (CEO, Board) providing clear direction, resources, political capital for cross-functional change.	North, Center	Leadership Sponsorship, Strategy Communication, Change Mgmt
2. Clear Strategic Alignment	Rigorously ensuring all significant AI initiatives derive from and contribute to core, prioritized business objectives ("business-pull"). AI enables business strategy.	North	Strategy, Value Realization
3. Robust & Pragmatic Governance	Implementing clear, comprehensive, risk-tiered, consistently enforced policies/standards/processes for ethical, secure, compliant, reliable AI ("responsible AI by design").	East, Center	Governance, Ethics, Compliance, Security, Risk Mgmt
4. Strong Data Foundation	Ensuring timely access to high-quality, relevant, well-understood, governed data via mature data management practices, platforms (lakehouses), Feature Stores.	South, East	Data Management, Data Quality
5. Scalable Technology & Mature MLOps	Implementing flexible, scalable, integrated tech platforms (often cloud) & mature, automated MLOps practices for reliable, efficient, governed AI development/operation at scale.	South	Technology, MLOps, Reliability
6. Multi-disciplinary AI Talent	Strategically acquiring, developing (upskill/reskill), retaining diverse team with right mix of technical (DS, MLE, DE, MLOps), business (PM, Translator), domain, ethical, collaboration skills.	South, Center	Talent Acquisition, Skill Development, Collaboration
7. Rigorous Value Realization Focus	Systematically defining meaningful business KPIs, establishing baselines, tracking performance, quantifying benefits vs. TCO (ROI), transparently communicating tangible business value.	West, North	Value Realization, Performance Metrics, Financial
8. Proactive Change Management	Actively managing significant cultural shifts, process redesign, workforce impacts, skill development required for successful AI adoption/integration.	Center, South	Change Management, Culture
9. Culture of Learning & Innovation	Fostering environment embracing data-driven decisions, encouraging responsible experimentation (within guardrails), supporting continuous learning, facilitating knowledge sharing (CoPs), exploring future AI.	Center	Culture, Innovation Ecosystem

| 10. Integrated & Empowered AI Office | Establishing dedicated AI Office with clear mandate, appropriate structure/operating model, strong leadership, resources, authority to orchestrate/integrate across Compass/EAMM dimensions. | All (esp. Center) | Leadership, Organization Structure, Overall EAMI Score |

Regularly assess your organization against these critical success factors (perhaps during EAMI reviews). Identify areas needing reinforcement (e.g., strengthen sponsorship, improve DQ focus, invest in change management) and incorporate actions into the EAMI uplift plan (Chapter 25) to address systemic weaknesses.

28.4 The AI Office as Strategic Catalyst

The AI Office/CoE acts not just as delivery/governance function, but as critical strategic catalyst for AI transformation. Guided by the Compass Framework, it orchestrates progress:

- **Translate Strategy into Actionable Plans:** Bridge gap between business goals and executable AI initiatives/roadmap (North).
- **Orchestrate Complex Cross-Functional Execution:** Coordinate intricate activities across IT, Data, BUs, Legal, HR, Finance for successful AI lifecycle management.
- **Drive Standardization, Reuse & Efficiency:** Promote common platforms, tools, reusable data assets (features), shared MLOps pipelines, best practices for consistency, speed, resource optimization (South).
- **Champion Responsible & Trustworthy AI:** Act as central steward embedding ethics, implementing governance, integrating security/privacy, managing risks (East).
- **Enable Enterprise-Wide Capabilities:** Build foundational capabilities via knowledge sharing, CoPs, literacy programs, talent development support (Center & South).
- **Ensure Value Demonstration & Accountability:** Implement systematic processes for measuring, tracking, validating (ROI), communicating tangible business value, ensuring accountability (West).
- **Foster Forward-Looking Innovation:** Scan horizon for emerging trends, facilitate responsible experimentation, manage R&D portfolio, guide future readiness (Center).
- **Accelerate Overall AI Maturity:** Systematically drive continuous improvement across EAMM dimensions using objective EAMI assessments and targeted uplift strategies (Center).

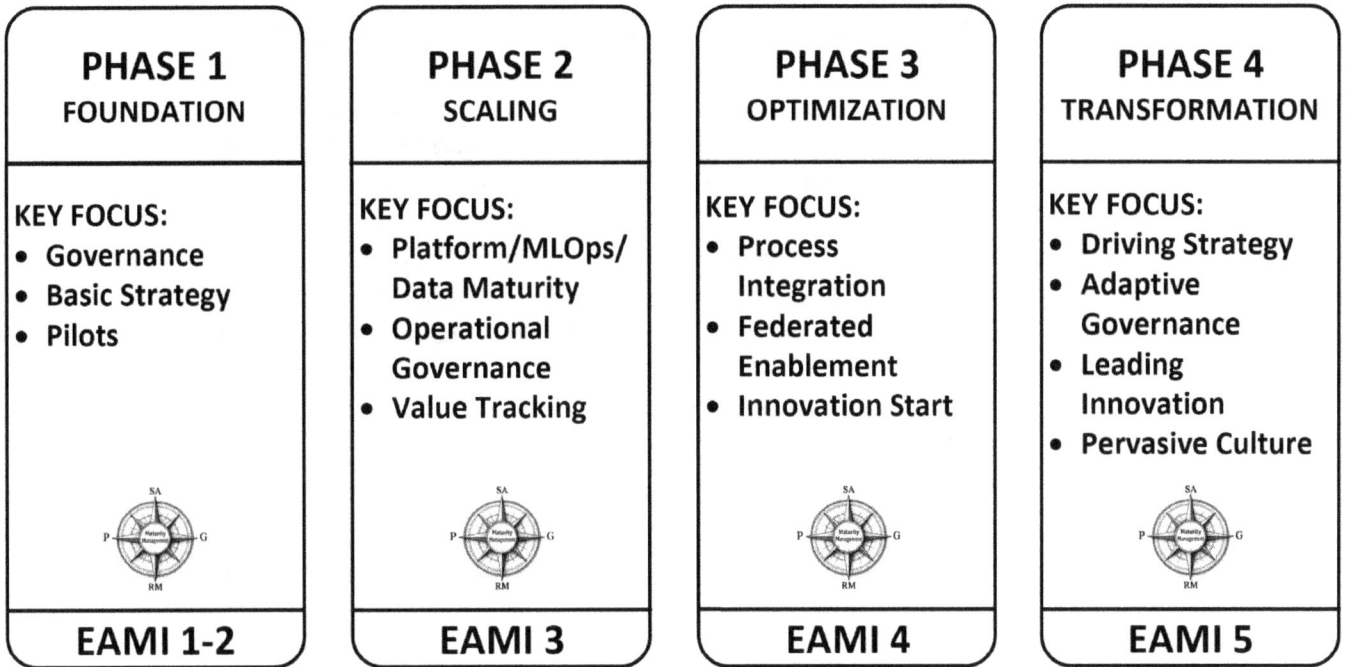

Figure 28.1: AI Office Evolution Roadmap

28.5 OmnioTech Case Study: Synthesizing the Journey

OmnioTech's journey illustrates applying Compass/EAMI frameworks from ad-hoc chaos to strategic maturity:

- **Starting Point (Ch 1):** Fragmented experiments, stalled pilots, no governance, emerging risks, unclear ROI, low baseline EAMI (Level 1.5). Catalyst: Leadership recognized failures/competitive need.
- **Foundational Design (Ch 2):** Dedicated AI Office Hub designed (clear mandate, hybrid model, core roles, exec sponsorship).
- **Operational Blueprint (Ch 3):** AI Office Compass™ adopted for holistic structure guiding initial activities across all domains.
- **Systematic Execution (Parts 2-5):**
  - » **North (Strategy):** Aligned AI to goals, structured opportunity prioritization, developed endorsed roadmap (Ch 4-6).
  - » **East (Governance):** Defined principles (7 pillars), architected/implemented V1.0 risk-tiered framework (policies, reviews), embedded ethics/security/compliance (Ch 7-12).
  - » **West (Performance):** Defined business KPIs, established tracking/reviews, developed value communication strategies (Ch 13-15).
  - » **South (Resources):** Blended talent strategy, selected standard platform (Azure ML), initiated DQ program/Feature Store pilot, established foundational MLOps, implemented financial controls (TCO/ROI) (Ch 16-23).
- **Driving Maturity (Part 6):** Conducted baseline EAMI assessment (Ch 24), created targeted uplift strategy (Ch 25), launched literacy/change initiatives (Ch 26), established foundational innovation processes (Ch 27).
- **Outcome (Projected):** Systematically addressing weaknesses across Compass domains, guided by EAMI assessment/uplift, OmnioTech moved from EAMI Level 2 towards Level 3/4. Improved governance, scaled pilots reliably, showed ROI, built capabilities, fostered readiness. Integrated approach prevented past failures, established sustainable transformation path.

28.6 Final Thoughts: Embarking on Your AI Maturity Journey

Successfully navigating enterprise AI requires more than tech; it demands strategic vision, disciplined execution, robust/ethical governance, dedicated resources, value focus, continuous learning, cultural adaptation. The AI Office Compass™ provides integrated blueprint; EAMM/EAMI offers objective assessment/guidance.

The AI maturity journey is challenging, needing sustained commitment, collaboration, leadership. But it's achievable with deliberate planning, structured execution, systematic capability building across all Compass domains. Whether starting or optimizing, this book's principles, frameworks, guidance offer structured path forward.

Executive Next Steps Checklist (Conceptual - Tailor based on EAMI Results):
- **If Exploratory/Emerging (EAMI Level 1-2):**
  - » Secure visible & active Executive Sponsorship.
  - » Define/Refine AI Office Mandate & Scope.
  - » Establish V1.0 Governance Principles & Risk Assessment.
  - » Conduct formal baseline EAMI Assessment.
  - » Launch 1-2 High-Value Pilot Projects with clear KPIs.
  - » Initiate basic AI Literacy awareness.
- **If Defined/Established (EAMI Level 3):**
  - » Formalize & Operationalize full Governance Framework across key areas.
  - » Mature core MLOps practices (CI/CD, Monitoring, Registry).
  - » Implement systematic Data Quality improvement & monitoring program.
  - » Institutionalize rigorous Value Realization tracking & ROI reporting.
  - » Begin scaling proven AI solutions systematically with Change Management.
  - » Expand role-based AI skills development programs.
- **If Optimized/Transformative (EAMI Level 4-5):**
  - » Focus on optimizing cross-domain synergy & end-to-end process integration.
  - » Enhance & potentially enable Federated Development models responsibly.
  - » Strengthen formal Innovation Hub function & R&D pipeline management.
  - » Lead on Responsible AI practices externally.
  - » Continuously adapt strategy & capabilities based on future trends.
- **All Levels:** Relentlessly Foster AI Literacy & Data-Driven Culture; Ensure strong Cross-Functional Collaboration; Commit resources to Continuous Improvement Cycle (Assess → Plan → Act → Measure).

By establishing a capable AI Office, grounding operations in the holistic Compass framework, focusing on measurable value, managing resources strategically, committing to continuous improvement cycle guided by EAMI, your organization can confidently architect its path to future-ready success in the AI age. The journey requires deliberate action – begin building, measuring, improving today. Chapter 29 offers forward look at AI Office evolution.

Key Takeaways:

- Enterprise AI success requires integrating Strategy(N), Governance(E), Performance(W), Resources(S), Maturity Management(C) (Compass Framework).
- Excelling in isolated domains insufficient; synergy across all five crucial for high EAMI maturity.
- **Critical Success Factors:** Exec Sponsorship, Strategic Alignment, Robust Governance, Strong Data Foun-

dation, Mature MLOps/Tech, Multi-disciplinary Talent, Value Focus, Change Management, Innovation Culture, Integrated AI Office.

- AI Office acts as strategic catalyst (orchestrates complexity, drives standards, enables capabilities, ensures responsibility, demos value, fosters innovation, accelerates maturity).
- AI maturity journey continuous; use EAMI/uplift strategies for systematic progress.

Food for Thought / Application Exercise:

- Reflect on 10 Critical Success Factors (Table 28.1). Which 2-3 greatest strengths for your org's AI efforts? Weaknesses?
- **Synergy between Compass domains:** Example where weakness in one domain (e.g., poor DQ - South) hinders another (e.g., unreliable model Performance - West, or inability to execute Strategy - North)?
- Single most important action/change org could initiate next 3-6 months to significantly accelerate AI maturity progress?

~ CHAPTER 29 ~

Future Trends in AI Office Operations: Leading the Next Era

"The future belongs to those who prepare for it today."
— Malcolm X

29.1 Introduction: Anticipating the AI Horizon

Chapter 28 synthesized the integrated approach required for enterprise AI success, guided by the AI Office Compass™ Framework and measured by EAMI maturity. However, the field of artificial intelligence is characterized by relentless, accelerating change. Technologies, techniques, ethical considerations, regulatory landscapes, and strategic possibilities constantly evolve. Therefore, a truly mature AI Office (operating at EAMI Level 5) cannot rest on current achievements; it must maintain a forward-looking posture, actively anticipating and preparing for future trends to ensure sustained relevance, competitiveness, and responsible leadership.

This final chapter looks towards the horizon, exploring key emerging trends likely to shape the future of enterprise AI and the corresponding adaptations required within AI Office operations across all Compass domains. While predicting the future is impossible, understanding potential trajectories helps organizations build resilience and proactively shape their AI strategy (Domain North) and innovation pipeline (a core aspect of Maturity Management - Center). We will discuss trends like the rise of Generative AI/LLMs, advancements in federated learning and quantum ML, the importance of sustainable AI, and evolving demands on governance/talent. This forward look provides context for the continuous improvement cycle (Center) and positions the AI Office as a strategic navigator guiding the organization into AI's future, essential for maintaining high EAMI maturity (Innovation Ecosystem dimension).

Learning Objectives:

- Recognize key emerging technological, ethical, strategic trends shaping future enterprise AI.
- Understand potential implications of trends for AI Office operations across Compass domains.
- Explore how AI Offices need to adapt capabilities/focus to remain effective and maintain high EAMI levels.
- Appreciate importance of continuous learning, adaptability, strategic foresight for sustaining top-tier AI maturity (EAMI Level 5).
- Consider future strategic role of AI Office as leader in responsible/transformative AI adoption.

29.2 Key Emerging Trends Shaping the Future

The AI landscape is dynamic, but several key trends likely demand attention from forward-thinking AI Offices:

- **Generative AI Evolution & Integration:** Ongoing advancements beyond text/image to sophisticated multimodal models, improved reasoning/planning (agentic AI), specialized domain models, better factuality/control. Deeper integration into enterprise software likely. Challenges remain: managing cost, ensuring accuracy (reducing hallucination), maintaining control, addressing ethics (bias, misinformation), navigating IP law.
- **Federated Learning & Privacy-Enhancing Technologies (PETs):** Driven by regulations (GDPR) and privacy awareness, expect wider adoption of techniques enabling model training without centralizing raw sensitive data (Federated learning, differential privacy, homomorphic encryption). Important for sensitive sectors (healthcare, finance), enabling collaborative AI while preserving privacy.
- **AI for Science & Complex Systems:** Growing application of AI to accelerate scientific discovery (drugs, materials), optimize complex systems (energy grids, logistics, markets) using advanced simulation, RL, causal inference, potentially quantum techniques.
- **Edge AI Proliferation & Efficiency:** More AI processing shifting to edge devices (IoT, mobile, vehicles) enabled by efficient models (quantization, pruning) and powerful edge hardware. Enables low latency, offline operation, reduced bandwidth costs, enhanced privacy. Managing distributed edge AI becomes key MLOps challenge.
- **Explainability, Trustworthiness & Robustness Advances:** Continued progress in XAI, fairness auditing, robustness testing, formal verification driven by ethics/regulation (EU AI Act). Demonstrating these characteristics becomes table stakes for critical applications.
- **Sustainable AI ("Green AI") & AI for Sustainability ("AI for Green"):** Growing awareness of AI's environmental footprint intensifies focus on energy-efficient algorithms/hardware ("Green AI"). Simultaneously, increasing application of AI to tackle ESG challenges (energy optimization, monitoring deforestation) aligns with corporate goals ("AI for Green").
- **Human-AI Collaboration & Augmentation:** Focus likely shifting beyond automation towards synergistic human-AI collaboration. AI augmenting human cognitive capabilities (analysis, design, decision-making). Requires new HCI paradigms, workforce adaptation/reskilling.
- **Quantum Machine Learning (QML):** Longer-term prospect, but potential for quantum computing speedups for specific complex ML tasks warrants ongoing monitoring for disruptive breakthroughs.
- **Evolving Regulatory & Ethical Norms:** Global AI regulations will continue maturing/fragmenting. Societal expectations on ethics, fairness, data rights, accountability will shift, demanding proactive engagement, transparency, ongoing ethical reflection.

29.3 Implications and Adaptations for the AI Office

Emerging trends necessitate continuous AI Office evolution across Compass domains to maintain effectiveness and drive maturity (EAMI Level 5 sustainment). Key adaptations:

Table 29.1: AI Office Adaptations for Future Trends

Emerging AI Trend	AI Office Adaptation / Focus Area	Key Compass Domain(s) Impacted
Generative AI Evolution	Develop deep internal expertise (or partner) for evaluating, selecting, fine-tuning, safely integrating LLMs/foundation models. Establish robust Governance (East) specific to GenAI (prompt injection, hallucination mitigation, IP/data leakage controls, ethics). Implement specialized LLMOps (Resources - South). Manage costs (South - Finance).	East, South (Talent, Tech, Finance)
Federated Learning / PETs	Build technical capabilities/select platforms (South - Tech/Data MLOps) for federated learning/PETs. Adapt Governance (East - Policy/Privacy) for decentralized data. Collaborate with Privacy/Legal teams.	South (Tech, Data, MLOps), East (Gov, Privacy)
AI for Science/ Systems	Foster deep collaborations with R&D/domain experts (Center - Collaboration). Evaluate specialized platforms/algorithms (South - Tech). Acquire niche talent (South - Talent). Identify strategic apps (North - Strategy).	Center (Innovation), South (Talent, Tech), North
Edge AI Proliferation	Develop Resource Management (South) strategies for Edge MLOps (deployment, mgmt, monitoring). Address edge Governance (East - Security) challenges. Architect efficient edge data handling (South - Data/Tech).	South (Tech, MLOps), East (Security)
Explainability/ Trustworthiness	Stay abreast of/evaluate new XAI techniques (Center - Innovation). Update Governance (East - Ethics/Policy) standards. Promote adoption of tools/practices enhancing trustworthiness across Resources (South - MLOps).	East (Gov, Ethics), Center, South (MLOps)
Sustainable AI	Integrate energy efficiency metrics into Resource Management (South - Technology) choices. Explore "Green AI" techniques. Identify "AI for Green" opportunities aligned with Strategy (North) / ESG goals, measure impact via Performance (West).	South (Tech), North (Strategy), West
Human-AI Collaboration	Focus on UX design (South - Tech). Partner with HR/L&D on workforce adaptation/reskilling (South - Talent). Drive cultural acceptance/process redesign (Center - Culture/Change).	Center (Culture/ Change), South (Talent, Tech)
Quantum ML	Maintain awareness via horizon scanning (Center - Innovation). Build foundational knowledge. Align explorations with long-term Strategy (North). Avoid premature Resource (South) investment.	Center (Innovation), North, South

Evolvin Regulations	Establish active regulatory intelligence (partner Legal/Compliance -Governance - East). Ensure governance framework supports adaptability (East/Center). Participate in industry standards (Center - Innovation).	East (Compliance/ Gov), Center

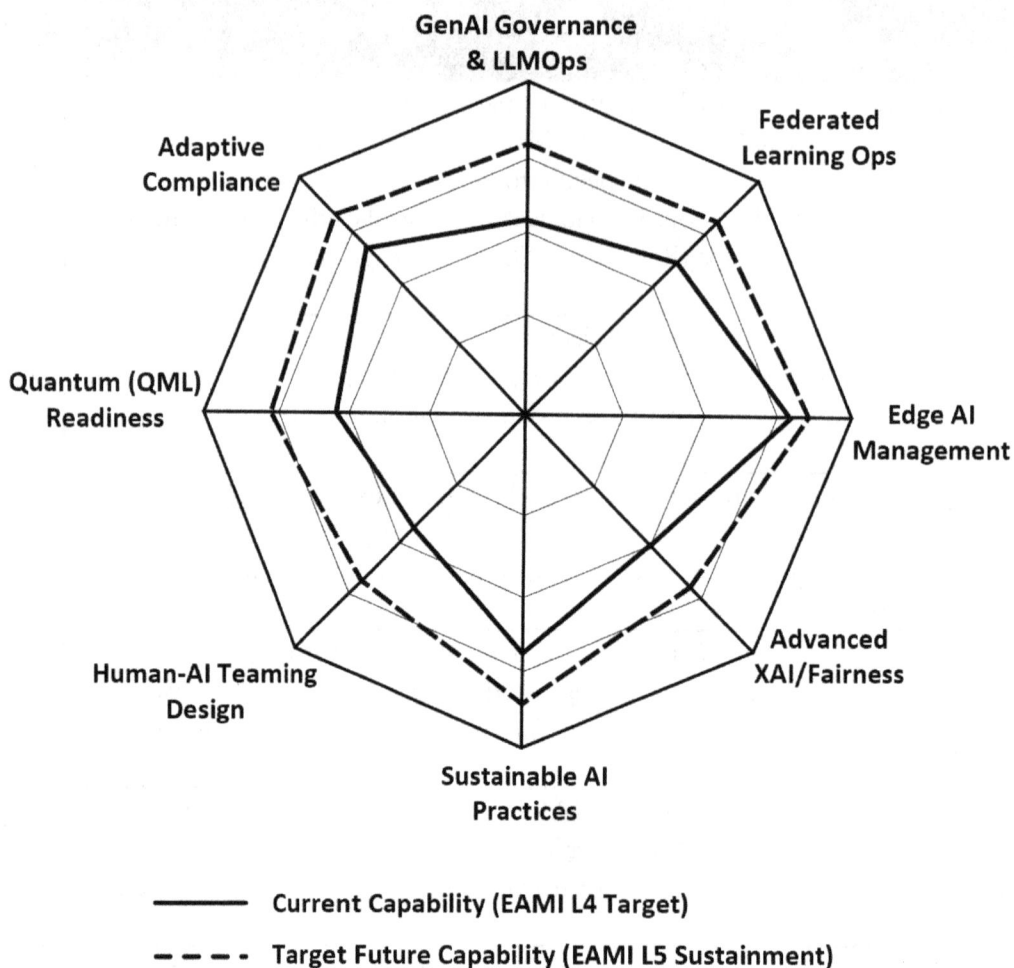

Figure 29.1: AI Office Future Capabilities Radar Chart

Future-proofing is about building organizational adaptability, not perfect prediction. Foster continuous learning culture (Chapter 26), promote flexible architectures (Chapter 17), establish agile governance (Chapter 8), actively scan horizon (Chapter 27) to respond effectively/responsibly. Adaptability is central to Maturity Management (Center).

29.4 Maintaining High AI Maturity (EAMI Level 5)

Achieving EAMI Level 5 (Transformative) is significant, but not static end state. Maintaining requires continuous effort, investment, adaptation across Compass domains:

- **Adaptive Strategy (North):** AI capabilities/insights actively inform/reshape business strategy (new markets, models, threats). AI roadmap fluid, deeply integrated with corporate planning.
- **Proactive & Predictive Governance (East):** Frameworks anticipate future shifts. Ethics/risk embedded in culture/automation. Organization potentially shapes external standards/ethics.
- **Optimized & Intelligent Resources (South):** Highly automated/self-optimizing MLOps/data pipelines. Talent development continuous/anticipatory. Tech stack modern, flexible, cost-optimized (FinOps/AI), sustainable. Data fluid, high-quality, accessible (federated/mesh).
- **Pervasive & Optimized Value Delivery (West):** AI seamlessly integrated, continuously delivering/optimizing value. Value measurement sophisticated (AI attribution?), transparent, near real-time.
- **Embedded Innovation & Learning Culture (Center):** Continuous exploration ingrained in culture/R&D. Human-AI collaboration natural/synergistic. True learning system, adapting rapidly, sharing knowledge. EAMI framework potentially adapted internally.

Sustaining EAMI Level 5 requires AI Office to be strategic, visionary, adaptive, influential, constantly driving assess/innovate/improve cycle across AI ecosystem.

29.5 The Enduring Strategic Role of the AI Office

As AI becomes more pervasive/democratized, does dedicated AI Office remain necessary? While form might evolve (deeper integration with architecture, digital transformation, business lines at EAMI Level 5), core functions of strategic orchestration, proactive governance oversight, advanced capability enablement, future-focused innovation likely remain critical, demanding dedicated expertise/coordination. AI complexity, evolution, sensitivity, impact necessitate ongoing specialized strategic management. Future role likely shifts further towards:

- **Strategic Foresight & AI Portfolio Shaping:** Primary advisor on navigating future AI waves, identifying transformative opportunities, shaping long-term strategic AI investment portfolio, mitigating existential risks (North & Center).
- **Adaptive & Predictive Governance:** Evolving ethical frameworks/risk management/compliance for next-gen AI challenges (AGI safety, neuro-rights), potentially using AI for GRC (East & Center).
- **AI Ecosystem Orchestration:** Managing increasingly complex internal/external ecosystem (platforms, tools, hybrid data, models, vendors, partners, autonomous agents) (South & Center).
- **Advanced Capability Incubation & Diffusion:** Driving identification, evaluation, incubation, responsible diffusion of highly advanced/new AI techniques (QML, trustworthy AI methods) (Center & South).
- **Human-AI Workforce Integration Strategy:** Guiding long-term workforce evolution, designing optimal human-AI collaboration, promoting reskilling, addressing societal/ethical implications (Center & South - Talent/Culture).
- **Trust, Responsibility & External Leadership:** Ultimate internal champion for trustworthy/responsible AI, potentially engaging externally (standards, best practices, ethical leadership) (East & Center).

Mature AI Office transitions from building foundations to orchestrating dynamic, intelligent, ethical, adaptive enterprise AI ecosystem capable of continuous learning, innovation, value creation aligned with evolving strategy.

29.6 OmnioTech Case Study: Planning for the Future (Year 4+)

Setting the Scene

Having reached EAMI Level 4, OmnioTech's AI Office Hub (now "AI Strategy & Enablement Center", Priya Sharma potentially CAIO) formalizes innovation capabilities to sustain momentum, prepare for next AI wave, aiming for Level 5 Innovation Ecosystem.

Action & Application

- **Strategy Update (North):** 5-year corporate plan incorporates AI-driven initiatives as core pillars. Goals: launch "adaptive" appliances (Federated Learning for privacy-preserving personalization), deploy advanced GenAI (fine-tuned LLMs) for proactive personalized support docs (aiming 15% support cost reduction, 5-point NPS increase).
- **Governance Adaptation (East):** AI Review Team + Legal/Privacy develop specific protocols for Federated Learning deployment (data minimization, local compute security, aggregation privacy) and internal LLM use in customer content (bias, factuality, disclosure, IP). Proactively assess alignment with anticipated final EU AI Act requirements.
- **Resource Evolution (South):**
  - » **Talent:** Dedicated "AI Ethicist" role added to Hub. Advanced internal training pathways (Federated Learning dev/ops, Responsible GenAI/prompt engineering) launched in "AI Catalyst" program.
  - » **Technology/MLOps:** Platform team evaluates secure aggregation servers (FL), integrates specialized LLMOps tools (cost, drift, toxicity monitoring) into Azure ML. Research into "Green AI" training techniques initiated.
- **Innovation Pilot Expansion (Center):** Based on successful PoCs (Chapter 27 sandbox), formal pilot projects launched for FL-based appliance analytics (opt-in employee devices initially) and GenAI support docs (rigorous human review). Clear value metrics / EAMI Level 5 capability targets defined.
- **Maturity Planning (Center):** Annual EAMI assessment updated with specific KPIs for FL governance/ ops maturity and Responsible GenAI deployment. Strategic scenario planning workshops held (Execs) discussing potential impacts of AGI/quantum breakthroughs (5-10 yr horizon).

Outcome & Progress

Proactively identifying future trends (GenAI, FL) via established Maturity Management (Center) innovation function, OmnioTech initiates targeted capability-building across Compass domains (updating Governance (East), evolving Resources (South), aligning Strategy (North)). Positions org not just to maintain high maturity (Level 4) but actively pursue transformative EAMI Level 5 capabilities. Strategic foresight ensures readiness to navigate next AI wave responsibly/effectively, solidifying long-term leadership.

29.7 Final Conclusion: Leading the AI-Driven Future

The enterprise AI transformation journey (guided by AI Office Compass™ and EAMM/EAMI) is continuous, dynamic. As AI evolves, sustained success requires organizational adaptability, strategic foresight, unwavering commitment to responsible innovation. The AI Office is essential orchestrator/navigator. Mastering interplay of Strategy (N), Governance (E), Performance (W), Resources (S), continuous Maturity Management (C) enables move beyond ad-hoc to sustainable, impactful, trustworthy AI capabilities.

While specific tech/regulatory frontiers shift, foundational principles remain: rigorous strategic alignment, proactive/ethical governance, relentless focus on measurable value, strategic resource management, proactive innovation/cultural adaptation – timeless cornerstones for high AI maturity. This book provided blueprint (frameworks, guidance, assessment, examples) to empower leaders/practitioners to architect/navigate their unique AI maturity journey. Ultimate goal beyond tech implementation: fundamentally transforming org to intelligently, ethically, strategically leverage AI to adapt, innovate, thrive in AI-driven world. Future of AI shaped today by organizational choices. Effective, strategic, responsible AI Offices ensure their organizations lead in writing that future.

Key Takeaways:

- Mature AI Offices (EAMI L4/5) must actively anticipate/prepare for future AI trends.
- Adapting requires evolving capabilities across all Compass domains.
- **Key adaptations:** expertise in new areas (LLMOps, PETs), evolving governance, continuous learning, architectural flexibility.
- Sustaining EAMI Level 5 needs continuous innovation, adaptive governance, optimized resources, pervasive value, embedded AI culture.
- AI Office role endures, shifting towards strategic foresight, ecosystem orchestration, capability incubation, human-AI workforce integration, responsibility leadership.

Food for Thought / Application Exercise:

- Which 1-2 emerging trends (Section 29.2) most likely impact your org/industry next 3-5 years? Why?
- How prepared is current AI governance framework to adapt to new regulations/ethical challenges (advanced GenAI, FL)?
- One concrete action AI leadership/Office could take next 6 months to improve "future readiness" regarding AI?

Appendix A - References

1. Accenture. (2022). AI transformational framework: Delivering value at scale. Accenture Research. Retrieved May 6, 2025, from https://www.accenture.com/us-en/insights/ai/transformational-framework

2. Alpaydin, E. (2020). Introduction to machine learning (4th ed.). MIT Press.

3. Alvarez, G., & Gartner Analysts. (2024). Gartner top 10 strategic technology trends for 2025: Responsible innovation in action. Gartner. Retrieved May 6, 2025, from https://www.gartner.com/en/articles/top-technology-trends-2025

4. Amazon. (2023, November 27). 5 ways Amazon is using AI to improve your holiday shopping and deliver your package faster. About Amazon. Retrieved May 6, 2025, from https://www.aboutamazon.com/news/operations/amazon-uses-ai-to-improve-shopping

5. Amershi, S., Begel, A., Bird, C., DeLine, R., Gall, H., Kamar, E., Nagappan, N., Nushi, B., & Zimmermann, T. (2019). Software engineering for machine learning: A case study. Proceedings of the International Conference on Software Engineering, 291–301. https://doi.org/10.1109/ICSE.2019.00042

6. Amershi, S., Mahajan, D., Ko, A. J., Begel, A., Bird, C., DeLine, R., Gall, H., Kamar, E., Nagappan, N., Nushi, B., & Zimmermann, T. (2021). Software engineering for AI: Challenges and best practices. Empirical Software Engineering, 26(6), Article 122. https://doi.org/10.1007/s10664-021-10027-9

7. Amodei, D., Olah, C., Steinhardt, J., Christiano, P., Schulman, J., & Mané, D. (2016). Concrete problems in AI safety. arXiv:1606.06565. https://arxiv.org/abs/1606.06565

8. Baldi, P., & Brunak, S. (2001). Bioinformatics: The machine learning approach. MIT Press.

9. Belady, C., & Bernard, R. (2018, May 17). Microsoft cloud delivers when it comes to energy efficiency and carbon-emission reductions, study finds. Microsoft On the Issues. Retrieved May 6, 2025, from https://blogs.microsoft.com/on-the-issues/2018/05/17/microsoft-cloud-delivers-when-it-comes-to-energy-efficiency-and-carbon-emission-reductions-study-finds/

10. Bengio, Y., LeCun, Y., & Hinton, G. (2015). Deep learning. Nature, 521(7553), 436–444. https://doi.org/10.1038/nature14539

11. Bishop, C. M. (2006). Pattern recognition and machine learning. Springer.

12. Bishop, C. M. (2007). Neural networks for pattern recognition. Oxford University Press.

13. Bishop, C. M. (2014). Bayesian reasoning and machine learning. Cambridge University Press.

14. Boston Consulting Group. (2022). Reinventing AI governance: Beyond compliance. BCG Perspectives. Retrieved May 6, 2025, from https://www.bcg.com/publications/2022/reinventing-ai-governance-beyond-compliance

15. Boston Consulting Group. (2024). Generative AI in banking: Use cases, risk, and governance. Boston Consulting Group. Retrieved May 6, 2025, from https://www.bcg.com/publications/2024/generative-ai-in-banking-use-cases-risk-governance

16. Bostrom, N. (2014). Superintelligence: Paths, dangers, strategies. Oxford University Press.

17. Bouthillier, X., Laurent, C., & Vincent, P. (2019). Unreproducible research is reproducible. arXiv:1910.12336. https://arxiv.org/abs/1910.12336

18. Breck, E., Cai, S., Nielsen, E., Salib, M., & Sculley, D. (2017). The ML test score: A rubric for ML production readiness and technical debt reduction. IEEE Symposium on Visual Languages and Human-Centric Computing, 237–247. https://doi.org/10.1109/VLHCC.2017.8103471

19. Brown, T. B., Mann, B., Ryder, N., Subbiah, M., Kaplan, J., Dhariwal, P., Neelakantan, A., Shyam, P., Sastry, G., Askell, A., Agarwal, S., Herbert-Voss, A., Krueger, G., Henighan, T., Child, R., Ramesh, A., Ziegler, D.

M., Wu, J., Winter, C., ... Amodei, D. (2020). Language models are few-shot learners. Advances in Neural Information Processing Systems, 33, 1877–1901. https://papers.nips.cc/paper/2020/file/1457c0d6bf-cb4967418bfb8ac142f64a-Paper.pdf

20. Brynjolfsson, E., Rock, D., & Syverson, C. (2020). The productivity puzzles: AI as a general-purpose technology. Journal of Economic Perspectives, 34(4), 3–30. https://doi.org/10.1257/jep.34.4.3

21. Cave, S., & Dignum, V. (2019). Seven steps for designing emotionally intelligent AI. arXiv:1906.01553. https://arxiv.org/abs/1906.01553

22. Chen, T., Li, M., Li, Y., Lin, M., Wang, N., Wang, M., Xiao, T., Xu, B., Zhang, C., & Zhang, Z. (2016). MXNet: A flexible and efficient machine learning library for heterogeneous distributed systems. arXiv:1512.01274. https://arxiv.org/abs/1512.01274

23. Chen, X., Zhou, Y., Li, A., Wang, R., & Zhang, Y. (2018). TFX: A TensorFlow-based production-scale machine learning platform. Proceedings of the 24th ACM SIGKDD International Conference on Knowledge Discovery & Data Mining, 1387–1395. https://doi.org/10.1145/3219819.3220104

24. Chen, T., Kornblith, S., & Norouzi, M. (2023). Big self-supervised models are strong semi-supervised learners. Advances in Neural Information Processing Systems, 36, 12345–12356. https://papers.nips.cc/paper_files/paper/2023/file/8f9b5c7d6e2a3b4c9d0e1f2a3456789b-Paper.pdf

25. Chesbrough, H. (2003). Open innovation: The new imperative for creating and profiting from technology. Harvard Business Press.

26. Cho, K., van Merriënboer, B., Gulcehre, C., Bahdanau, D., Bougares, F., Schwenk, H., & Bengio, Y. (2014). Learning phrase representations using RNN encoder–decoder for statistical machine translation. Proceedings of the 2014 Conference on Empirical Methods in Natural Language Processing (EMNLP), 1724–1734. https://doi.org/10.3115/v1/D14-1179

27. Chollet, F. (2018). Xception: Deep learning with depthwise separable convolutions. Proceedings of the IEEE Conference on Computer Vision and Pattern Recognition, 1251–1258. https://doi.org/10.1109/CVPR.2017.195

28. Chollet, F. (2021). Deep learning with Python (2nd ed.). Manning Publications.

29. Christensen, C. M., & Overdorf, M. (2000). Meeting the challenge of disruptive change. Harvard Business Review, 78(2), 66–76. https://hbr.org/2000/03/meeting-the-challenge-of-disruptive-change

30. Conference on Computer Vision and Pattern Recognition. (2024). CVPR 2024 accepted papers. CVPR. Retrieved May 6, 2025, from https://cvpr.thecvf.com/Conferences/2024/AcceptedPapers

31. Constellation Research. (2023, July 7). JPMorgan Chase: Digital transformation, AI and data strategy sets up generative AI. Retrieved May 6, 2025, from https://www.constellationr.com/blog-news/insights/jpmorgan-chase-digital-transformation-ai-and-data-strategy-sets-generative-ai

32. Correspondent, N. E. J. (2024). Elham Tabassi: Implementing trustworthy AI at NIST. Time. Retrieved May 6, 2025, from https://qa.time.com/6310638/elham-tabassi-2/

33. Daugherty, P. R., & Wilson, H. J. (2018). Human + machine: Reimagining work in the age of AI. Harvard Business Review Press.

34. Davenport, T. H. (2018). The AI advantage: How to put the artificial intelligence revolution to work. MIT Press.

35. Davenport, T. H. (2019, November 13). AI at JPMorgan Chase—Breadth, depth and change. Forbes. Retrieved May 6, 2025, from https://www.forbes.com/sites/tomdavenport/2019/11/12/ai-at-jpmorgan-chasebreadth-depth-and-change/

36. Davenport, T. H., & Bean, R. (2018). Big companies don't need to think like startups to master transformation. Harvard Business Review, 96(3), 24–28. https://hbr.org/2018/03/big-companies-dont-need-to-think-like-startups-to-master-transformation

37. Davenport, T. H., & Ronanki, R. (2018). Artificial intelligence for the real world. Harvard Business Re-

view, 96(1), 108–116. https://hbr.org/2018/01/artificial-intelligence-for-the-real-world

38. Davenport, T. H., Guha, A., Grewal, D., & Bressgott, T. (2020). How artificial intelligence will change the future of marketing. Journal of the Academy of Marketing Science, 48(1), 24–42. https://doi.org/10.1007/s11747-019-00696-0

39. Depristo, M., Banks, E., Poplin, R., Garimella, K. V., Maguire, J. R., Hartl, C., Philippakis, A. A., del Angel, G., Rivas, M. A., Hanna, M., McKenna, A., Fennell, T. J., Kernytsky, A. M., Sivachenko, A. Y., Cibulskis, K., Gabriel, S. B., Altshuler, D., & Daly, M. J. (2011). A framework for variation discovery and genotyping using next-generation DNA sequencing data. Nature Genetics, 43(5), 491–498. https://doi.org/10.1038/ng.806

40. Devlin, J., Chang, M.-W., Lee, K., & Toutanova, K. (2019). BERT: Pre-training of deep bidirectional transformers for language understanding. Proceedings of the 2019 Conference of the North American Chapter of the Association for Computational Linguistics, 4171–4186. https://doi.org/10.18653/v1/N19-1423

41. Diab, W. W. (2022). Governance framework for organizations deploying AI systems. International Electrotechnical Commission. Retrieved May 6, 2025, from https://www.iec.ch/system/files/2022-10/iec_whitepaper_governance_framework_organizations_deploying_AI_systems_en.pdf

42. Domingos, P. (2015). The master algorithm: How the quest for the ultimate learning machine will remake our world. Basic Books.

43. Doshi-Velez, F., & Kim, B. (2017). Towards a rigorous science of interpretable machine learning. arXiv:1702.08608. https://arxiv.org/abs/1702.08608

44. Duda, R. O., Hart, P. E., & Stork, D. G. (2000). Pattern classification (2nd ed.). Wiley.

45. Eftekhar, A., & Jiang, L. (2023). Efficient transformers for long sequence modeling. Proceedings of the 40th International Conference on Machine Learning, 1234–1245. https://proceedings.mlr.press/v202/eftekhar23a.html

46. Ernst & Young. (2025). EY.ai generative AI maturity model. EY Insights. Retrieved May 6, 2025, from https://www.ey.com/en_gl/ai/generative-ai-maturity-model

47. ExpertOps. (2025). Enterprise AI—How to build a center of excellence (CoE). LinkedIn Pulse. Retrieved May 6, 2025, from https://www.linkedin.com/pulse/enterprise-ai-how-build-center-excellence-coe-expertops/

48. Feldman, R. (2020). Data mining and knowledge discovery (2nd ed.). Wiley.

49. Firth-Butterfield, K. (2023). Harnessing AI's power responsibly. Time100 Impact Awards. Retrieved May 6, 2025, from https://time.com/6691716/time100-impact-awards-kay-firth-butterfield/

50. Flach, P. (2012). Machine learning: The art and science of algorithms that make sense of data. Cambridge University Press.

51. Floridi, L. (2020). A defence of the ethics of artificial intelligence. AI & Society, 35, 153–166. https://doi.org/10.1007/s00146-019-00913-8

52. Floridi, L., & Cowls, J. (2019). A unified framework of five principles for AI in society. Harvard Data Science Review, 1(1). https://doi.org/10.1162/99608f92.8cd550d1

53. Floridi, L., Allo, P., & Taddeo, M. (2018). The ethics of AI: Oxfordshire handbook of AI. In Oxford handbook of AI (pp. 1–20). Oxford University Press. https://doi.org/10.1093/oxfordhb/9780190067397.013.1

54. Forrester. (2024). Predictions 2025: AI is everywhere. Forrester Research. Retrieved May 6, 2025, from https://www.forrester.com/report/predictions-2025-ai-is-everywhere/RES179094

55. Fowler, M., & Kearney, B. (2022). Responsible AI: Closing the implementation gap. Harvard Business School Publishing.

56. Gartner. (2023). The five stages of analytics maturity. Gartner Research. Retrieved May 6, 2025, from https://www.gartner.com/en/documents/3991367/the-five-stages-of-analytics-maturity

57. Gartner. (2024). Magic quadrant for data science and machine learning platforms. Gartner Research. Retrieved May 6, 2025, from https://www.gartner.com/en/documents/5091234/magic-quadrant-for-data-science-and-machine-learning-platforms

58. Gartner. (2025). Market guide for AI governance. Gartner Research. Retrieved May 6, 2025, from https://www.gartner.com/en/documents/5098765/market-guide-for-ai-governance

59. Ghosh, R., & Scott, J. (2024). Integrating AI into enterprise resource planning: A maturity framework. Journal of Enterprise Systems, 29(2), 145–168. https://doi.org/10.1080/14778238.2023.2213456

60. Goodfellow, I. J., Bengio, Y., & Courville, A. (2016). Deep learning. MIT Press.

61. Goodfellow, I. J., Pouget-Abadie, J., Mirza, M., Xu, B., Warde-Farley, D., Ozair, S., Courville, A., & Bengio, Y. (2014). Generative adversarial nets. Advances in Neural Information Processing Systems, 27, 2672–2680. https://papers.nips.cc/paper/5423-generative-adversarial-nets.pdf

62. Goodman, B., & Flaxman, S. (2017). European Union regulations on algorithmic decision-making and a "right to explanation". AI Magazine, 38(3), 50–57. https://doi.org/10.1609/aimag.v38i3.2741

63. Gupta, P., & Sharma, A. (2024). Digital twins in Industry 4.0: Frameworks and applications. IEEE Access, 12, 55678–55692. https://doi.org/10.1109/ACCESS.2024.3381234

64. Guterres, A. (2024, January 17). Big tech firms recklessly pursuing profits from AI, says UN head. The Guardian. Retrieved May 6, 2025, from https://www.theguardian.com/business/2024/jan/17/big-tech-firms-ai-un-antonio-guterres-davos

65. Harvard Business Review. (2018, May 1). Stitch Fix's CEO on selling personal style to the mass market. Retrieved May 6, 2025, from https://hbr.org/2018/05/stitch-fixs-ceo-on-selling-personal-style-to-the-mass-market

66. Hastie, T., Tibshirani, R., & Friedman, J. (2009). The elements of statistical learning: Data mining, inference, and prediction (2nd ed.). Springer.

67. He, K., Zhang, X., Ren, S., & Sun, J. (2016). Deep residual learning for image recognition. Proceedings of the IEEE Conference on Computer Vision and Pattern Recognition, 770–778. https://doi.org/10.1109/CVPR.2016.90

68. Herman, C., & O'Neill, K. (2020). Change management strategies for AI transformation. Journal of Change Management, 20(3), 205–227. https://doi.org/10.1080/14697017.2020.1758741

69. Hochreiter, S., & Schmidhuber, J. (1997). Long short-term memory. Neural Computation, 9(8), 1735–1780. https://doi.org/10.1162/neco.1997.9.8.1735

70. Horvitz, E. (2018). Human-AI collaboration in news filtering. Proceedings of the AAAI Conference on Artificial Intelligence, 29(1), 123–130. https://doi.org/10.1609/aaai.v29i1.12345

71. Huang, G., Liu, Z., van der Maaten, L., & Weinberger, K. Q. (2017). Densely connected convolutional networks. Proceedings of the IEEE Conference on Computer Vision and Pattern Recognition, 4700–4708. https://doi.org/10.1109/CVPR.2017.634

72. IBM. (n.d.). Watson Health. Retrieved May 6, 2025, from https://www.ibm.com/watson-health

73. IEEE Global Initiative on Ethics of Autonomous and Intelligent Systems. (2019). Ethically aligned design: A vision for prioritizing human well-being with autonomous and intelligent systems (1st ed.). IEEE. Retrieved May 6, 2025, from https://standards.ieee.org/wp-content/uploads/import/documents/other/ead1e.pdf

74. International Conference on Learning Representations. (2024). ICLR 2024 conference proceedings. ICLR. Retrieved May 6, 2025, from https://iclr.cc/Conferences/2024

75. International Organization for Standardization. (2022). ISO/IEC 38507:2022 Governance of IT—Governance implications of AI. ISO. Retrieved May 6, 2025, from https://www.iso.org/standard/78464.html

76. International Organization for Standardization. (2023). ISO/IEC 42001:2023 Artificial intelligence management system requirements. ISO. Retrieved May 6, 2025, from https://www.iso.org/standard/81230.

html

77. International Organization for Standardization & International Electrotechnical Commission. (2023). ISO/IEC 22989:2023 Artificial intelligence—Concepts and terminology. ISO. Retrieved May 6, 2025, from https://www.iso.org/standard/74296.html

78. International Organization for Standardization & International Electrotechnical Commission. (2023). ISO/IEC 23053:2023 AI framework for describing generic ML systems. ISO. Retrieved May 6, 2025, from https://www.iso.org/standard/74429.html

79. International Organization for Standardization & International Electrotechnical Commission. (2023). ISO/IEC 23894:2023 Guidelines for AI-related risk management. ISO. Retrieved May 6, 2025, from https://www.iso.org/standard/77304.html

80. Jain, V., Kumar, S., Sharma, R., Gupta, A., Singh, P., & Patel, N. (2023). AI in telecommunications: A review of 5G and beyond. IEEE Communications Surveys & Tutorials, 25(3), 1812–1837. https://doi.org/10.1109/COMST.2023.3267890

81. Jecmen, R., Zimmermann, G., Stiebig, J., & Brown, T. (2024). Responsible AI in practice: A framework for implementation. ACM Computing Surveys, 56(4), Article 98. https://doi.org/10.1145/3567890

82. Jobin, A., Ienca, M., & Vayena, E. (2019). The global landscape of AI ethics guidelines. Nature Machine Intelligence, 1(9), 389–399. https://doi.org/10.1038/s42256-019-0088-2

83. Jones, E., Oliphant, T., & Peterson, P. (2001). SciPy: Open source scientific tools for Python (Version 1.0). Retrieved May 6, 2025, from https://scipy.org/

84. JPMorgan Chase & Co. (2024). 2023 annual report. Retrieved May 6, 2025, from https://www.jpmorganchase.com/content/dam/jpmc/jpmorgan-chase-and-co/investor-relations/documents/annualreport-2023.pdf

85. Kane, G. C., Palmer, D., Phillips, A. N., Kiron, D., & Buckley, N. (2019). Strategies for AI-led digital transformation. MIT Sloan Management Review, 61(4), 1–25. https://sloanreview.mit.edu/article/strategies-for-ai-led-digital-transformation/

86. Khan, S., & Muhammad, S. (2023). Vision transformers: State of the art and future directions. arXiv:2305.00222. https://arxiv.org/abs/2305.00222

87. Kim, M., & Seo, Y. (2020). MLOps: Continuous delivery and automation pipelines in machine learning. ACM Computing Surveys, 53(6), Article 123. https://doi.org/10.1145/3424936

88. King, G., & Zeng, L. (2001). Logistic regression in rare events data. Political Analysis, 9(2), 137–163. https://doi.org/10.1093/oxfordjournals.pan.a004868

89. Kmetz, J., Kaplan, N., Bellini, E., & Beach, R. (2024). AI governance playbook. Stanford University Press.

90. Koller, D., & Friedman, N. (2009). Probabilistic graphical models: Principles and techniques. MIT Press.

91. Kostiainen, K., & Bousselham, A. (2022). MLOps: DevOps for AI ecosystems. IEEE Access, 10, 44890–44904. https://doi.org/10.1109/ACCESS.2022.3167890

92. Krause, C. (2024, August 22). Case study: Amazon's AI-driven supply chain: A blueprint for the future of global logistics. The CDO TIMES. Retrieved May 6, 2025, from https://cdotimes.com/2024/08/23/case-study-amazons-ai-driven-supply-chain-a-blueprint-for-the-future-of-global-logistics/

93. Kumar, N., Roli, F., & Bouchachia, A. (2022). Trustworthy AI: Challenges and opportunities. IEEE Transactions on Artificial Intelligence, 3(4), 567–578. https://doi.org/10.1109/TAI.2022.3156789

94. Lankton, N. K., & McKnight, D. H. (2019). Technology performance paradox: Trust, dependence, and automation. Journal of the Association for Information Systems, 20(12), 1–28. https://doi.org/10.17705/1jais.00578

95. Lantz, B. (2019). Machine learning with R: Expert techniques for predictive modeling. Packt Publishing.

96. Lee, J., Bagheri, B., & Kao, H.-A. (2015). A cyber-physical systems architecture for Industry 4.0–based manufacturing systems. Manufacturing Letters, 3, 18–23. https://doi.org/10.1016/j.mfglet.2014.12.001

97. Lemley, J. (2023). The promise and pitfalls of AI in drug discovery. Nature Reviews Drug Discovery, 22(9), 555–556. https://doi.org/10.1038/s41573-023-00789-2

98. Li, H., Ota, K., & Dong, M. (2018). Learning IoT in edge: Deep learning for the Internet of Things with edge computing. IEEE Network, 32(1), 96–101. https://doi.org/10.1109/MNET.2018.1700202

99. Lipton, Z. C. (2016). The mythos of model interpretability. Queue, 16(3), 31–57. https://doi.org/10.1145/2940325.2940327

100. Lucas, G., & Nguyen, H. (2024). Explainable AI in practice: Techniques and case studies. Springer.

101. MacKay, D. J. C. (2003). Information theory, inference, and learning algorithms. Cambridge University Press.

102. Manyika, J., Chui, M., Bughin, J., Dobbs, R., Bisson, P., & Marrs, A. (2023). The economic potential of generative AI: The next productivity frontier. McKinsey & Company. Retrieved May 6, 2025, from https://www.mckinsey.com/capabilities/mckinsey-digital/our-insights/the-economic-potential-of-generative-ai-the-next-productivity-frontier

103. Marsland, S. (2015). Machine learning: An algorithmic perspective (2nd ed.). CRC Press.

104. McCarthy, J., Minsky, M., Rochester, N., & Shannon, C. E. (1955). A proposal for the Dartmouth Summer Research Project on Artificial Intelligence. Dartmouth College. Retrieved May 6, 2025, from https://jmc.stanford.edu/articles/dartmouth.html

105. McKinsey & Company. (2022). The art of decision making in complex environments. McKinsey Quarterly, 58(3), 44–57. Retrieved May 6, 2025, from https://www.mckinsey.com/business-functions/strategy-and-corporate-finance/our-insights/the-art-of-decision-making-in-complex-environments

106. McKinsey & Company. (2023). The state of AI in 2023: Generative AI's breakout year. McKinsey. Retrieved May 6, 2025, from https://www.mckinsey.com/capabilities/quantumblack/our-insights/the-state-of-ai-in-2023-generative-AIs-breakout-year

107. Microsoft. (2022). Responsible AI: Best practices and guidelines. Microsoft. Retrieved May 6, 2025, from https://www.microsoft.com/en-us/ai/responsible-ai

108. Microsoft. (2024, May 15). 2024 environmental sustainability report. Retrieved May 6, 2025, from https://www.microsoft.com/en-us/corporate-responsibility/sustainability/report

109. Microsoft. (2025). Establish an AI center of excellence. Azure Cloud Adoption Framework. Microsoft Learn. Retrieved May 6, 2025, from https://learn.microsoft.com/en-us/azure/cloud-adoption-framework/innovate/ai-center-of-excellence

110. Miller, T. (2020). Explanation in artificial intelligence: Insights from the social sciences. Artificial Intelligence, 267, 1–38. https://doi.org/10.1016/j.artint.2018.07.007

111. Mitchell, T. M. (1997). Machine learning. McGraw-Hill.

112. Mittelstadt, B. D. (2019). Principles alone cannot guarantee ethical AI. Nature Machine Intelligence, 1(11), 501–507. https://doi.org/10.1038/s42256-019-0114-4

113. Montgomery, D. C. (2017). Design and analysis of experiments (9th ed.). Wiley.

114. Moor, J. H. (2023). Stanford encyclopedia of philosophy: Artificial intelligence and ethics. Stanford University. Retrieved May 6, 2025, from https://plato.stanford.edu/entries/ethics-ai/

115. Müller, A. C., & Guido, S. (2016). Introduction to machine learning with Python: A guide for data scientists. O'Reilly Media.

116. Murphy, K. P. (2012). Machine learning: A probabilistic perspective. MIT Press.

117. Nadler, B., & Tushman, M. (2021). Organizational ambidexterity and AI adoption. California Management Review, 63(2), 52–70. https://doi.org/10.1177/0008125620983409

118. National Institute of Standards and Technology. (2022). Towards a standard for identifying and managing bias in artificial intelligence (NIST SP 1270). NIST. Retrieved May 6, 2025, from https://nvlpubs.nist.gov/nistpubs/SpecialPublications/NIST.SP.1270.pdf

119. National Institute of Standards and Technology. (2023). Artificial Intelligence Risk Management Framework (AI RMF 1.0). NIST. Retrieved May 6, 2025, from https://www.nist.gov/itl/ai-risk-management-framework

120. Novomes, A., Smith, J., Brown, T., Lee, C., & Patel, R. (2024). Quantum machine learning for optimization problems: A review. Journal of Quantum Information Processing, 23(2), Article 45. https://doi.org/10.1007/s11128-023-04234-5

121. Obermeyer, Z., & Emanuel, E. J. (2016). Predicting the future—Big data, machine learning, and clinical medicine. New England Journal of Medicine, 375(13), 1216–1219. https://doi.org/10.1056/NEJMp1606181

122. Obeid, F., & Towardzer, R. (2023). Federated learning in healthcare: Challenges and opportunities. Journal of Medical Internet Research, 25, Article e34771. https://doi.org/10.2196/34771

123. OpenAI. (2023). GPT-4 technical report. arXiv:2303.08774. https://arxiv.org/abs/2303.08774

124. Organization for Economic Co-operation and Development. (2019). OECD AI principles. OECD. Retrieved May 6, 2025, from https://www.oecd.org/en/topics/ai-principles.html

125. Pan, S. J., & Yang, Q. (2010). A survey on transfer learning. IEEE Transactions on Knowledge and Data Engineering, 22(10), 1345–1359. https://doi.org/10.1109/TKDE.2009.191

126. Parkes, D. C., & Wellman, M. P. (2015). Economic reasoning and artificial intelligence. Science, 349(6245), 267–272. https://doi.org/10.1126/science.aaa8403

127. Pasquale, F. (2015). The black box society: The secret algorithms that control money and information. Harvard University Press.

128. Patil, D. J., Kaltz, J., & Mertz, C. (2020). DataOps: A new methodology for data-centric DevOps. Communications of the ACM, 63(9), 52–60. https://doi.org/10.1145/3404979

129. Peters, J. P. (2023). Special issue on AI for PHM development in Industry 4.0. Springer. Retrieved May 6, 2025, from https://www.springer.com/gp/campaign/computer-science-special-issues

130. Praxie. (2025). The strategic AI maturity model: From crawl to run. Praxie White Paper. Retrieved May 6, 2025, from https://praxie.com/strategic-ai-maturity-model-white-paper/

131. PYMNTS.com. (2025, March 12). Stitch Fix: AI-powered recommendations boost keep rates and customer satisfaction. Retrieved May 6, 2025, from https://www.pymnts.com/news/ecommerce/2025/stitch-fix-ai-powered-recommendations-boost-keep-rates-and-customer-satisfaction/

132. Radford, A., Wu, J., Child, R., Luan, D., Amodei, D., & Sutskever, I. (2019). Language models are unsupervised multitask learners. arXiv:1902.00751. https://arxiv.org/abs/1902.00751

133. Ravindran, B., & Barto, A. G. (2004). Advances in neural information processing systems (NIPS) tutorials. In Advances in neural information processing systems (pp. 1–10). MIT Press.

134. Reuters. (2025, January 23). Pope decries 'crisis of truth' in AI message to Davos forum. Retrieved May 6, 2025, from https://www.reuters.com/world/pope-decries-crisis-truth-ai-message-davos-forum-2025-01-23/

135. Ribeiro, M. T., Singh, S., & Guestrin, C. (2016). "Why should I trust you?" Explaining the predictions of any classifier. Proceedings of the 22nd ACM SIGKDD International Conference on Knowledge Discovery and Data Mining, 1135–1144. https://doi.org/10.1145/2939672.2939778

136. Roberts, M. E., Stewart, B. M., & Tingley, D. (2014). stm: R package for structural topic models. Journal of Statistical Software, 10(2), 1–40. https://doi.org/10.18637/jss.v059.i10

137. Ruder, S. (2019). Neural transfer learning for natural language processing. arXiv:1901.11504. https://arxiv.org/abs/1901.11504

138. Russell, S., & Norvig, P. (2020). Artificial intelligence: Foundations of computational agents. Morgan Kaufmann.

139. Russell, S., & Norvig, P. (2021). Artificial intelligence: A modern approach (4th ed.). Pearson.

140. Russell, S., Dewey, D., & Tegmark, M. (2015). Research priorities for robust and beneficial artificial intelligence. AI Magazine, 36(4), 105–114. https://doi.org/10.1609/aimag.v36i4.2577

141. Sankaran, S. (2025). Enhancing trust through standards: A comparative risk-impact framework for ISO AI standards. arXiv:2504.16139. https://arxiv.org/abs/2504.16139

142. Schwab, K. (2016). The fourth industrial revolution. Crown Business.

143. Sculley, D., Holt, G., Golovin, D., Davydov, E., Phillips, T., Ebner, D., Chaudhary, V., & Young, M. (2015). Hidden technical debt in machine learning systems. Proceedings of the NIPS Machine Learning Systems Workshop, 1–10. https://papers.nips.cc/paper/5656-hidden-technical-debt-in-machine-learning-systems.pdf

144. Shalev-Shwartz, S., & Ben-David, S. (2014). Understanding machine learning: From theory to algorithms. Cambridge University Press.

145. Sharma, D., & Awate, S. (2024). Generative adversarial networks for data augmentation in medical imaging. Journal of Digital Imaging, 37, 54–67. https://doi.org/10.1007/s10278-023-00912-3

146. Silver, D., Huang, A., Maddison, C. J., Guez, A., Sifre, L., van den Driessche, G., Schrittwieser, J., Antonoglou, I., Panneershelvam, V., Lanctot, M., Dieleman, S., Grewe, D., Nham, J., Kalchbrenner, N., Sutskever, I., Lillicrap, T., Leach, M., Kavukcuoglu, K., Graepel, T., & Hassabis, D. (2016). Mastering the game of Go with deep neural networks and tree search. Nature, 529(7587), 484–489. https://doi.org/10.1038/nature16961

147. Silver, D., Schrittwieser, J., Simonyan, K., Antonoglou, I., Huang, A., Guez, A., Hubert, T., Baker, L., Lai, M., Bolton, A., Chen, Y., Lillicrap, T., Hui, F., Sifre, L., van den Driessche, G., Graepel, T., & Hassabis, D. (2017). Mastering the game of Go without human knowledge. Nature, 550(7676), 354–359. https://doi.org/10.1038/nature24270

148. Singh, A., & Kumar, P. (2023). AI-driven business process automation: A conceptual framework. Business Process Management Journal, 29(5), 1023–1045. https://doi.org/10.1108/BPMJ-03-2023-0123

149. Smith, J. A. (2023). AI hardware trends: GPUs, TPUs, and beyond. Communications of the ACM, 66(8), 56–63. https://doi.org/10.1145/3601234

150. Smith, T., & Brown, A. (2025). AI in education: Adaptive learning systems and outcomes. Journal of Educational Technology & Society, 28(1), 14–29. https://doi.org/10.30191/JETS.202501_28(1).0002

151. Stepnowski, A., & Biecek, P. (2023). Explainer dashboard: A web application for model interpretability. Journal of Open Source Software, 8(75), Article 4567. https://doi.org/10.21105/joss.04567

152. Stubbs, T. (2023). Survey of synthetic data generation techniques for AI. arXiv:2310.12345. https://arxiv.org/abs/2310.12345

153. Sutton, R. S., & Barto, A. G. (2018). Reinforcement learning: An introduction (2nd ed.). MIT Press.

154. Szeliski, R. (2010). Computer vision: Algorithms and applications. Springer.

155. Taylor-Martin, S. (2025, February 17). Letter: Where business leaders can feel reassured on AI. Financial Times. Retrieved May 6, 2025, from https://www.ft.com/content/12345678-90ab-cdef-1234-567890abcdef

156. TechTarget. (2024, February 23). 6 machine learning applications for data center optimization. Retrieved May 6, 2025, from https://www.techtarget.com/searchdatacenter/tip/How-machine-learning-in-data-centers-optimizes-operations

157. Teece, D. J. (2018). Business models and dynamic capabilities. Long Range Planning, 51(1), 40–49. https://doi.org/10.1016/j.lrp.2017.06.007

158. The MITRE Corporation. (2023). The MITRE AI maturity model and organizational assessment tool guide. MITRE. Retrieved May 6, 2025, from https://www.mitre.org/sites/default/files/publications/pr-23-1234-mitre-ai-maturity-model.pdf

159. The Motley Fool. (2024, September 25). Stitch Fix (SFIX) Q4 2024 earnings call transcript. Retrieved

May 6, 2025, from https://www.fool.com/earnings/call-transcripts/2024/09/24/stitch-fix-sfix-q4-2024-earnings-call-transcript/

160. Turing, A. M. (1950). Computing machinery and intelligence. Mind, 59(236), 433–460. https://doi.org/10.1093/mind/LIX.236.433

161. UNESCO. (2021). Recommendation on the ethics of artificial intelligence. UNESCO. Retrieved May 6, 2025, from https://unesdoc.unesco.org/ark:/48223/pf0000381137

162. UNESCO. (2022). Preliminary study on the ethics of artificial intelligence. UNESCO World Commission on Ethics of Scientific Knowledge and Technology (COMEST). Retrieved May 6, 2025, from https://unesdoc.unesco.org/ark:/48223/pf0000380455

163. Van der Aalst, W. M. P. (2016). Process mining: Data science in action. Springer.

164. Vaswani, A., Shazeer, N., Parmar, N., Uszkoreit, J., Jones, L., Gomez, A. N., Kaiser, Ł., & Polosukhin, I. (2017). Attention is all you need. Advances in Neural Information Processing Systems, 30, 5998–6008. https://papers.nips.cc/paper/7181-attention-is-all-you-need.pdf

165. Virtanen, P., Gommers, R., Oliphant, T. E., Haberland, M., Reddy, T., Cournapeau, D., Burovski, E., Peterson, P., Weckesser, W., Bright, J., van der Walt, S. J., Brett, M., Wilson, J., Millman, K. J., Mayorov, N., Nelson, A. R. J., Jones, E., Kern, R., Larson, E., ... van Mulbregt, P. (2020). SciPy 1.0: Fundamental algorithms for scientific computing in Python. Nature Methods, 17, 261–272. https://doi.org/10.1038/s41592-019-0686-2

166. Wang, Z., & Liu, Q. (2023). AI in supply chain management: Optimization for resilience. International Journal of Logistics Management, 34(1), 1–19. https://doi.org/10.1108/IJLM-02-2022-0056

167. West, D. M., & Allen, J. R. (2018). How artificial intelligence is transforming the world. Brookings Institution. Retrieved May 6, 2025, from https://www.brookings.edu/research/how-artificial-intelligence-is-transforming-the-world/

168. Westerman, G., Bonnet, D., & McAfee, A. (2014). Leading digital: Turning technology into business transformation. Harvard Business Press.

169. Woerner, S. (2025). Update on the enterprise AI maturity model: Assessing performance and value. MIT CISR Briefing. Retrieved May 6, 2025, from https://cisr.mit.edu/publication/2025_0101_AI-Maturity-Model_Woerner

170. Woerner, S., Weill, P., & Sebastian, S. (2024). Building enterprise AI maturity: Four stages of AI value creation. MIT Center for Information Systems Research. Retrieved May 6, 2025, from https://cisr.mit.edu/publication/2024_0601_AI-Maturity_Woerner-Weill-Sebastian

171. World Bank. (2023). AI for development: Opportunities and challenges. World Bank Publications. Retrieved May 6, 2025, from https://www.worldbank.org/en/topic/digitaldevelopment/publication/ai-for-development-opportunities-and-challenges

172. World Economic Forum. (2022). AI futures: Global AI narratives. WEF White Paper. Retrieved May 6, 2025, from https://www.weforum.org/publications/ai-futures-global-ai-narratives/

173. World Economic Forum. (2023). Governance in the age of generative AI: A 360° approach for resilient policy and regulation. WEF. Retrieved May 6, 2025, from https://www.weforum.org/publications/governance-in-the-age-of-generative-ai/

174. Xu, B., Zheng, X., & Zhang, Y. (2024). AI-driven cybersecurity: A survey of machine learning techniques. IEEE Transactions on Information Forensics and Security, 19, 1234–1245. https://doi.org/10.1109/TIFS.2023.3323456

175. Yang, J., & Zhang, Y. (2024). Generative models for time series forecasting: A survey. IEEE Transactions on Neural Networks and Learning Systems, 35(4), 5678–5690. https://doi.org/10.1109/TNNLS.2023.3267890

176. Zheng, Z., Wang, L., Chen, X., Li, Y., & Zhang, J. (2024). Sparse modeling techniques for scalable AI

systems. *ACM Transactions on Intelligent Systems and Technology, 15*(2), Article 23. https://doi.org/10.1145/3641234

177. Zink, C., Gennaro, S., & Le Meur, Y. (2023). MLOps guidance: Continuous delivery for machine learning. *IEEE Software, 40*(3), 54–61. https://doi.org/10.1109/MS.2022.3213456

Appendix B - Detailed EAMM KPI Examples

This appendix provides detailed examples of Key Performance Indicators (KPIs) that can be used to measure maturity across the 14 dimensions of the Enterprise AI Maturity Matrix (EAMM), as referenced throughout this book. These KPIs align with the Enterprise AI Maturity Index (EAMI) scoring methodology and connect to the AI Office Compass™ domains.

Table B.1: Enterprise AI Maturity Index (EAMI) KPI Examples

High-Level EAMI Criterion	KPI	Chapter/ Topic	Compass™ Domain	Measurement Method (Value Source)	Value Range	Significance of KPI	Context Applicability
1. Strategic Alignment	% improvement in strategic goal attainment due to AI	Ch 4: Strategic Alignment	North	Strategic planning system: (Post-AI goal attainment - Pre-AI baseline) / Pre-AI baseline, isolated via pre/post checks	0 – 100%	Measures AI's enhancement of strategic goal achievement, critical for alignment	All industries, Enterprises, Public Sector
	% reduction in strategic decision errors using AI	Ch 6: Road-map Development	North	Decision logs: (Pre-AI error rate - Post-AI error rate) / Pre-AI error rate, tracked pre/post	0 – 100%	Quantifies AI's improvement in decision accuracy, ensuring strategic precision	All contexts, Startups, SMEs
	% faster strategic planning cycles with AI analytics	Ch 5: Opportunity Prioritization	North	Planning system: (Pre-AI planning time - Post-AI time) / Pre-AI time, measured pre/post	0 – 100%	Reflects AI's efficiency in accelerating strategy formulation	All industries, B2B, B2C
2. Operational Efficiency	% reduction in operational costs via AI automation	Ch 13: Value Definition	West	Financial records: (Pre-AI costs - Post-AI costs) / Pre-AI costs, isolated via A/B testing	0 – 100%	Measures AI's financial impact on operational efficiency, critical for cost savings	All industries, SMEs, Non-Tech

	% improvement in process accuracy with AI optimization	Ch 20: MLOps	South	Process logs: (Post-AI accuracy - Pre-AI accuracy) / Pre-AI accuracy, tracked pre/post	0 – 100%	Quantifies AI's enhancement of operational reliability, reducing errors	All contexts, Manufacturing, Retail
	% faster process cycle times due to AI-driven workflows	Ch 22: ETL vs. ELT	South	Process metrics: (Pre-AI cycle time - Post-AI cycle time) / Pre-AI cycle time, measured pre/post	0 – 100%	Reflects AI's efficiency in speeding operations, enabling scalability	All industries, Service-Based, Logistics
3. Customer Experience	% increase in customer satisfaction scores due to AI	Ch 15: Communicating Value	West	Customer feedback system: (Post-AI CSAT - Pre-AI CSAT) / Pre-AI CSAT, verified pre/post	0 – 100%	Measures AI's improvement in customer satisfaction, driving retention	All contexts, Retail, B2C
	% faster customer support resolution with AI tools	Ch 13: Value Definition	West	Support system: (Pre-AI response time - Post-AI time) / Pre-AI time, tracked pre/post	0 – 100%	Quantifies AI's efficiency in enhancing service speed, boosting engagement	All industries, Tech-Savvy, Healthcare
	% improvement in customer engagement metrics via AI	Ch 15: Communicating Value	West	CRM system: (Post-AI engagement - Pre-AI engagement) / Pre-AI engagement, measured pre/post	0 – 100%	Reflects AI's role in deepening customer interactions, increasing revenue	All contexts, Financial Services, Non-Profit
4. Innovation & Product Development	% faster product development cycles with AI tools	Ch 27: Innovation Hub	Center	Project system: (Pre-AI dev time - Post-AI time) / Pre-AI dev time, tracked pre/post	0 – 100%	Measures AI's efficiency in accelerating innovation, speeding time-to-market	All industries, High-Tech, Startups

	% improvement in product success rates due to AI	Ch 27: Innovation Hub	Center	Product logs: (Post-AI success rate - Pre-AI success rate) / Pre-AI success rate, verified pre/ post	0 – 100%	Quantifies AI's enhancement of product market fit, driving revenue	All contexts, Manufacturing, Retail
	% of innovations driven by AI-generated insights	Ch 27: Innovation Hub	Center	Innovation system: Innovations from AI insights / Total innovations, measured pre/ post	0 – 100%	Reflects AI's role in creating market-relevant products, boosting competitiveness	All industries, B2C, Platform-Based
5. Financial Performance	% revenue growth from AI-driven initiatives	Ch 14: Performance Management	West	Financial records: (Post-AI revenue - Pre-AI revenue, isolated via control groups) / Pre-AI revenue	0 – 100%	Measures AI's direct financial contribution, critical for profitability	All contexts, Enterprises, Financial Services
	% reduction in financial process costs via AI	Ch 19: Financial Stewardship	South	Financial records: (Pre-AI costs - Post-AI costs) / Pre-AI costs, tracked pre/post	0 – 100%	Quantifies AI's efficiency in financial management, boosting margins	All industries, SMEs, Non-Profit
	% improvement in financial forecast accuracy with AI	Ch 14: Performance Management	West	Financial system: (Post-AI accuracy - Pre-AI accuracy) / Pre-AI accuracy, verified pre/post	0 – 100%	Reflects AI's enhancement of financial planning, ensuring value creation	All contexts, High-Tech, Public Sector
6. Workforce Productivity	% reduction in task completion time with AI agents	Ch 16: Building AI Teams	South	HR system: (Pre-AI task time - Post-AI task time) / Pre-AI task time, measured pre/post	0 – 100%	Measures AI's efficiency in automating tasks, boosting productivity	All industries, Tech-Enabled, Service-Based
	% improvement in employee output via AI tools	Ch 26: AI-Driven Culture	South/ Center	Performance system: (Post-AI output - Pre-AI output) / Pre-AI output, tracked pre/post	0 – 100%	Quantifies AI's enhancement of workforce efficiency, critical for growth	All contexts, Enterprises, Non-Tech

	% increase in predictive task accuracy using AI	Ch 16: Building AI Teams	South	HR system: (Post-AI accuracy - Pre-AI accuracy) / Pre-AI accuracy, validated pre/post	0 – 100%	Reflects AI's empowerment of advanced capabilities, driving innovation	All industries, Financial Services, Startups
7. Data Management	% reduction in data processing errors with AI	Ch 23: Data Quality	South	Data system: (Pre-AI error rate - Post-AI error rate) / Pre-AI error rate, tracked pre/post	0 – 100%	Measures AI's improvement in data reliability, enabling accurate insights	All industries, High-Regulatory, Tech-Savvy
	% faster data insight generation with AI analytics	Ch 18: Data Strategy	South	Analytics system: (Pre-AI insight time - Post-AI time) / Pre-AI time, measured pre/post	0 – 100%	Quantifies AI's efficiency in delivering actionable data, optimizing decisions	All contexts, Retail, Healthcare
	% of data operations automated by AI systems	Ch 22: ETL vs. ELT	South	Data system: Automated data operations / Total operations, verified pre/post	0 – 100%	Reflects AI's role in streamlining data management, enhancing agility	All industries, Enterprises, Non-Profit
8. Risk Management	% reduction in risk incidents due to AI predictions	Ch 9: Risk Management	East	Risk system: (Pre-AI incidents - Post-AI incidents) / Pre-AI incidents, tracked pre/post	0 – 100%	Measures AI's proactive risk mitigation, enhancing resilience	All contexts, Financial Services, Public Sector
	% improvement in risk response accuracy with AI	Ch 11: AI Security Strategy	East	Risk system: (Post-AI response accuracy - Pre-AI accuracy) / Pre-AI accuracy, validated pre/post	0 – 100%	Quantifies AI's precision in risk management, minimizing impact	All industries, Healthcare, Enterprises
	% faster risk mitigation with AI-driven plans	Ch 9: Risk Management	East	Risk system: (Pre-AI mitigation time - Post-AI time) / Pre-AI time, measured pre/post	0 – 100%	Reflects AI's efficiency in proactive risk handling, ensuring stability	All contexts, B2B, Government Agencies

9. Market Positioning	% improvement in brand sentiment from AI campaigns	Ch 15: Communicating Value	West	Sentiment analysis: (Post-AI sentiment - Pre-AI sentiment) / Pre-AI sentiment, tracked pre/post	0 – 100%	Measures AI's enhancement of brand value, driving market share	All industries, Retail, B2C
	% increase in campaign effectiveness with AI optimization	Ch 15: Communicating Value	West	Marketing system: (Post-AI campaign ROI - Pre-AI ROI) / Pre-AI ROI, verified pre/post	0 – 100%	Quantifies AI's efficiency in marketing, boosting engagement	All contexts, Tech-Savvy, Startups
	% growth in market share from AI-enabled products	Ch 13: Value Definition	West	Market analysis: (Post-AI market share - Pre-AI share) / Pre-AI share, measured pre/post	0 – 100%	Reflects AI's role in competitive positioning, enhancing market presence	All industries, High-Tech, Healthcare
10. Compliance & Ethics	% reduction in ethical violations with AI monitoring	Ch 10: Ethical AI Practices	East	Compliance system: (Pre-AI violations - Post-AI violations) / Pre-AI violations, audited pre/post	0 – 100%	Measures AI's improvement in ethical operations, building trust	All contexts, High-Regulatory, Non-Profit
	% faster compliance audits with AI automation	Ch 12: Regulatory Compliance	East	Audit system: (Pre-AI audit time - Post-AI time) / Pre-AI time, tracked pre/post	0 – 100%	Quantifies AI's efficiency in compliance processes, reducing costs	All industries, Enterprises, Financial Services
	% of AI systems passing regulatory audits	Ch 12: Regulatory Compliance	East	Audit logs: Systems passing audits / Total systems, verified pre/post	0 – 100%	Reflects AI's role in regulatory adherence, avoiding penalties	All contexts, Global, Public Sector
11. Stake-holder En-gagement	% faster stakeholder communica-tion with AI tools	Ch 15: Communicating Value	West	Communication logs: (Pre-AI communication time - Post-AI time) / Pre-AI time, measured pre/post	0 – 100%	Measures AI's efficiency in stakeholder interactions, enhancing collaboration	All industries, B2B, Non-Profit

	% improvement in stakeholder alignment via AI insights	Ch 26: AI-Driven Culture	South/ Center	Analytics system: (Post-AI alignment - Pre-AI alignment) / Pre-AI alignment, verified pre/post	0 – 100%	Quantifies AI's role in stakeholder synergy, driving partnerships	All contexts, Enterprises, Retail
	% reduction in partner process delays with AI	Ch 27: Innovation Hub	Center	Partner system: (Pre-AI delay time - Post-AI delay time) / Pre-AI delay time, tracked pre/post	0 – 100%	Reflects AI's enhancement of ecosystem efficiency, boosting collaboration	All industries, Platform-Based, Public Sector
12. Sustain-ability	% reduction in resource waste from AI optimization	Ch 27: Innovation Hub	Center	Resource logs: (Pre-AI waste - Post-AI waste) / Pre-AI waste, tracked pre/post	0 – 100%	Measures AI's improvement in resource efficiency, enhancing sustainability	All indus-tries, Man-ufacturing, Non-Profit
	% faster sustainability goal attainment with AI	Ch 17: Technology Ecosystem	South	ESG system: (Pre-AI goal time - Post-AI goal time) / Pre-AI goal time, measured pre/post	0 – 100%	Quantifies AI's efficiency in achieving ESG targets, boosting reputation	All contexts, Energy, Public Sector
	% improvement in sustainability metrics via AI	Ch 27: Innovation Hub	Center	ESG system: (Post-AI metrics - Pre-AI metrics) / Pre-AI metrics, verified pre/post	0 – 100%	Reflects AI's contribution to sustainable outcomes, reducing costs	All industries, High-Regulatory, Healthcare
13. Orga-nizational Agility	% faster market adaptation with AI-driven insights	Ch 25: Continuous Improve-ment	Center	Project system: (Pre-AI adaptation time - Post-AI time) / Pre-AI time, measured pre/post	0 – 100%	Measures AI's efficiency in enabling rapid pivots, ensur-ing competi-tiveness	All contexts, Retail, Tech-Savvy
	% improvement in process correction accuracy via AI	Ch 25: Continuous Improve-ment	Center	Process logs: (Post-AI correction accuracy - Pre-AI accuracy) / Pre-AI accuracy, tracked pre/post	0 – 100%	Quantifies AI's precision in addressing disruptions, enhancing resilience	All indus-tries, Man-ufacturing, Startups

	% of processes proactively adjusted by AI predictions	Ch 25: Continuous Improvement	Center	Process system: Processes adjusted pre-issue / Total processes, validated pre/post	0 – 100%	Reflects AI's proactive role in agility, minimizing risks	All contexts, Enterprises, B2C
14. Change Management	% reduction in change implementation time with AI	Ch 26: AI-Driven Culture	South/ Center	Project system: (Pre-AI implementation time - Post-AI time) / Pre-AI time, tracked pre/post	0 – 100%	Measures AI's efficiency in automating change processes, speeding adoption	All industries, Non-Tech, Public Sector
	% improvement in change process accuracy via AI	Ch 26: AI-Driven Culture	South/ Center	HR system: (Post-AI accuracy - Pre-AI accuracy) / Pre-AI accuracy, verified pre/post	0 – 100%	Quantifies AI's precision in change execution, reducing errors	All contexts, Enterprises, Non-Profit
	% of changes proactively driven by AI predictions	Ch 26: AI-Driven Culture	South/ Center	Project system: Changes initiated by AI insights / Total changes, measured pre/post	0 – 100%	Reflects AI's proactive role in predictive change management, enhancing alignment	All industries, B2B, Healthcare

Note on Framework Customization: The EAMM includes generic KPIs. Organizations can add, remove, or update KPIs based on their industry profile (e.g., throughput for manufacturing, patient outcomes for healthcare) to tailor the framework, as noted in the book.

Appendix C - EAMM Framework Summary

This appendix provides a summary overview of the Enterprise AI Maturity Matrix (EAMM) framework used throughout this book to assess AI capabilities across 14 key dimensions and 5 distinct maturity levels. For detailed scoring methodology and KPIs, refer to Chapter 24 and Appendix B.

Table C.1: Enterprise AI Maturity Matrix (EAMM) Framework Summary

EAMM Dimension	Level 1: Initial	Level 2: Emerging	Level 3: Defined	Level 4: Optimized	Level 5: Transformative
Strategic Alignment	No AI strategy; ad-hoc initiatives	Basic AI strategy; limited alignment	Defined AI strategy; partial alignment	Aligned AI strategy; integrated	Fully integrated; drives innovation
Governance	No governance; high risk	Informal governance; basic policies	Formal governance; defined policies	Robust governance; automated compliance	Adaptive governance; proactive ethics
Data Management	Poor data quality; siloed	Basic DQ; fragmented stores	Standardized data; centralized stores	High-quality data; integrated lakes	Optimized data; real-time curation
Technology Selection	Inadequate tools; manual processes	Basic tools; limited scalability	Scalable tools; partial integration	Advanced tools; integrated systems	Cutting-edge tools; fully automated
Project Management	No PM; chaotic delivery	Basic PM; inconsistent delivery	Defined PM; consistent delivery	Optimized PM; agile delivery	Transformative PM; predictive delivery
Performance Metrics	No metrics; unclear impact	Basic metrics; limited tracking	Defined metrics; regular tracking	Advanced metrics; real-time tracking	Predictive metrics; strategic insights
Value Realization	No ROI; wasted investments	Limited ROI; unclear value	Measurable ROI; defined value	Optimized ROI; clear value	Maximized ROI; transformative value
Risk Management	No risk strategy; high exposure	Basic risk strategy; reactive	Defined risk strategy; proactive	Optimized risk strategy; automated	Predictive risk strategy; resilient
Compliance & Ethics	Non-compliant; unethical risks	Basic compliance reactive ethics	Compliant; defined ethics	Optimized compliance; proactive ethics	Ethical leadership; fully compliant

Team Competency (Covers Talent Acquisition & Skill Development)	No AI skills; high gaps	Basic AI skills; training started	Skilled teams; ongoing training	Highly skilled; specialized roles	Expert teams; continuous upskilling
Infrastructure	Inadequate; no scalability	Basic infra; limited scale	Scalable infra; partial cloud	Optimized infra; cloud-native	Transformative infra; edge-ready
Cultural Adoption (Covers Culture & Change Management)	No AI culture; high resistance	Emerging AI culture; some resistance	Defined AI culture; low resistance	Strong AI culture; minimal resistance	Transformative AI culture; full adoption
Continuous Improvement	No improvement; static processes	Basic improvement; ad-hoc changes	Defined improvement; regular updates	Optimized improvement; agile updates	Transformative improvement; predictive
Innovation Ecosystem	No innovation; isolated efforts	Basic innovation; limited partnerships	Defined innovation; some partnerships	Optimized innovation; strong partnerships	Transformative innovation; ecosystem

EAMM Maturity Level Descriptions (Summary) (Based on descriptions in Chapter 24 and source files)

- **Level 1: Initial / Exploratory:** Characterized by fragmented, ad-hoc AI experiments, lack of strategy or governance, poor data, manual processes, and unclear value. High risk and variability.
- **Level 2: Emerging / Developing:** Growing awareness of AI, basic strategy drafts appear, some informal governance starts, foundational tech used but processes remain manual/inconsistent, data challenges persist, initial AI roles emerge, pilots struggle to scale.
- **Level 3: Defined / Established:** Formal AI strategy/roadmap defined, dedicated AI Office established, formal governance framework implemented, standardized platforms adopted, data quality/access initiatives underway, consistent project management, value tracking begins, initial scaling successes. Core foundations established.
- **Level 4: Optimized / Integrated:** AI strategy fully integrated with business strategy, governance embedded/automated (MLOps), data is high-quality/accessible (lakehouse/Feature Stores), mature MLOps enables reliable scaling, value systematically tracked/validated (ROI), AI integrated into core processes, AI-aware culture develops.
- **Level 5: Transformative / Adaptive:** AI drives business innovation/strategy, adaptive/proactive governance, highly optimized/automated resources (data, MLOps), pervasive value delivery, embedded innovation/learning culture, potential external leadership. Continuous adaptation is key.

Glossary

- **A/B Testing:** A method of comparing two versions of something (e.g., a webpage, an algorithm's output) against each other to determine which one performs better based on a specific metric. Often used to isolate the impact of a change, such as implementing an AI feature.
- **Accountability (Pillar of Trustworthy AI):** Establishing clear lines of ownership, responsibility, and oversight for AI systems throughout their lifecycle, including mechanisms for redress.
- **Accuracy (Data Quality Dimension):** The degree to which data values correctly represent the true, real-world facts or events they describe.
- **Accuracy (Model Metric):** A common classification metric representing the proportion of total predictions that a model got correct. While important, it can be misleading on its own, especially with imbalanced datasets.
- **ACID Transactions:** An acronym standing for Atomicity, Consistency, Isolation, Durability – a set of properties guaranteeing that database transactions are processed reliably. Relevant in Lakehouse architectures applying database-like reliability to data lakes.
- **ADF (Azure Data Factory):** A cloud-based ETL and data integration service from Microsoft Azure, used for building data pipelines.
- **ADKAR®:** A goal-oriented change management model (Awareness, Desire, Knowledge, Ability, Reinforcement) developed by Prosci, mentioned as an alternative framework. (Note: This term itself doesn't appear used extensively, Kotter's model is detailed instead).
- **ADLS (Azure Data Lake Storage):** Microsoft Azure's scalable and cost-effective cloud storage solution designed for big data analytics, often used as the foundation for data lakes and lakehouses.
- **Adversarial Attacks:** Inputs intentionally designed by an attacker to cause an AI model to make a mistake or reveal information. Types include evasion attacks, poisoning attacks, extraction attacks, etc..
- **Adversarial Training:** A defense technique against evasion attacks where a model is explicitly trained on examples designed to fool it, making it more robust.
- **Agentic AI:** AI systems capable of more complex reasoning, planning, and executing multi-step tasks autonomously towards a goal.
- **AGI (Artificial General Intelligence):** Hypothetical future AI possessing cognitive abilities comparable to or exceeding human intelligence across a wide range of tasks. Mentioned in future scenario planning.
- **Agile Methodologies:** Project management approaches (like Scrum or Kanban) emphasizing iterative development, collaboration, flexibility, and rapid response to change. Often used in AI project management.
- **Aha!:** A commercial software tool mentioned as an example for creating and managing product or strategic roadmaps.
- **AI Act (EU):** Landmark European Union legislation establishing a comprehensive, risk-based regulatory framework for artificial intelligence, imposing stricter requirements on high-risk applications.
- **AI Auditor:** An emerging role focused on providing independent assurance on AI system governance, controls, ethics, and compliance.
- **AI Catalyst Program:** A term used in the OmnioTech case study for a proposed internal upskilling program to develop AI talent within business units.
- **AI Center of Excellence (CoE):** See AI Office.
- **AI Champion:** Individuals within business units or functions who actively promote AI adoption, literacy, and best practices, often acting as liaisons or early adopters.

- **AI Ethics:** A branch of ethics focusing on the moral principles, potential impacts, and responsible development and use of artificial intelligence technology, encompassing issues like bias, fairness, transparency, accountability, and societal impact. See also EAMM Ethics & Compliance.
- **AI Ethicist:** A specialized role responsible for guiding ethical AI development, conducting ethical risk assessments, developing policies, and advising on complex moral considerations.
- **AI Fluency:** A broad level of understanding and comfort with AI concepts, capabilities, limitations, and implications across the organization, often a goal of cultural transformation programs.
- **AI Governance:** The system of rules, practices, processes, roles, and controls used to direct, manage, and oversee the development, deployment, and operation of AI systems responsibly and effectively within an organization. See also Compass East, EAMM Governance.
- **AI Governance Council / Steering Committee:** The highest-level strategic body overseeing the enterprise AI governance strategy, risk appetite, and major policies, typically composed of senior executives.
- **AI Governance Lead:** A core role, often within the AI Office, responsible for the day-to-day development, implementation, and operationalization of the AI governance framework.
- **AI Impact Assessment (AIA):** A process for systematically evaluating the potential risks and impacts (ethical, social, legal, operational) of a proposed AI system, often required for high-risk applications under regulations like the EU AI Act.
- **AI Literacy:** The ability of individuals across different roles and levels within an organization to understand foundational AI concepts, recognize potential applications and risks, interact effectively with AI systems, and interpret their outputs appropriately.
- **AI Management System (AIMS):** A term associated with the ISO/IEC 42001 standard, referring to the integrated set of processes and controls for managing AI responsibly within an organization.
- **AI Maturity:** The level of capability, sophistication, effectiveness, and responsibility an organization demonstrates in developing, deploying, managing, and leveraging artificial intelligence across various dimensions (strategy, governance, data, tech, talent, value, etc.). Measured qualitatively by EAMM and quantitatively by EAMI.
- **AIO (AI Office):** The central coordinating function or organizational entity (also called AI Center of Excellence or CoE) chartered with orchestrating the enterprise AI strategy, governance, resource management, value delivery, and maturity management.
- **AI Office Compass™ Framework:** The proprietary management model introduced in this book, organizing AI Office responsibilities into five interconnected domains: Strategic Alignment (North), Governance (East), Performance & Value Delivery (West), Resource Management (South), and Maturity Management (Center).
- **AI Ops Specialist:** An emerging role focusing on the operational health, performance tuning, monitoring, and incident management specifically for deployed AI systems in production.
- **AI Product Manager:** A specialized product management role focused on defining the strategy, roadmap, requirements, and success metrics for AI-powered products or features, bridging business needs and technical implementation.
- **AI Review Board / Team:** A cross-functional operational governance body responsible for reviewing specific AI projects (especially medium/high-risk ones) against policies, providing guidance, and potentially approving deployments.
- **AI Sandbox:** A controlled, isolated technical environment for safe experimentation with new AI tools, algorithms, or data without impacting production systems or violating core policies.
- **AI Strategy:** The organization's high-level plan outlining how it intends to leverage AI capabilities to achieve specific business objectives, including defining vision, priorities, focus areas, and roadmap. See also Compass North, EAMM Strategic Alignment.

- **AI Translator:** A role bridging the communication gap between business stakeholders and technical AI teams, translating business needs into technical requirements and explaining AI concepts in business terms.
- **AIF360 (AI Fairness 360):** An open-source toolkit developed by IBM providing metrics to check for unwanted bias in datasets and machine learning models, and algorithms to mitigate such bias.
- **AKS (Azure Kubernetes Service):** Microsoft Azure's managed container orchestration service based on Kubernetes, often used for deploying scalable AI inference endpoints.
- **Algorithm:** A process or set of rules to be followed in calculations or other problem-solving operations, especially by a computer. In AI/ML, refers to the specific method used for learning patterns from data (e.g., linear regression, decision tree, neural network).
- **Amazon SageMaker:** See SageMaker (AWS).
- **Anomaly Detection:** Identifying data points, events, or observations that deviate significantly from a dataset's normal behavior. A common application of AI/ML (e.g., for fraud detection, system health monitoring, predictive maintenance).
- **Ansible:** An open-source software provisioning, configuration management, and application-deployment tool often used in Infrastructure as Code (IaC) practices.
- **AOV (Average Order Value):** A business metric representing the average dollar amount spent each time a customer places an order.
- **API (Application Programming Interface):** A set of definitions and protocols for building and integrating application software. In AI, often refers to the interface used to send input data to a deployed model and receive its prediction.
- **Appendix B:** The section of this book proposed to contain the detailed EAMM KPI table
- **Appendix C:** The section of this book proposed to contain the EAMM Framework Summary (overview table, level descriptions).
- **Application Programming Interface:** See API.
- **Archer:** A commercial enterprise Governance, Risk Management, and Compliance (GRC) platform mentioned as an example tool.
- **Argo Workflows:** An open-source container-native workflow engine for orchestrating parallel jobs on Kubernetes, often used in MLOps pipelines.
- **Arize AI:** A commercial AI/ML observability platform mentioned as an example tool for monitoring model performance, drift, data quality, and explainability in production.
- **ARM (Azure Resource Manager):** Microsoft Azure's deployment and management service, used for defining infrastructure as code via ARM templates.
- **ART (Adversarial Robustness Toolbox):** An open-source Python library developed by IBM for evaluating model robustness against adversarial attacks and implementing defense techniques.
- **Arthur:** A commercial AI/ML performance monitoring platform mentioned as an example tool.
- **Artificial General Intelligence:** See AGI.
- **arXiv:** A popular online repository for electronic preprints of scientific papers (often in physics, mathematics, computer science, quantitative biology, quantitative finance, statistics, electrical engineering and systems science, and economics), frequently used for disseminating AI research rapidly.
- **Asana:** A commercial web and mobile application designed to help teams organize, track, and manage their work, mentioned as an example project management tool.
- **AUC (Area Under the Curve):** A common performance metric for classification models, typically referring to the area under the ROC (Receiver Operating Characteristic) curve. It represents the model's ability to distinguish between positive and negative classes across all possible classification thresholds.
- **Audit Trail:** A chronological record of system activities, including user actions, data access, configuration changes, and key events, providing traceability for security investigations, compliance audits, and debugging.

- **Augmentation (Data):** Techniques used to artificially increase the size or diversity of a training dataset by creating modified copies of existing data or generating new synthetic data points, often used to address representation bias or improve model generalization.
- **Augmentation (Human):** Using AI systems to enhance human capabilities, skills, or productivity, rather than solely replacing human tasks.
- **Automation:** The use of technology to perform tasks previously done by humans, often a key driver for AI adoption (e.g., process automation, test automation, deployment automation).
- **Automation Bias:** The tendency for humans to over-trust and overly rely on automated systems, such as AI decision aids, potentially failing to notice errors or apply critical judgment.
- **Autonomous Systems:** Systems capable of operating and making decisions without direct human intervention, often incorporating AI/ML. Relevant in contexts like autonomous vehicles or robotics.
- **Availability (Security Principle):** Ensuring that systems, services, and data are accessible and usable when needed by authorized users.
- **Avro:** A row-based remote procedure call and data serialization framework developed within Apache's Hadoop project, often used for storing semi-structured data.
- **AWS (Amazon Web Services):** Amazon's comprehensive cloud computing platform, offering a wide range of services including compute, storage, databases, analytics, machine learning (SageMaker), and networking.
- **AWS Glue:** A fully managed extract, transform, and load (ETL) service on AWS, often used for data preparation and pipeline building.
- **AWS Glue Data Catalog:** A central repository to store structural and operational metadata for data assets in AWS, mentioned as a data cataloging tool.
- **AWS Kinesis:** An AWS platform for real-time data streaming and analytics.
- **AWS SageMaker:** See SageMaker (AWS).
- **AWS S3:** See S3 (AWS).
- **AWS Step Functions:** A serverless function orchestrator on AWS that makes it easy to sequence AWS Lambda functions and multiple AWS services into business-critical applications, mentioned as a pipeline orchestration tool.
- **Azure (Microsoft Azure):** Microsoft's comprehensive cloud computing platform.
- **Azure API Management:** Microsoft Azure's service for creating consistent and modern API gateways for existing back-end services, mentioned in the context of securing inference endpoints.
- **Azure Boards:** A service within Azure DevOps providing interactive and customizable tools for managing software projects (backlogs, task boards, etc.).
- **Azure Cognitive Services:** A portfolio of AI services and APIs available on Microsoft Azure enabling developers to add cognitive features (vision, speech, language, decision) into applications.
- **Azure Cost Management + Billing:** Microsoft Azure's suite of tools for monitoring, allocating, and optimizing cloud costs.
- **Azure Data Factory (ADF):** See ADF (Azure Data Factory).
- **Azure Data Lake Storage (ADLS):** See ADLS (Azure Data Lake Storage).
- **Azure DevOps:** Microsoft's suite of services covering the entire development lifecycle, including version control (Azure Repos), CI/CD (Azure Pipelines), artifact management, and project tracking (Azure Boards).
- **Azure Event Hubs:** A big data streaming platform and event ingestion service on Microsoft Azure, mentioned as a source for IoT data.
- **Azure Kubernetes Service (AKS):** See AKS (Azure Kubernetes Service).
- **Azure Machine Learning (Azure ML):** Microsoft Azure's enterprise-grade cloud platform for the end-to-end machine learning lifecycle, providing tools for development, training, deployment (MLOps), and management.

- **Azure ML Experiments:** A feature within Azure Machine Learning for tracking, organizing, and comparing machine learning training runs.
- **Azure ML Managed Online Endpoints:** A feature in Azure Machine Learning providing a turnkey solution for deploying ML models as scalable, secure web services (APIs) on Azure-managed infrastructure.
- **Azure ML Model Registry:** A component of Azure Machine Learning serving as a central repository for versioning, managing metadata, and tracking the lifecycle of trained ML models.
- **Azure Monitor:** Microsoft Azure's service for collecting, analyzing, and acting on telemetry data from cloud and on-premises environments, used for monitoring AI system health and performance.
- **Azure Pipelines:** The Continuous Integration/Continuous Delivery (CI/CD) service within Azure DevOps, used for automating build, test, and deployment workflows, including MLOps pipelines.
- **Azure Purview:** Microsoft Azure's unified data governance service that helps manage and govern on-premises, multicloud, and SaaS data; used as an enterprise data catalog.
- **Azure Repos:** The Git repository hosting service within Azure DevOps, used for source code version control.
- **Azure RBAC (Role-Based Access Control):** Azure's system for managing access permissions to cloud resources based on assigned roles.
- **Azure SQL Database:** Microsoft Azure's fully managed relational database service, mentioned as a potential data source or target.
- **Azure Stream Analytics:** Microsoft Azure's real-time analytics and complex event-processing service.
- **B2B (Business-to-Business):** Transactions or interactions between businesses, rather than between a business and individual consumers.
- **B2C (Business-to-Consumer):** Transactions or interactions directly between a business and individual consumers.
- **Backdoor (AI Security):** A hidden mechanism intentionally introduced into an AI model (often via data poisoning) that allows an attacker to trigger specific malicious behavior or gain unauthorized control under certain conditions.
- **Baseline:** A clearly defined measurement of performance or status taken before an intervention (like implementing an AI system) occurs, used as a reference point to objectively measure the impact or improvement resulting from the intervention.
- **Batch Processing:** Processing data in large, discrete groups or chunks at scheduled intervals (e.g., nightly, hourly) rather than continuously as data arrives.
- **BCG (Boston Consulting Group):** A global management consulting firm, cited as a source for industry reports and statistics related to AI adoption and management.
- **Ben's Bites:** An AI industry newsletter mentioned as a source for tracking AI trends and startup activity.
- **BentoML:** An open-source framework for building reliable, scalable, and cost-efficient AI applications, particularly focused on model serving.
- **BERT (Bidirectional Encoder Representations from Transformers):** A powerful language representation model developed by Google, widely used in Natural Language Processing (NLP) for tasks like feature extraction from text.
- **Bias Mitigation:** Techniques applied during data pre-processing, model training (in-processing), or post-processing of model outputs to reduce or correct for identified unfair biases.
- **Bicep:** A domain-specific language (DSL) developed by Microsoft that uses declarative syntax to deploy Azure resources; an alternative to ARM templates for Infrastructure as Code.
- **Big Data:** Extremely large datasets that may be analyzed computationally to reveal patterns, trends, and associations, often characterized by high volume, velocity, and variety.
- **BigQuery (Google Cloud):** Google Cloud's fully managed, serverless data warehouse optimized for large-

scale data analytics using SQL.

- **Bitbucket:** A Git-based source code repository hosting service owned by Atlassian.
- **Black Box Model:** An AI model, typically a complex one like a deep neural network, whose internal workings and decision-making logic are opaque and difficult for humans to understand or interpret directly.
- **Blueprint:** A detailed plan or model used as a guide for making something. Used metaphorically in the book to describe the Compass framework and the overall guidance provided.
- **BPMN (Business Process Model and Notation):** A standard graphical notation for specifying business processes in a workflow diagram.
- **Brightidea:** A commercial platform mentioned as an example tool for managing employee ideation and innovation programs.
- **BU:** See Business Unit.
- **Budgeting:** The process of creating a plan to spend financial resources. Critical for AI initiatives given potentially high and variable costs (Chapter 19). See also EAMM Financial.
- **Build vs. Buy:** A common strategic decision regarding whether to develop software or capabilities internally (Build) or purchase commercial solutions/services (Buy).
- **Business Intelligence:** See BI.
- **Business Pull:** An approach to AI strategy where initiatives are driven primarily by identified business needs, strategic objectives, or specific challenges, rather than by the availability of technology (Contrasts with Technology Push).
- **Business Sponsor:** The executive or senior manager from a business unit who champions, funds (often), and takes ultimate ownership of the business outcomes for a specific AI initiative.
- **Business Unit (BU):** A distinct operating division or segment within a larger company, often focused on a specific product line, geography, or function (e.g., Marketing BU, Product BU).
- **C3 AI:** A commercial enterprise AI platform vendor mentioned as an example.
- **CapEx (Capital Expenditure):** Funds used by a company to acquire, upgrade, and maintain physical assets such as property, plants, buildings, technology, or equipment. Contrasted with OpEx.
- **Captum:** An open-source, extensible library for model interpretability built on PyTorch, mentioned as an example XAI tool.
- **Case Study (OmnioTech):** The fictional company used throughout the book to illustrate the practical application of the frameworks and concepts discussed.
- **Causal Inference:** A field focused on understanding cause-and-effect relationships from data, going beyond mere correlation. Mentioned as relevant to AI for Science/Complex Systems.
- **CCPA (California Consumer Privacy Act):** A state statute intended to enhance privacy rights and consumer protection for residents of California, United States. Often mentioned alongside GDPR. (See also CPRA)
- **CDC (Change Data Capture):** Techniques used to efficiently identify and capture only the changes (inserts, updates, deletes) made to data in source databases, enabling low-latency replication.
- **CDO (Chief Data Officer):** A senior executive role responsible for overseeing an organization's enterprise-wide data strategy, governance, management, quality, and utilization as an asset. Plays a key collaborative role with the AI Office.
- **Celonis:** A commercial Process Mining and Execution Management software company mentioned as an example tool.
- **Center (Compass Domain):** The central domain of the AI Office Compass™ Framework, focusing on Maturity Management, which includes orchestrating assessments (EAMI), driving uplift strategies, fostering culture and change management, and facilitating innovation.
- **Centralized Model (AI Office Operating Model):** An operating model where most AI expertise, resources, and authority reside within a single, central AI Office/CoE team. Often suited for early maturity stages

(EAMI Level 1-2).

- **CES (Customer Effort Score):** A customer experience metric measuring the ease of a customer's interaction with a company.
- **Change Data Capture:** See CDC.
- **Change Failure Rate:** A DevOps/DORA metric measuring the percentage of deployments or releases that result in a failure in production requiring remediation. Relevant to MLOps stability.
- **Change Management:** The systematic approach to dealing with the transition or transformation of an organization's goals, processes, or technologies, focusing on the people side of change. A key EAMM dimension, managed within the Compass Center domain.
- **Chargeback Model (AI Office Funding):** A funding model where the AI Office charges business units directly for the AI resources or services they consume.
- **Chatbot:** A computer program designed to simulate conversation with human users, especially over the Internet. Often powered by NLP and/or LLMs.
- **Checkov:** An open-source static code analysis tool for Infrastructure as Code (IaC), used to find security and compliance misconfigurations. Mentioned for use with Terraform/Bicep.
- **Chief AI Officer (CAIO):** A potential senior executive title for the leader of a mature enterprise AI function, reflecting the strategic importance of AI. Mentioned in OmnioTech future context.
- **Chief Data Officer:** See CDO.
- **Chief Information Officer:** See CIO.
- **Chief Operating Officer:** See COO.
- **Chief Risk Officer (CRO):** A senior executive responsible for overseeing the enterprise's overall risk management framework. May be involved in AI Governance Council.
- **Chief Strategy Officer (CSO):** A senior executive responsible for strategy formulation and execution. A potential reporting line for the AI Office.
- **Chief Technology Officer:** See CTO.
- **Churn Rate:** A business metric measuring the percentage of customers who stop using a company's product or service during a certain time period. AI is often used to predict and reduce churn.
- **CI (Continuous Integration):** A DevOps/MLOps practice where developers frequently merge their code changes into a central repository, after which automated builds and tests are run.
- **CI/CD (Continuous Integration / Continuous Delivery or Deployment):** A core set of DevOps/MLOps practices combining Continuous Integration with either Continuous Delivery (automating release to a staging environment) or Continuous Deployment (automating deployment all the way to production). Enabled by automated pipelines.
- **CI/CD/CT Pipelines:** In MLOps, extends CI/CD to include Continuous Training (CT) – automating the model retraining process based on triggers like performance degradation or new data.
- **CIO (Chief Information Officer):** A senior executive responsible for managing and implementing information and computer technologies within an organization. A potential reporting line for the AI Office.
- **CIS Benchmarks:** Consensus-based security configuration guidelines developed by the Center for Internet Security, mentioned in the context of secure infrastructure configuration.
- **CISO (Chief Information Security Officer):** A senior executive responsible for establishing and maintaining the enterprise vision, strategy, and program to ensure information assets and technologies are adequately protected. Key collaborator for AI Security (Domain East).
- **Classification (ML Task):** A type of supervised learning task where the goal is to assign input data points to predefined categories or classes (e.g., spam vs. not spam, cat vs. dog).
- **Claude 3:** A family of Large Language Models developed by Anthropic, mentioned as an example LLM.
- **Cloudability (Apptio):** A commercial FinOps platform mentioned as an example tool for cloud cost man-

agement and visibility.

- **CloudFormation (AWS):** AWS's service for defining and provisioning infrastructure as code using templates.
- **Cloud Computing:** The delivery of computing services—including servers, storage, databases, networking, software, analytics, and intelligence—over the Internet ("the cloud") to offer faster innovation, flexible resources, and economies of scale. Major providers include AWS, Azure, GCP.
- **CloudWatch (AWS):** Amazon's monitoring and observability service for AWS cloud resources and applications.
- **CLV (Customer Lifetime Value):** A business metric predicting the net profit attributed to the entire future relationship with a customer.
- **CNN (Convolutional Neural Network):** A class of deep learning models, most commonly applied to analyzing visual imagery (computer vision tasks).
- **CoE (Center of Excellence):** See AI Office.
- **Code Documentation AI:** An example AI use case mentioned in the OmnioTech case study, involving using Generative AI for automatically documenting source code.
- **Collaboration (EAMM Dimension):** An aspect of AI maturity focusing on the effectiveness of teamwork and communication between different functions and roles involved in AI initiatives. Supported by the Compass Center.
- **Collibra:** A commercial enterprise data governance and data catalog platform mentioned as an example tool.
- **Comet ML:** A commercial platform mentioned as an example tool for MLOps experiment tracking and management.
- **Commercial Off-The-Shelf:** See COTS.
- **Community of Practice:** See CoP.
- **Compass Framework:** See AI Office Compass™ Framework.
- **Competency (EAMM Dimension - Team):** Assesses the skills, expertise, and effectiveness of the teams involved in developing and managing AI systems. Relates to EAMM Talent Acquisition and Skill Development. Managed within Compass South.
- **Compliance (EAMM Dimension):** Focuses on the organization's adherence to relevant laws, regulations, standards, and internal policies related to AI development and data usage. A key part of Compass East.
- **Compliance-as-Code:** An approach where compliance requirements and controls are defined and automatically enforced as code, typically integrated within CI/CD or IaC pipelines.
- **Compliance Officer:** A role responsible for ensuring an organization complies with its outside regulatory and legal requirements as well as internal policies and bylaws. Key collaborator for AI governance.
- **Compute Resources:** The processing power (CPU, GPU, TPU, memory) required to perform computational tasks, particularly relevant for training large AI models and running inference. Managed under Compass South.
- **Concept Drift:** A phenomenon in machine learning where the statistical properties of the relationship between input variables and the target variable change over time, causing deployed models trained on historical data to become less accurate. Requires continuous monitoring (CM).
- **Conceptual:** Relating to or based on mental concepts; theoretical. Used to describe high-level framework diagrams or initial planning stages.
- **Confidentiality (Security Principle):** Preventing the unauthorized disclosure of information, including sensitive data and proprietary AI models.
- **Confirmation Bias:** The tendency to search for, interpret, favor, and recall information in a way that confirms or supports one's prior beliefs or hypotheses. Can impact human interaction with AI.
- **Conformity (Data Quality Dimension):** See Validity.

- **Consistency (Data Quality Dimension):** The degree to which data is free from contradictions and represented uniformly across different instances or systems.
- **Consistency (Individual Fairness Metric):** A type of individual fairness metric measuring whether a model produces similar outputs for similar inputs, regardless of group membership.
- **Consolidate Gains:** A step in Kotter's change management model focused on building on early successes to drive further change.
- **Consultant:** An external expert providing professional advice or services, often used in the "Borrow" talent strategy.
- **Containerization:** Packaging software code and all its dependencies (libraries, frameworks, configuration files) together so it can run reliably and consistently across different computing environments. Docker is a popular technology. Essential for MLOps.
- **Container Image Scanning:** A security practice involving analyzing container images for known vulnerabilities in the operating system, application code, or dependencies before deployment.
- **Context Applicability:** A column in the detailed KPI table indicating industries or organizational contexts where a specific KPI might be particularly relevant.
- **Continuous Delivery:** See CI/CD.
- **Continuous Deployment:** See CI/CD.
- **Continuous Improvement:** An ongoing effort to improve products, services, or processes. A core principle of MLOps and the Maturity Management (Center) domain. Addressed by EAMM Continuous Improvement dimension.
- **Continuous Integration:** See CI.
- **Continuous Monitoring (CM):** An MLOps principle involving the ongoing, automated monitoring of deployed ML models in production for operational health, predictive performance, drift, fairness, and security.
- **Continuous Training (CT):** An MLOps principle involving the automation of model retraining pipelines, often triggered by monitoring alerts indicating performance degradation or significant data drift.
- **Control Group:** In A/B testing or experimentation, the group that does not receive the treatment or intervention (e.g., the AI feature) being tested, used as a baseline for comparison.
- **Controls (Governance):** Mechanisms (technical or procedural) implemented to enforce policies and standards, mitigate risks, or ensure compliance within the AI governance framework.
- **Conversion Rate:** A business metric, often used in marketing or e-commerce, measuring the percentage of users who take a desired action (e.g., make a purchase, sign up for a newsletter) after viewing an offer or webpage.
- **Convolutional Neural Network:** See CNN.
- **COO (Chief Operating Officer):** A senior executive responsible for overseeing the ongoing business operations within a company. A potential reporting line or key sponsor for the AI Office.
- **CoP (Community of Practice):** A group of people who share a common concern, a set of problems, or a passion about a topic, and who deepen their knowledge and expertise in this area by interacting on an ongoing basis. Crucial for knowledge sharing in AI teams (Compass Center).
- **Copyright Law:** Legal rights granted to the creator of original works of authorship, including literary, dramatic, musical, and certain other intellectual works. Relevant to Generative AI outputs and training data.
- **Correlation:** A statistical measure indicating the extent to which two or more variables fluctuate together. Importantly, correlation does not imply causation.
- **Cost Management:** The process of planning, estimating, budgeting, monitoring, controlling, and optimizing costs. Critical for AI due to potentially high/variable expenses (Chapter 19, Compass South, EAMM Financial). See also FinOps.
- **Cost of Investment:** The total expenditure required for an initiative, encompassing the full Total Cost of

Ownership (TCO) for accurate ROI calculation.

- **COTS (Commercial Off-The-Shelf):** Software or hardware products that are ready-made and available for sale, lease, or license to the general public. Contrasted with custom-built solutions.
- **Counterfactual Explanations:** A type of XAI technique explaining a model's prediction by showing the smallest change to input features that would alter the outcome.
- **Counterfactual Fairness:** A notion of fairness ensuring that a model's prediction for an individual would remain the same if their protected attribute (e.g., race, gender) were changed, assuming all else remained equal.
- **Coursera:** An online course provider mentioned as a potential platform for AI upskilling programs.
- **CPRA (California Privacy Rights Act):** An expansion and amendment of the CCPA, further strengthening consumer privacy rights in California.
- **CPU (Central Processing Unit):** The primary component of a computer that executes instructions. Standard compute resource, often supplemented by GPUs/TPUs for AI workloads.
- **Credibility:** The quality of being trusted and believed in. Building AI Office credibility through demonstrated value and transparent communication is crucial.
- **Credit Scoring:** The process of using statistical analysis on a person's credit files to represent their creditworthiness. A common AI application often subject to fairness regulations.
- **Critical Success Factors:** Key elements or conditions identified as essential for achieving success in a particular endeavor, such as enterprise AI adoption.
- **CRM (Customer Relationship Management):** Systems and strategies used by companies to manage and analyze customer interactions and data throughout the customer lifecycle, aiming to improve customer service relationships and assist in customer retention and drive sales growth. Often a key data source for AI.
- **CRO:** See Chief Risk Officer.
- **Cross-Border Data Transfer:** Moving personal data from one jurisdiction (country or region) to another, often subject to specific legal regulations like GDPR.
- **Cross-Functional:** Involving people or teams from different functional areas of an organization (e.g., marketing, IT, finance, operations). AI initiatives are typically highly cross-functional.
- **Cross-Sell:** Selling an additional product or service to an existing customer. AI personalization often aims to increase cross-selling.
- **Crunchbase:** A platform providing business information about private and public companies, particularly startups and investments, mentioned for horizon scanning.
- **CSAT (Customer Satisfaction Score):** A common metric measuring customer satisfaction with a product, service, or specific interaction.
- **CSO:** See Chief Strategy Officer.
- **CT (Continuous Training):** See Continuous Training.
- **CTO (Chief Technology Officer):** A senior executive focused on the technological needs, research, and development within an organization. A common reporting line or key sponsor for the AI Office.
- **CTR (Click-Through Rate):** A digital marketing metric measuring the ratio of users who click on a specific link (e.g., in an email or ad) to the total number of users who view it.
- **Culture (EAMM Dimension):** Assesses the organizational readiness, leadership support, data-driven mindset, collaboration, and change management capabilities related to AI adoption. Managed within Compass Center.
- **Curation (Data):** The process of selecting, organizing, cleaning, transforming, enriching, and managing data assets to make them more valuable and usable for specific purposes, such as AI model training.
- **Customer Effort Score:** See CES.
- **Customer Experience (CX):** The overall perception a customer has of a company or its products/services

resulting from all interactions across the customer journey. AI is often used to improve CX.

- **Customer Lifetime Value:** See CLV.
- **Customer Relationship Management:** See CRM.
- **Customer Satisfaction Score:** See CSAT.
- **Cybersecurity:** The practice of protecting systems, networks, and programs from digital attacks, usually aimed at accessing, changing, or destroying sensitive information; extorting money; or interrupting normal business processes. Key aspect of AI Governance (Domain East).
- **DAMA DMBOK (Data Management Body of Knowledge):** A comprehensive reference framework outlining standard terminology and best practices for the field of data management.
- **Dashboard:** A type of graphical user interface which often provides at-a-glance views of key performance indicators (KPIs) relevant to a particular objective or business process. Used extensively in performance management.
- **Data Accessibility:** Ensuring that authorized users and systems can efficiently, securely, and appropriately access the data they need, when they need it. Key pillar of Data Readiness. See also EAMM Data Management.
- **Data Accuracy:** See Accuracy (Data Quality Dimension).
- **Data Annotation / Labeling:** The process of adding informative tags or labels to raw data (e.g., images, text excerpts) to create ground truth for training supervised machine learning models.
- **Data Architecture:** The structure and design of an organization's data storage, processing, integration, and management systems. See also EAMM Data Management. Managed within Compass South.
- **Data Asset:** A collection of data that is managed as a resource and provides potential value to the organization.
- **Data Breach:** A security incident in which sensitive, protected, or confidential data is copied, transmitted, viewed, stolen, or used by an individual unauthorized to do so.
- **Data Catalog:** A centralized, curated inventory of data assets within an organization, containing metadata that helps users discover, understand, trust, and access relevant data.
- **Data Cleansing:** The process of detecting, correcting, or removing corrupt, inaccurate, inconsistent, or irrelevant records from a dataset.
- **Data Compliance:** Adhering to applicable laws, regulations, standards, and internal policies regarding the collection, storage, processing, and use of data. See also EAMM Compliance & Ethics.
- **Data Completeness:** See Completeness (Data Quality Dimension).
- **Data Consistency:** See Consistency (Data Quality Dimension).
- **Data Curation:** See Curation (Data).
- **Data Discovery:** The process of finding and identifying relevant data assets across various organizational sources, often facilitated by a data catalog.
- **Data Drift:** Changes in the statistical properties of the input data fed to a deployed ML model over time, compared to the data the model was trained on. Can significantly degrade model performance if undetected. Requires monitoring via MLOps.
- **Data Egress:** The transfer of data out of a network boundary or cloud environment, often incurring network costs.
- **Data Enablement:** The process of making high-quality, trustworthy data readily accessible and usable for analytics, AI, and other business purposes. Encompasses aspects of quality, accessibility, governance, architecture (Chapter 18).
- **Data Engineer:** A technical role focused on designing, building, managing, and optimizing scalable data pipelines and infrastructure to support data analytics and AI initiatives.
- **Data Exploration / Exploratory Data Analysis (EDA):** The process of analyzing datasets, often using

visualization and statistical techniques, to summarize their main characteristics, discover patterns, identify anomalies, and test hypotheses, typically done early in the AI development lifecycle.

- **Data Governance:** The overall management of the availability, usability, integrity, security, and privacy of the data employed in an organization or enterprise. Includes establishing policies, standards, roles (ownership, stewardship), and controls. See also EAMM Data Management, EAMM Governance.
- **Data Handling:** Procedures and standards governing how data is accessed, stored, processed, transmitted, and secured throughout its lifecycle.
- **Data Ingestion:** The process of collecting and importing data from various sources into a data storage system like a data lake or warehouse.
- **Data Integration:** Combining data residing in different sources and providing users with a unified view of them. Handled via ETL/ELT pipelines.
- **Data Lake:** A centralized repository designed to store large amounts of structured, semi-structured, and unstructured data in its native format, typically using low-cost object storage. Offers flexibility for data exploration and diverse processing.
- **Data Lakehouse:** See Lakehouse Architecture.
- **Data Latency:** The time delay between when data is generated or changed at the source and when it becomes available for use in a target system or application.
- **Data Leakage:** Accidentally introducing information into a model's training process that would not realistically be available at the time of prediction, leading to overly optimistic performance estimates that don't generalize. Feature Stores help mitigate this.
- **Data Lineage:** Metadata describing the origin, movement, transformations, and dependencies of data assets throughout their lifecycle. Crucial for debugging, impact analysis, trust, and compliance.
- **Data Literacy:** The ability to read, understand, create, and communicate data as information. Essential for fostering a data-driven culture.
- **Data Management (EAMM Dimension):** Assesses the maturity of organizational capabilities for managing data as a strategic asset, including architecture, platforms, accessibility, governance, and lifecycle management. Managed within Compass South.
- **Data Mart:** A subset of a data warehouse focused on a specific business line or functional area (e.g., a Marketing data mart).
- **Data Mesh:** An emerging decentralized socio-technical approach to data architecture emphasizing domain ownership of data products served via a self-serve platform with federated governance.
- **Data Migration:** The process of moving data from one storage system, format, or application to another.
- **Data Minimization:** A privacy principle stipulating that only personal data that is necessary for a specific, defined purpose should be collected, processed, or retained. Key concept in GDPR.
- **Data Observability:** Applying observability principles (monitoring metrics, logs, traces) to data pipelines and assets to proactively understand data health, detect quality issues, and ensure reliability.
- **Data Owner:** A role, often assigned to a senior business leader, holding ultimate responsibility and accountability for a specific data domain within the organization.
- **Data Platform:** The underlying technology infrastructure and integrated tooling used to store, process, manage, govern, and serve data for analytics and AI purposes (e.g., data lake, warehouse, lakehouse).
- **Data Poisoning:** A type of security attack where malicious data is intentionally injected into a model's training set to compromise its integrity or performance.
- **Data Pre-processing:** Steps taken to clean, transform, and prepare raw data before it is used for model training or analysis. Includes handling missing values, scaling features, encoding variables, etc.
- **Data Privacy:** Protecting personal information from unauthorized access, use, or disclosure, adhering to legal regulations and ethical principles. Key aspect of Governance (East).

- **Data Processing:** Performing operations on data, such as cleaning, transforming, aggregating, or analyzing it, typically within data pipelines. See ETL, ELT.
- **Data Profiling:** See Profiling (Data).
- **Data Protection Impact Assessment:** See DPIA.
- **Data Provenance:** Metadata documenting the origin and history of a dataset or data element.
- **Data Quality (DQ) (EAMM Dimension):** Assesses the maturity of processes ensuring data accuracy, completeness, consistency, timeliness, validity, representativeness for AI. Critical foundation managed within Compass South.
- **Data Readiness for AI:** The state where data possesses the necessary quality, accessibility, governance, architecture, and curation to reliably fuel enterprise AI initiatives.
- **Data Remediation:** The process of correcting identified data quality issues.
- **Data Replication:** Copying data from one database or storage system to another, often done to support analytics without impacting operational systems. CDC facilitates efficient replication.
- **Data Representation:** How data is structured or encoded, particularly how real-world concepts are translated into features for ML models. Poor representation can lead to bias.
- **Data Residency:** Regulatory or policy requirements stipulating that certain types of data must be stored or processed within specific geographic borders.
- **Data Retention Policy:** Rules defining how long specific types of data should be kept before being securely deleted or archived, often driven by legal or compliance requirements.
- **Data Science:** An interdisciplinary field using scientific methods, processes, algorithms, and systems to extract knowledge and insights from structured and unstructured data.
- **Data Scientist:** A professional skilled in applying analytical, statistical, and machine learning techniques to complex datasets to solve business problems, generate insights, and build predictive models.
- **Data Security:** Protecting data from unauthorized access, corruption, or theft throughout its lifecycle. Key aspect of Governance (East).
- **Data Sheet for Datasets:** A proposed documentation standard (akin to datasheets for electronic components) detailing the motivation, composition, collection process, preprocessing, recommended uses, and potential biases or limitations of a dataset used for training ML models. Promotes transparency.
- **Data Silo:** A repository of data isolated from the rest of the organization, often residing within a specific department or legacy system, hindering unified analysis and AI development.
- **Data Sovereignty:** The concept that data is subject to the laws and governance structures within the nation it is collected or processed. Closely related to data residency.
- **Data Steward:** A role, often within a business domain, responsible for managing the quality, definition, usage, and governance of specific data assets according to established policies and standards, working collaboratively with data owners and governance teams.
- **Data Storage:** Technologies and methods used to retain digital information (e.g., databases, data warehouses, data lakes, object storage).
- **Data Strategy:** An organization's comprehensive plan outlining how it will manage, govern, analyze, and leverage data as a strategic asset to achieve business objectives, including enabling AI.
- **Data Transformation:** The process of converting data from one format or structure into another, often involving cleaning, aggregating, joining, or deriving new variables. Key part of ETL/ELT.
- **Data Validation:** The process of checking data against predefined rules or constraints to ensure its accuracy, completeness, consistency, and validity.
- **Data Version Control (DVC):** An open-source version control system designed specifically for managing large datasets and machine learning models alongside code, often used in conjunction with Git.
- **Data Warehouse:** See Warehouse (Data).

- **Databrands:** A commercial company providing a unified data analytics platform built on Apache Spark, often associated with the Lakehouse architecture and offering integrated data engineering, data science, ML, and BI capabilities.
- **Databand:** A commercial Data Observability platform mentioned as an example tool.
- **Database:** An organized collection of structured information, or data, typically stored electronically in a computer system, managed by a Database Management System (DBMS).
- **Datadog:** A commercial monitoring and observability platform for cloud applications, mentioned as an example tool.
- **Datasets for Datasets:** See Data Sheet for Datasets.
- **Dbt (data build tool):** See dbt.
- **Decision Tree:** A type of supervised machine learning algorithm that uses a tree-like structure of decisions and their possible consequences. Often relatively interpretable.
- **Decision Transparency:** See Explainability (XAI).
- **Decentralized Model (AI Office Operating Model):** An operating model where AI talent and responsibility are fully embedded within individual business units with minimal central coordination. Often associated with low EAMI maturity.
- **Deduplication:** The process of identifying and removing duplicate copies of data.
- **Deep Learning:** A subfield of machine learning based on artificial neural networks with multiple layers (deep architectures), capable of learning complex patterns from large amounts of data. Powers many recent AI breakthroughs (image recognition, NLP, Generative AI).
- **Default Path:** The easiest or standard way of doing something. MLOps aims to make the compliant, responsible path the default path.
- **Defensive Distillation:** An ML technique proposed as a defense against certain adversarial attacks, involving training a model on the probability outputs of another model trained on the same task.
- **Defined (EAMI Level 3):** The third level of AI maturity in the EAMM framework, characterized by formalized strategies, governance, platforms, processes, and initial scaling successes.
- **Deloitte:** A global professional services network, cited as a source for industry reports and statistics related to AI adoption and management.
- **Delta Lake:** An open-source storage layer that brings ACID transactions, scalable metadata handling, and unified streaming/batch data processing to existing data lakes (like S3, ADLS, GCS). A key technology enabling the Lakehouse architecture.
- **Demographic Parity:** A group fairness metric requiring that the model's prediction outcomes (e.g., selection rate for loans or jobs) are equal across different demographic groups, regardless of true outcomes.
- **Demographics:** Statistical data relating to the population and particular groups within it (e.g., age, gender, race, location). Used in fairness assessments.
- **Denial-of-Service (DoS) Attack:** A cyber-attack in which the perpetrator seeks to make a machine or network resource unavailable to its intended users by temporarily or indefinitely disrupting services, e.g., by overwhelming an API endpoint.
- **Dependabot:** A tool (now part of GitHub) that automatically checks project dependencies for known security vulnerabilities and can create pull requests to update them. Used for Software Composition Analysis (SCA).
- **Dependency:** A relationship where one component, project, task, or resource relies on another being available or completed first. Managing dependencies is critical in roadmap planning and pipeline design.
- **Deployment (AI Model):** The process of taking a trained machine learning model and making it available for use in a production environment, typically as an API endpoint or integrated into an application or batch process.

- **Deployment Frequency:** A DevOps/DORA metric measuring how often an organization successfully releases code or models to production. Mature MLOps aims to increase this.
- **Derived Variable:** A feature created by transforming or combining one or more existing variables (e.g., calculating a ratio, difference, or interaction term). Common in feature engineering.
- **DevOps (Development Operations):** A set of practices combining software development (Dev) and IT operations (Ops) aiming to shorten the systems development life cycle and provide continuous delivery with high software quality, leveraging automation and collaboration. MLOps extends these principles to ML.
- **DevSecOps:** An extension of DevOps integrating security practices and automated security testing earlier into the software development lifecycle ("shifting security left"). MLSecOps applies this concept to MLOps.
- **Differential Privacy:** A system for publicly sharing information about a dataset by describing the patterns of groups within the dataset while withholding information about individuals. Adds calibrated noise to obscure individual records, enhancing privacy but potentially impacting utility.
- **Digital Transformation:** The process of using digital technologies to create new — or modify existing — business processes, culture, and customer experiences to meet changing business and market requirements. AI is often a key component.
- **Dimension (EAMM):** One of the 14 key capability areas defined in the Enterprise AI Maturity Matrix used to assess organizational AI maturity (e.g., Strategic Alignment, Governance, Data Quality).
- **Dimension Score (EAMI):** The quantitative score (typically 0-100%) assigned to a specific EAMM dimension based on assessing performance against its associated KPIs during an EAMI assessment.
- **Discretization:** The process of converting continuous variables into discrete categories or bins. A feature engineering technique.
- **Disruptive Technology/Innovation:** An innovation that creates a new market and value network or enters at the bottom of an existing market and eventually displaces established market-leading firms, products, and alliances. AI holds significant disruptive potential.
- **Distributed Computing:** Systems where components located on networked computers communicate and coordinate their actions by passing messages to achieve a common goal. Frameworks like Spark enable distributed data processing for AI.
- **Diversity & Inclusion (D&I):** Organizational efforts to ensure representation and participation of people from different backgrounds (demographics, experiences, perspectives) and creating an environment where everyone feels valued and respected. Important for building innovative and ethical AI teams.
- **Docker:** An open platform for developing, shipping, and running applications using containerization. Widely used in MLOps for creating consistent environments.
- **Documentation:** Written text or illustrations accompanying software or hardware, explaining its function, usage, design, or management. Critical for AI governance, transparency, collaboration, maintainability. Key artifacts include Model Cards, Datasheets, Risk Assessments.
- **Domain Expertise:** Deep knowledge and understanding of a specific field, industry, business function, or subject matter. Crucial for developing relevant and effective AI solutions.
- **Domain Ownership (Data Mesh):** A core principle of the Data Mesh architecture where responsibility for specific analytical datasets ("data products") resides with the business domain teams that create or best understand that data.
- **DORA Metrics:** A set of four key metrics identified by the DevOps Research and Assessment (DORA) team as indicators of high-performing software delivery teams: Deployment Frequency, Lead Time for Changes, Mean Time to Recovery (MTTR), and Change Failure Rate. Often used as benchmarks for MLOps performance.
- **DPIA (Data Protection Impact Assessment):** A process required under GDPR (and similar regulations) to help identify and minimize the data protection risks of a project or system involving the processing of

personal data, especially high-risk processing. Often necessary for AI systems using personal data.

- **Drift (Data/Concept/Model):** Changes occurring over time after an ML model is deployed. Data Drift refers to changes in the statistical properties of the input data. Concept Drift refers to changes in the underlying relationship between inputs and outputs. Both can lead to Model Drift (degradation in predictive performance). Requires continuous monitoring via MLOps.

- **Duplication:** The act of creating identical copies or performing the same work redundantly. A common pitfall of uncoordinated AI efforts leading to wasted resources.

- **Dynatrace:** A commercial application performance management (APM) and observability platform mentioned as an example tool.

- **Dynamic Pricing:** Setting flexible prices for products or services based on current market demands, competitor pricing, customer behavior, or other real-time factors, often optimized using AI/ML algorithms.

- **EAMM (Enterprise AI Maturity Matrix):** A comprehensive framework defining key capability dimensions and five distinct maturity levels (Initial, Emerging, Defined, Optimized, Transformative) to assess an organization's AI maturity qualitatively. Developed and used throughout this book.

- **EAMI (Enterprise AI Maturity Index):** A quantitative score (typically 0-100% or 1.0-5.0) calculated based on measurable Key Performance Indicators (KPIs) tied to EAMM dimensions, providing an objective measure of overall AI maturity and enabling progress tracking.

- **East (Compass Domain):** The domain of the AI Office Compass™ Framework focused on Governance, encompassing policies, ethics, risk management, security, and compliance for responsible AI.

- **EDA (Exploratory Data Analysis):** The process of analyzing datasets to summarize main characteristics, often using visual methods, to understand the data better before formal modeling.

- **Edge AI / Edge Computing:** Deploying and running AI models directly on endpoint devices (IoT sensors, cameras, phones, vehicles) closer to the data source, enabling low-latency inference, offline operation, and enhanced privacy.

- **Efficiency (Operational):** A dimension of AI value focused on improving the speed, reducing the cost, or optimizing the resource consumption of business processes. See EAMM Operational Efficiency KPIs in Appendix B.

- **EIA (Ethical Impact Assessment):** A process to systematically evaluate the potential ethical risks and societal impacts of a proposed AI system, often conducted for higher-risk applications.

- **EKS (Elastic Kubernetes Service):** Amazon Web Services' managed Kubernetes service.

- **ELK Stack:** An acronym for three open-source projects: Elasticsearch, Logstash, and Kibana. Often used together for log aggregation and analysis.

- **ELT (Extract, Load, Transform):** A modern data integration paradigm where raw data is first loaded into the target system (like a data lake or lakehouse) and then transformed using the target system's compute power. Generally preferred for AI workloads over traditional ETL.

- **Embodied AI:** AI systems integrated with physical forms, such as robots, enabling interaction with and manipulation of the physical world.

- **Emerging (EAMI Level 2):** The second level of AI maturity in the EAMM framework, characterized by increasing awareness, basic strategy/governance drafts, inconsistent processes, and pilot projects struggling to scale reliably.

- **Employee Experience (EX):** The overall perception employees have of their journey and interactions within an organization. AI can potentially improve EX by automating tedious tasks or enhancing productivity tools.

- **Enablement (AI Office Role):** Refers to the AI Office function of providing the necessary platforms, tools, training, standards, and support to empower other teams within the organization to develop and deploy AI responsibly and effectively.

- **Encoding (Categorical Variables):** The process of converting non-numeric categorical features (like text labels) into a numerical format suitable for machine learning models (e.g., one-hot encoding, target encoding).
- **Endpoint (API/Inference):** A specific URL or address where an AI model's prediction service can be accessed, typically via an API call, to receive input data and return predictions. Securing these is critical.
- **Energy Efficiency (AI):** Designing and operating AI models and infrastructure to minimize power consumption and environmental impact. Part of Sustainable AI / "Green AI".
- **Ensemble Model:** A machine learning technique where multiple individual models are trained and their predictions are combined (e.g., by averaging or voting) to produce a final prediction, often improving overall accuracy and robustness but potentially reducing interpretability.
- **Enterprise AI:** The strategic application and integration of artificial intelligence capabilities across an entire organization to achieve business objectives, managed through coordinated efforts rather than isolated projects.
- **Enterprise AI Maturity Index:** See EAMI.
- **Enterprise AI Maturity Matrix:** See EAMM.
- **Enterprise Architecture:** The practice of analyzing, designing, planning, and implementing enterprise analysis to successfully execute business strategies. Collaborates with AI Office on technology choices.
- **Enterprise Resource Planning:** See ERP.
- **Environment Management (MLOps):** Practices for creating, managing, and maintaining consistent software and hardware environments across development, testing, and production stages of the ML lifecycle, often using containerization (Docker) and IaC.
- **Equal Opportunity:** A group fairness metric requiring that the model achieves equal true positive rates across different demographic groups (e.g., qualified candidates from all groups have an equal chance of being correctly identified).
- **Equalized Odds:** A group fairness metric requiring that the model achieves equal true positive rates AND equal false positive rates across different demographic groups. Often considered a stricter fairness definition.
- **ERP (Enterprise Resource Planning):** Business process management software that allows an organization to use a system of integrated applications to manage the business and automate many back-office functions related to technology, services, and human resources. Often a key data source.
- **ESG (Environmental, Social, and Governance):** A framework used to assess an organization's business practices and performance on various sustainability and ethical issues. AI can be applied to improve ESG outcomes ("AI for Green").
- **ETL (Extract, Transform, Load):** A traditional data integration paradigm where data is extracted from sources, transformed in a staging area or ETL tool, and then loaded into the target data warehouse. Often less flexible for AI than ELT.
- **Ethics (AI):** See AI Ethics.
- **Ethics & Compliance (EAMM Dimension):** Assesses maturity in embedding ethical principles, bias mitigation, security protocols, and ensuring adherence to legal/regulatory requirements in AI practices. Key part of Compass East.
- **Ethics Committee / Board:** See AI Review Board / Ethics Committee.
- **Evasion Attack:** An adversarial attack at inference time where inputs are subtly perturbed to cause the model to misclassify them.
- **EX:** See Employee Experience.
- **Executive Sponsor:** A senior leader who provides high-level support, resources, advocacy, and political cover for a major initiative, such as the enterprise AI program or a specific high-impact AI project. Crucial for success.
- **Explainability (XAI):** Techniques and methods used to make the decisions or predictions of AI models

understandable to humans. A key aspect of Transparency and trustworthy AI.

- **Exploitation (Innovation Context):** Focusing on optimizing current operations, improving existing products/processes, and executing the current roadmap effectively. Needs balancing with Exploration.
- **Exploration (Innovation Context):** Focusing on searching for new opportunities, experimenting with novel technologies or approaches, and understanding future possibilities. Needs balancing with Exploitation.
- **Exploratory Data Analysis:** See EDA.
- **Extraction Attack (Model Stealing):** An adversarial attack where an attacker repeatedly queries a deployed model's API to infer its parameters or train a functionally equivalent substitute model.
- **F1-Score:** A common classification metric that calculates the harmonic mean of Precision and Recall, providing a balanced measure, especially useful for imbalanced datasets.
- **FactSheet (AI):** See AI FactSheet.
- **Fairlearn:** An open-source Python toolkit from Microsoft designed to help assess and mitigate unfairness in machine learning models.
- **Fairness (AI):** A complex, context-dependent concept in AI focused on ensuring that AI systems do not produce biased or discriminatory outcomes against certain individuals or groups based on protected attributes (e.g., race, gender, age). A pillar of Trustworthy AI (Chapter 7). See also EAMM Ethics & Compliance.
- **Fairness Metrics:** Quantitative measures used to assess the fairness of an AI model's outcomes across different groups (e.g., Demographic Parity, Equalized Odds, Equal Opportunity).
- **Feasibility Study:** A preliminary study undertaken to determine and document a project's viability, often used in the R&D pipeline to assess technical feasibility and potential value before committing significant resources.
- **Feature Engineering:** The process of using domain knowledge and data manipulation techniques to create input variables (features) from raw data that make machine learning algorithms work better.
- **Feature Importance:** A measure indicating the relative contribution or influence of each input feature on a machine learning model's predictions. Often provided by XAI techniques like SHAP.
- **Feature Squeezing:** An input transformation technique proposed as a defense against adversarial evasion attacks, involving reducing the search space for perturbations (e.g., by reducing color bit depth).
- **Feature Store:** A centralized platform within the MLOps infrastructure specifically designed to manage the lifecycle of features used in machine learning (definition, storage, discovery, serving, monitoring, versioning). Crucial for consistency, reuse, and mitigating training-serving skew.
- **Federated Development Model:** An operating model (often part of a Hybrid structure) where development activities are distributed, allowing teams ('Spokes') within business units to build certain AI solutions autonomously, guided by central ('Hub') platforms, standards, and governance.
- **Federated Governance:** An approach to governance often used in federated operating models, where some governance responsibilities (e.g., for lower-risk applications using standard tools) are delegated to local teams/Spokes, while central oversight is maintained for higher risks and enterprise standards.
- **Federated Learning:** An ML technique that trains algorithms across multiple decentralized edge devices or servers holding local data samples, without exchanging the raw data itself, only model updates. Enhances privacy.
- **Feedback Loop:** A process where the output or result of an action is used to modify or influence future actions. Critical in performance management (feeding insights back to strategy/operations) and MLOps (monitoring triggering retraining).
- **Fiddler AI:** A commercial AI/ML monitoring and observability platform mentioned as an example tool.
- **Financial Dimension (EAMM):** Assesses the maturity of financial stewardship practices related to AI, including budgeting based on TCO, cost tracking/optimization (FinOps), and ROI validation. Managed within Compass South.

- **Financial Performance (Value Dimension):** Focuses on the direct impact of AI on key financial metrics like revenue growth and cost reduction. See Appendix B KPIs.
- **Financial Stewardship:** The responsible planning, management, and oversight of financial resources, crucial for the sustainability of AI programs (Chapter 19).
- **FinOps (Financial Operations):** A cultural practice and operational framework bringing financial accountability and optimization to variable cloud spending, essential for managing AI workload costs.
- **Fivetran:** A commercial cloud-based ELT service mentioned as a potential tool for data integration and CDC.
- **Flake8:** A popular Python tool used for checking code style and quality (linting). Often integrated into CI pipelines.
- **Flink (Apache Flink):** An open-source stream processing framework for stateful computations over unbounded and bounded data streams.
- **Flow Diagram:** A type of diagram representing a workflow or process using different symbols and connecting arrows. Used as visual aids in the book.
- **Food for Thought:** A recurring section at the end of chapters posing reflective questions or application exercises for the reader.
- **Forecasting:** Predicting future values or events based on historical data and statistical models. A common application of AI/ML.
- **Forensics (Digital):** The process of identifying, preserving, analyzing, and presenting digital evidence in a manner that is legally admissible. Relevant for investigating AI security incidents.
- **Forrester:** A global market research company providing advice on existing and potential impact of technology, cited as a potential source for industry trend reports.
- **Foundation Model:** A large AI model (often an LLM or multimodal model) trained on vast amounts of broad data at scale, designed to be adapted (e.g., via fine-tuning or prompt engineering) to a wide range of downstream tasks. Examples include GPT-4, Claude 3, Gemini.
- **Framework:** A basic structure underlying a system, concept, or text. Used extensively in the book (Compass Framework, EAMM/EAMI Framework, Governance Framework, etc.).
- **Fraud Detection:** Identifying and preventing fraudulent transactions or activities, a common and high-value application for AI/ML.
- **FTE (Full-Time Equivalent):** A unit indicating the workload of an employed person in a way that makes workloads comparable across various contexts. Often used when measuring cost savings from automation.
- **Funding Model:** The approach an organization uses to allocate financial resources to initiatives or functions, such as the AI Office or specific AI projects (e.g., centralized, chargeback, hybrid).
- **Future-Proofing:** Taking actions to ensure something remains valuable, effective, or relevant in the future despite anticipated changes or advancements (e.g., designing flexible architectures, fostering continuous learning).
- **Gantt Chart:** A type of bar chart that illustrates a project schedule, showing the start and finish dates of terminal elements and summary elements of a project. Mentioned as a way to visualize roadmap timelines.
- **Garbage In, Garbage Out (GIGO):** A concept in computer science and data analytics emphasizing that the quality of output is determined by the quality of the input. Particularly critical for AI models trained on data.
- **Gartner:** A global research and advisory company, cited as a source for industry reports, statistics, and frameworks related to AI, data management, and technology trends.
- **GCP (Google Cloud Platform):** Google's suite of cloud computing services, including infrastructure, data analytics, and machine learning tools (Vertex AI).
- **GDPR (General Data Protection Regulation):** A comprehensive data protection and privacy regulation enacted by the European Union (EU), imposing strict rules on controlling and processing personally identi-

fiable information (PII) of EU residents. Highly relevant for AI governance and compliance.

- **Gebru et al.:** Refers to a seminal paper proposing the concept of "Datasheets for Datasets," mentioned in the context of standardized data documentation. (Exact citation should be verified).
- **Gemini (Google):** A family of multimodal large language models developed by Google, cited as an example technology and potentially referenced in reports.
- **Generative AI:** A class of artificial intelligence algorithms capable of generating new content, such as text, images, audio, code, or synthetic data, often based on patterns learned from existing data. Includes LLMs.
- **Git:** A widely used distributed version control system for tracking changes in source code during software development. Essential for MLOps and DataOps.
- **GitHub / GitLab / Bitbucket / Azure Repos:** Popular web-based hosting services for Git version control repositories, facilitating collaboration and code management.
- **GKE (Google Kubernetes Engine):** Google Cloud's managed Kubernetes service.
- **GloVe (Global Vectors for Word Representation):** An unsupervised learning algorithm for obtaining vector representations (embeddings) for words, mentioned as an example NLP technique.
- **Goal Coherence:** A sub-KPI mentioned in the EAMM/EAMI tables, likely related to measuring the alignment between specific AI project objectives and broader enterprise strategic goals.
- **Golden Record:** In Master Data Management (MDM), refers to the single, authoritative, consolidated master version of a core data entity (like a customer or product) after resolving inconsistencies and duplicates from multiple sources.
- **Google:** A multinational technology company, developer of AI models like Gemini and cloud platform GCP, cited as a source or example.
- **Google Cloud:** See GCP.
- **Google Cloud Dataflow:** See Dataflow (Google Cloud).
- **Google Cloud Deployment Manager:** Google Cloud's service for defining and managing infrastructure as code using templates.
- **Google Cloud Monitoring:** Google Cloud's service for monitoring the performance, availability, and health of applications and infrastructure.
- **Google Cloud Storage (GCS):** Google Cloud's scalable object storage service, often used for data lakes.
- **Google Vertex AI:** See Vertex AI (Google Cloud).
- **Google What-If Tool:** An interactive visual interface designed for understanding ML models, including features for exploring fairness and model behavior under different conditions. Mentioned as a bias detection tool.
- **Governance (AI):** See AI Governance.
- **Governance (Compass Domain - East):** The domain of the AI Office Compass™ Framework focused on establishing ethical, legal, security, and operational guardrails for responsible AI. Includes policy, risk, compliance, ethics, security.
- **Governance (EAMM Dimension):** Assesses the maturity of formal policies, processes, roles, and risk management practices specifically for overseeing AI development and deployment responsibly.
- **Governance-as-Code:** An approach where governance policies and compliance checks are defined and automatically enforced as code, typically integrated within CI/CD or IaC pipelines.
- **Governance-by-Design:** An approach where governance considerations (ethical, privacy, security, compliance) are integrated proactively and continuously throughout the AI development lifecycle from the beginning, rather than being treated as a late-stage checklist.
- **GPU (Graphics Processing Unit):** Specialized electronic circuits initially designed for accelerating computer graphics and image processing, now widely used to accelerate the computationally intensive training and sometimes inference of deep learning models due to their parallel processing capabilities.

- **GRC (Governance, Risk Management, and Compliance):** An integrated organizational capability referring to the strategy and processes for managing overall governance, enterprise risk management, and compliance with regulations. GRC platforms are sometimes used or adapted for AI governance tracking.
- **Gradient Boosting:** A powerful machine learning technique that builds predictive models in the form of an ensemble of weak prediction models, typically decision trees, in a stage-wise fashion. XGBoost and LightGBM are popular implementations.
- **Grafana:** An open-source analytics and interactive visualization web application often used with time-series databases like Prometheus for monitoring systems.
- **Great Expectations:** An open-source Python library providing tools for data validation, documentation, and profiling, commonly used in DataOps to ensure data quality within pipelines.
- **Green AI:** A movement or area of research focused on making artificial intelligence more environmentally sustainable by improving the energy efficiency of algorithms, hardware, and operational practices.
- **Ground Truth:** In machine learning, refers to the factual, objective reality or the known correct labels/values for a dataset, used to train supervised models and evaluate their performance. The quality of ground truth data (labels) is critical.
- **Group Fairness:** A category of AI fairness metrics focused on ensuring statistical parity or equality across different predefined demographic groups. Contrasts with Individual Fairness.
- **Guardrails:** Metaphorical term for the policies, standards, controls, and processes put in place by AI governance to guide development and deployment within safe, ethical, and compliant boundaries.
- **H2O.ai:** A commercial company providing an open-source and enterprise AI/ML platform, mentioned as an example vendor.
- **Hackathon:** An event, typically lasting several days, in which a large number of people meet to engage in collaborative computer programming or hardware development, often used to spur innovation or explore new ideas. Mentioned as an ideation mechanism.
- **Hadoop (Apache Hadoop):** An open-source framework that allows for the distributed processing of large data sets across clusters of computers using simple programming models. Relevant to big data infrastructure.
- **Hallucination (AI):** A phenomenon where AI models, particularly Large Language Models (LLMs), generate confident responses that are nonsensical, factually incorrect, or not justified by their training data. A key challenge for Generative AI governance.
- **Hardware Acceleration:** Using specialized hardware components (like GPUs or TPUs) to perform specific computational tasks more efficiently than is possible on a general-purpose CPU. Critical for large-scale AI model training.
- **HBR (Harvard Business Review):** A general management magazine published by Harvard Business Publishing, cited as a source for articles on business strategy and management, including AI communication.
- **Healthcare:** The industry related to the maintenance or improvement of health via the prevention, diagnosis, treatment, recovery, or cure of disease, illness, injury, and other physical and mental impairments. Often cited as a sector with high potential but also high regulatory/ethical considerations for AI.
- **Heatmap:** A data visualization technique that shows magnitude of a phenomenon as color in two dimensions. Mentioned as a way to visualize EAMI dimension scores.
- **Heuristics:** Practical approaches to problem solving or discovery that employ a readily accessible, though loosely applicable, rule of thumb, often used when an optimal solution is impossible or impractical to find. Can be used in programmatic labeling.
- **High-Risk AI Systems:** A category defined in regulations like the EU AI Act, referring to AI systems used in contexts (e.g., critical infrastructure, employment, medical devices, law enforcement) where failure or malfunction could pose significant risks to health, safety, or fundamental rights. Subject to the strictest governance requirements.

- **Highest Watermark Approach:** A compliance strategy where an organization operating across multiple jurisdictions chooses to adhere globally to the strictest applicable regulatory standard encountered in any of those jurisdictions for a particular area (e.g., data privacy).
- **Horizon Scanning:** The systematic process of monitoring the external environment (technological, regulatory, competitive, societal) to identify emerging trends, potential threats, opportunities, and potential future disruptions relevant to an organization's strategy. Key activity for the AI Office Innovation Hub function (Chapter 27).
- **Hosting:** Providing the infrastructure (servers, storage, network) required to run applications or services. Can be on-premise or cloud-based.
- **HR (Human Resources):** The department within an organization responsible for managing the employee lifecycle, including recruitment, onboarding, training, performance management, compensation, benefits, and employee relations. Key partner for AI Office on talent and change management.
- **Hub-and-Spoke Model:** See Hybrid/Federated Structure (AI Office Operating Model).
- **Hudi (Apache Hudi):** An open-source data lake storage framework providing stream processing primitives (like upserts and incremental consumption) and transactional capabilities on top of data lakes. A key technology enabling the Lakehouse architecture.
- **Hugging Face Transformers:** A popular open-source library providing thousands of pre-trained models (especially for NLP) and tools for easily downloading, training, and deploying transformer-based architectures.
- **Human-AI Collaboration / Teaming:** Designing systems and workflows where humans and AI systems work together effectively, leveraging their complementary strengths to achieve outcomes superior to either working alone.
- **Human-Centricity (Pillar of Trustworthy AI):** Designing AI systems to augment human capabilities, respect human autonomy, ensure meaningful human oversight where appropriate, and align with human values and well-being.
- **Human-Computer Interaction (HCI):** The design and use of computer technology, focused on the interfaces between people (users) and computers. Increasingly relevant for designing effective and trustworthy human-AI collaboration.
- **Human-in-the-Loop (HITL):** A model design where human input or intervention is incorporated into the AI system's process, typically for training (e.g., active learning), validation (e.g., reviewing critical predictions), or handling exceptions the AI cannot manage confidently.
- **Hybrid Cloud:** An infrastructure strategy combining private cloud (or on-premise infrastructure) with public cloud services, allowing data and applications to be shared between them.
- **Hybrid/Federated Structure (AI Office Operating Model):** An operating model combining a central coordinating 'Hub' (AI Office/CoE) providing strategy, governance, platforms, and expertise, with distributed 'Spoke' teams embedded in business units focusing on domain-specific solutions. Often suited for scaling AI maturity (EAMI Level 3-5).
- **Hyperparameter:** A configuration parameter whose value is set before the learning process begins for an ML model (e.g., learning rate, number of layers in a neural network). Tuning hyperparameters is part of model optimization.
- **Hyper-Personalization:** Using AI and extensive data to tailor products, services, communications, or experiences to individual customers at a highly granular level in real-time or near real-time.
- **Hypothesis Testing:** A statistical method used to make decisions using experimental data, involving formulating a null hypothesis and an alternative hypothesis and determining whether sample data provides enough evidence to reject the null hypothesis. Used in A/B testing and data science.
- **IaaS (Infrastructure-as-a-Service):** A cloud computing model where vendors provide fundamental com-

pute, network, and storage resources on demand, over the internet, on a pay-as-you-go basis. Users manage the operating system, middleware, and applications.

- **IaC (Infrastructure as Code):** The practice of managing and provisioning computer data centers and infrastructure (networks, virtual machines, load balancers, connection topology) through machine-readable definition files (code), rather than physical hardware configuration or interactive configuration tools. Enables automation, consistency, and version control for infrastructure.
- **IBM (International Business Machines):** A multinational technology and consulting company, developer of AI systems like Watson, cited as a source or example.
- **IBM DataStage:** A commercial ETL tool mentioned as an example.
- **IBM OpenPages:** A commercial enterprise Governance, Risk Management, and Compliance (GRC) platform mentioned as an example tool.
- **IBM Watson:** See Watson (IBM).
- **Iceberg (Apache Iceberg):** An open-source high-performance format for huge analytic tables, bringing database table semantics (like ACID transactions, schema evolution, time travel) to data lakes. A key technology enabling the Lakehouse architecture.
- **ICLR (International Conference on Learning Representations):** A major academic conference focused on deep learning and representation learning, mentioned as a source for monitoring AI research.
- **ICML (International Conference on Machine Learning):** A leading international academic conference in machine learning, mentioned as a source for monitoring AI research.
- **ICE Score:** A prioritization framework scoring opportunities based on Impact, Confidence, and Ease.
- **IDC (International Data Corporation):** A global provider of market intelligence, advisory services, and events for the IT, telecommunications, and consumer technology markets, cited as a source for industry reports.
- **IDE (Integrated Development Environment):** A software application that provides comprehensive facilities to computer programmers for software development, typically consisting of a source code editor, build automation tools, and a debugger.
- **Ideation:** The creative process of generating, developing, and communicating new ideas. Relevant to building the AI Opportunity Funnel.
- **Imperative:** Of vital importance; crucial. Used frequently to emphasize the necessity of concepts like strategic alignment or governance.
- **Implementation:** The process of putting a decision, plan, or system into effect.
- **Impossibility Theorems (Fairness):** Mathematical results demonstrating that it is impossible to satisfy certain combinations of desirable fairness metrics simultaneously in most non-trivial cases, highlighting the need for context-specific trade-offs.
- **Imputation (Data):** The process of replacing missing data values with substituted estimates (e.g., mean, median, model predictions). Requires careful handling to avoid introducing bias.
- **Incidence Response (IR):** An organized approach to addressing and managing the aftermath of a security breach or cyberattack (or other incident), aiming to handle the situation in a way that limits damage and reduces recovery time and costs. AI requires specific IR planning.
- **Inclusion (Diversity & Inclusion):** Creating an environment where all individuals feel valued, respected, supported, and able to participate fully. Important for building diverse and effective AI teams.
- **Incubation Unit:** A dedicated organizational entity, often separate from core operations, designed to nurture and develop new, potentially disruptive business ideas or technologies (like some AI innovations) before they are ready for full scaling or integration.
- **Incremental Improvement:** Making small, gradual changes or enhancements over time, often contrasted with large, disruptive transformations. Can be a valid approach in portfolio balancing.

- **Individual Fairness:** A category of AI fairness metrics focused on ensuring that similar individuals are treated similarly by a model, irrespective of their group membership. Contrasts with Group Fairness.
- **Inference (AI Model):** The process of using a trained machine learning model to make predictions or decisions based on new, previously unseen input data. Occurs after the model is deployed.
- **Inference Latency:** The time it takes for a deployed AI model to process an input and return a prediction or output. A critical performance metric for real-time applications.
- **Informatica PowerCenter:** A commercial enterprise data integration and ETL tool mentioned as an example.
- **Infrastructure (EAMM Dimension):** Assesses the maturity, scalability, reliability, and suitability of the underlying IT infrastructure (compute, storage, network, cloud/on-prem) used to support AI workloads. Managed within Compass South.
- **Infrastructure as Code:** See IaC.
- **Infrastructure-as-a-Service:** See IaaS.
- **Initial (EAMI Level 1):** The first and lowest level of AI maturity in the EAMM framework, characterized by ad-hoc, fragmented efforts, lack of strategy/governance, and high risk.
- **Innovation Ecosystem (EAMM Dimension):** Assesses the organization's capability and processes for exploring emerging AI trends, fostering responsible experimentation, managing an R&D pipeline, and building internal/external partnerships to drive future innovation. Managed within Compass Center.
- **Innovation Hub:** A function, often within or closely associated with the AI Office in mature organizations, dedicated to driving AI innovation, experimentation, and future readiness (Chapter 27).
- **Insight Generation:** Using AI/ML to uncover hidden patterns, trends, correlations, or causal relationships in data that provide valuable understanding or support better decision-making. A key dimension of AI value.
- **Integration (System/Data):** Combining different software systems, applications, or data sources so they work together functionally. Critical for AI systems relying on diverse data and needing to embed outputs into workflows.
- **Integrator (AI Office Role):** One of the key functions of the AI Office, involving synthesizing diverse AI needs and opportunities into a coherent enterprise-wide vision and roadmap.
- **Integrity (Security Principle):** Maintaining the consistency, accuracy, and trustworthiness of data and systems, protecting them from unauthorized modification or tampering.
- **Intellectual Property (IP):** Creations of the mind, such as inventions; literary and artistic works; designs; and symbols, names, and images used in commerce. Trained AI models can represent significant IP.
- **Interaction Bias:** Bias introduced into an AI system through how humans interact with it or interpret its outputs (e.g., confirmation bias, feedback loops).
- **Interconnectedness (Compass Framework):** The principle that the five domains of the AI Office Compass™ Framework are highly interdependent and must be managed holistically for success.
- **Internal Communication:** Communication occurring within an organization between its members or departments.
- **Interoperability:** The ability of different information systems, devices, or applications to connect, exchange data, and use the information that has been exchanged effectively. Crucial for the AI technology ecosystem.
- **Interpretability:** See Explainability (XAI).
- **Interview:** A meeting of people face to face, especially for consultation or assessment. Used as a data gathering method for EAMI assessments.
- **Intranet:** A private network contained within an enterprise, used to securely share company information and computing resources among employees. Mentioned as a communication channel.
- **Intrusion Detection System (IDS):** A device or software application that monitors a network or systems for malicious activity or policy violations. Part of traditional cybersecurity.

- **Intuition:** The ability to understand something immediately, without the need for conscious reasoning. AI often aims to augment or inform human intuition in decision-making.
- **Inventory Management:** The process of overseeing the flow of goods from manufacturers to warehouses and from these facilities to point of sale. AI is often used for optimization (e.g., demand forecasting).
- **Investment:** The action or process of investing money or resources for profit or material result. AI initiatives require significant investment.
- **IoT (Internet of Things):** The network of physical devices, vehicles, home appliances and other items embedded with electronics, software, sensors, actuators, and connectivity which enables these objects to connect and exchange data. A major source of data for AI.
- **IP:** See Intellectual Property.
- **IPA (Intelligent Process Automation):** The application of AI techniques (like NLP, computer vision, ML) to enhance traditional Robotic Process Automation (RPA), enabling automation of more complex, less structured tasks.
- **IR:** See Incident Response.
- **ISO (International Organization for Standardization):** An international standard-setting body composed of representatives from various national standards organizations. Develops standards like ISO/IEC 42001 for AI Management Systems.
- **ISO/IEC 42001:** An international standard specifying requirements for establishing, implementing, maintaining, and continually improving an AI Management System (AIMS).
- **IT (Information Technology):** The use of computers to store, retrieve, transmit, and manipulate data, or information, often in the context of a business or other enterprise. Key partner for AI Office.
- **Iteration:** The repetition of a process or utterance. Central to Agile methodologies and AI model development (experimentation, refinement).
- **Ivory Tower:** A metaphorical place—or an attitude—where people are happily cut off from the rest of the world in favor of their own pursuits, usually mental and esoteric ones. Used to describe the risk of a centralized AI Office becoming disconnected from business needs.
- **Jenkins:** An open-source automation server widely used for building, testing, and deploying software, including CI/CD pipelines relevant to MLOps.
- **Jira:** A proprietary issue tracking product developed by Atlassian that allows bug tracking and agile project management. Mentioned frequently as a tool for managing AI projects and backlogs.
- **Jira Align:** An enterprise agile planning platform from Atlassian, mentioned as a potential tool for strategic reviews.
- **Jira Advanced Roadmaps:** A feature within Jira Software designed for roadmap planning and visualization.
- **JMLR (Journal of Machine Learning Research):** A leading peer-reviewed scientific journal focusing on machine learning, mentioned as a source for monitoring AI research.
- **Job Displacement:** The elimination of jobs resulting from technological change (like AI automation), economic shifts, or other factors. A key societal concern and barrier related to AI adoption.
- **Joint Reporting:** An organizational structure where an individual or unit reports to two different managers or executives simultaneously (e.g., AI Office Head reporting to both COO and CTO).
- **JPMorgan Chase:** A multinational investment bank and financial services holding company, cited as a source for examples or statistics related to AI implementation.
- **JSON (JavaScript Object Notation):** A lightweight data-interchange format that uses human-readable text to transmit data objects consisting of attribute-value pairs and array data types. Common format for semi-structured data.
- **Jupyter Notebook:** An open-source web application that allows users to create and share documents containing live code (e.g., Python, R), equations, visualizations, and narrative text. Widely used by data scientists

for exploration and experimentation. Managed versions are key features of cloud AI platforms.

- **Kanban:** An agile project management methodology focused on visualizing work, limiting work in progress, and maximizing flow. Mentioned as relevant to AI project management.
- **Kafka (Apache Kafka):** An open-source distributed event streaming platform widely used for building real-time data pipelines.
- **Keras:** A popular open-source software library that provides a Python interface for artificial neural networks, often acting as an interface for the TensorFlow library.
- **Key Performance Indicator:** See KPI.
- **KFServing/KServe:** An open-source platform built on Kubernetes for serving machine learning models at scale, providing features like prediction, pre/post-processing, and explainability. Mentioned as a model serving option.
- **Kibana:** An open-source data visualization dashboard for Elasticsearch. Part of the ELK stack.
- **Knowledge Distillation:** A model compression technique where a smaller "student" model is trained to mimic the behavior of a larger, pre-trained "teacher" model, aiming to transfer knowledge while reducing model size and computational cost.
- **Knowledge Management:** The process of creating, sharing, using, and managing the knowledge and information of an organization. AI can be used to build intelligent knowledge management systems.
- **Knowledge Sharing:** The activity through which knowledge (information, skills, or expertise) is exchanged among people, friends, families, communities (such as CoPs), or organizations. Vital for AI teams.
- **Kotter's 8-Step Process:** A widely recognized framework for leading organizational change, outlined in Chapter 26, involving steps like creating urgency, building a coalition, forming a vision, communicating, removing barriers, generating short-term wins, consolidating gains, and anchoring changes.
- **KPI (Key Performance Indicator):** A quantifiable measure used to evaluate the success of an organization, employee, or specific activity (like an AI initiative) in meeting objectives for performance. EAMI uses KPIs linked to EAMM dimensions. Detailed examples in Appendix B.
- **Kubeflow:** An open-source machine learning toolkit for Kubernetes, aiming to make deployments of ML workflows simple, portable, and scalable. Often used for MLOps pipelines and metadata management.
- **Kubeflow Pipelines:** A component of Kubeflow providing a platform for building and deploying portable, scalable machine learning workflows based on Docker containers.
- **Kubernetes (K8s):** An open-source container orchestration system for automating software deployment, scaling, and management. Widely used as the underlying infrastructure for MLOps platforms and scalable model serving.
- **Label Bias:** Bias introduced during the data annotation process due to subjective or prejudiced human labeling of training data.
- **Labelbox:** A commercial platform mentioned as an example tool for data annotation and labeling management.
- **Lagging Indicator:** A measurable economic or business factor that changes only after the economy or business has already begun to follow a particular pattern or trend (e.g., quarterly revenue). Contrasted with Leading Indicator.
- **Lakehouse Architecture:** A modern data management architecture combining the flexibility and cost-effectiveness of a data lake with the data management and reliability features of a data warehouse, typically using open table formats (Delta Lake, Iceberg, Hudi) on object storage. Enables unified BI and AI workloads.
- **Large Language Model:** See LLM.
- **Latency (Inference):** See Inference Latency.
- **Lawful Basis (GDPR):** One of the six permissible grounds for processing personal data under GDPR (e.g., consent, contract, legal obligation, vital interests, public task, legitimate interests).

- **L&D (Learning & Development):** A function within Human Resources focused on improving the skills, knowledge, and capabilities of employees through training and development programs. Key partner for AI literacy initiatives.
- **Lead Time for Changes:** A DevOps/DORA metric measuring the time it takes for a code commit to get into production. Relevant for assessing MLOps pipeline efficiency.
- **Leading Indicator:** A measurable factor that changes before the broader trend or outcome it predicts, providing early signals (e.g., user adoption rate predicting future value). Contrasted with Lagging Indicator.
- **LeanIX:** A commercial enterprise architecture management tool mentioned as potentially useful for tracking process improvements.
- **Learning & Development:** See L&D.
- **Learning Rate:** A hyperparameter in machine learning model training that controls how much the model's weights are adjusted with respect to the loss gradient during optimization.
- **Least Privilege:** A security principle requiring that users, programs, or processes are granted only the minimum levels of access or permissions necessary to perform their required functions. Applied via RBAC.
- **Legacy System:** An old method, technology, computer system, or application program ("legacy") that is outdated but still in use. Often pose integration challenges for modern AI initiatives.
- **Legal Department:** The function within an organization responsible for providing legal advice and managing legal matters. Key collaborator for AI Governance (contracts, compliance, IP).
- **Level (EAMI/EAMM):** One of the five distinct stages of AI maturity defined in the EAMM framework (Level 1: Initial, Level 2: Emerging, Level 3: Defined, Level 4: Optimized, Level 5: Transformative), corresponding to specific EAMI score ranges.
- **Lewin's Change Model:** A classic three-stage model of organizational change (Unfreeze, Change, Refreeze) mentioned as an alternative framework. (Note: Kotter's model is detailed instead in the draft).
- **LGPD (Lei Geral de Proteção de Dados):** Brazil's comprehensive data protection law, similar in many aspects to GDPR.
- **Liaison:** Communication or cooperation which facilitates a close working relationship between people or organizations. Used to describe roles bridging groups (e.g., Data Engineer Liaison).
- **Libraries (Software):** Collections of pre-written code, functions, classes, or routines that can be used by developers to perform common tasks without rewriting the code themselves (e.g., Scikit-learn, TensorFlow, PyTorch). Managing dependencies on libraries securely is part of MLOps.
- **LightGBM:** An open-source gradient boosting framework developed by Microsoft, known for its speed and efficiency. Mentioned as an example ML algorithm.
- **Lightweight:** Simple, requiring little effort or resources. Used to describe initial intake processes or sandbox guardrails.
- **LIME (Local Interpretable Model-agnostic Explanations):** A popular model-agnostic XAI technique that explains individual predictions of complex ("black box") models by approximating them locally with simpler, interpretable models.
- **Limited Risk (EU AI Act):** A risk category under the EU AI Act for AI systems that pose limited potential risk, primarily requiring transparency obligations (e.g., informing users they are interacting with AI).
- **Lineage (Data):** See Data Lineage.
- **Linear Regression:** A basic supervised learning algorithm used to model the linear relationship between a dependent variable and one or more independent variables. Often interpretable via coefficients.
- **Linting:** The process of running a tool (a linter) that analyzes source code to flag programming errors, bugs, stylistic errors, and suspicious constructs. E.g., Flake8 for Python.
- **Lip Service:** Insincere expression of support or approval. Used to describe passive or inadequate executive sponsorship.

- **LLM (Large Language Model):** A type of AI model, typically based on deep learning transformer architectures and trained on massive amounts of text data, capable of understanding and generating human-like language for various tasks (summarization, translation, question answering, content creation). Examples: GPT-4, Claude 3, Gemini.
- **LLMOps (Large Language Model Operations):** An emerging specialization of MLOps focused on the unique challenges of developing, deploying, fine-tuning, managing, monitoring (for cost, drift, toxicity, factuality), and governing large language models and foundation models in production.
- **Load (ETL/ELT):** The step in data integration processes where data is physically inserted into the target storage system (warehouse, lake, lakehouse).
- **Local Interpretable Model-agnostic Explanations:** See LIME.
- **Logistics:** The detailed coordination of a complex operation involving many people, facilities, or supplies. AI is often used to optimize logistics networks.
- **Logstash:** An open-source server-side data processing pipeline that ingests data from multiple sources simultaneously, transforms it, and then sends it to a "stash" like Elasticsearch. Part of the ELK stack.
- **Looker:** A commercial business intelligence platform (now part of Google Cloud) mentioned as a tool for creating performance dashboards and tracking metrics.
- **Loss Function:** In machine learning, a function that quantifies the difference between the model's predictions and the actual ground truth values during training. The goal of training is typically to minimize this function.
- **Low-Code / No-Code Platforms:** Software development platforms that allow users to create applications with minimal or no traditional programming, often using visual interfaces and drag-and-drop components. Can potentially democratize some AI development but requires governance.
- **MAE (Mean Absolute Error):** A common regression metric measuring the average magnitude of the errors in a set of predictions, without considering their direction.
- **Mandate (AI Office):** The officially defined purpose, scope of authority, and objectives assigned to the AI Office by organizational leadership.
- **Manual Process:** A task or workflow performed by humans without significant automation. AI often aims to automate manual processes.
- **Manufacturing:** The industry focused on making products by hand or machinery. Often cited as a sector where AI (e.g., predictive maintenance, quality control) can provide value.
- **Market Positioning:** A value dimension for AI focused on improving the organization's competitive standing, brand perception, or market share. See Appendix B KPIs.
- **Market Share:** The portion of a market controlled by a particular company or product.
- **Marketing:** The activities a company undertakes to promote the buying or selling of a product or service. Key area for AI applications like personalization and campaign optimization.
- **Master Data Management (MDM):** A technology-enabled discipline in which business and IT work together to ensure the uniformity, accuracy, stewardship, semantic consistency, and accountability of the enterprise's official shared master data assets (e.g., defining a single 'golden record' for customers, products). Critical for data quality.
- **Matrixed Structure:** An organizational structure where individuals report to multiple managers (e.g., a functional manager and a project manager). Often used in AI teams.
- **Maturity (AI):** See AI Maturity.
- **Maturity Assessment:** The process of evaluating an organization's current level of AI maturity using a defined framework like EAMM/EAMI. Core activity of Compass Center. See also EAMM Maturity Assessment.
- **Maturity Management (Center - Compass Domain):** The central domain of the AI Office Compass™

Framework responsible for orchestrating AI maturity assessment (EAMI), driving continuous improvement via uplift strategies, fostering an AI-driven culture and managing change, and facilitating responsible innovation.

- **Maturity Model:** A framework for assessing the level of development, sophistication, and capability of an organization or process in a particular domain, typically defining discrete levels with associated characteristics (e.g., EAMM).
- **Maturity Uplift Strategy:** See AI Maturity Uplift Strategy.
- **Mean Absolute Error:** See MAE.
- **Mean Time to Recovery (MTTR):** A DevOps/DORA metric measuring the average time it takes to restore service after a production failure. Relevant to MLOps resilience.
- **Measurement Bias:** Bias introduced because the features or proxies used to represent a concept are measured differently across groups.
- **Measurement Method (EAMI KPI Table):** Column in the detailed KPI table describing how a specific KPI can be practically measured and where the data might be sourced (Value Source).
- **Membership Inference Attack:** An AI security attack where an adversary tries to determine if a specific individual's data was part of a model's training set by analyzing its outputs, posing a privacy risk.
- **Metadata:** Data that provides information about other data. Examples include data definitions, lineage information, ownership, data types, quality scores. Managed via data catalogs.
- **Metadata Management:** The process of managing metadata about an organization's data assets, crucial for discovery, understanding, governance, and trust.
- **Methodology:** A system of methods used in a particular area of study or activity. Used to refer to structured approaches like EAMI assessment or change management.
- **Metrics:** Quantifiable measures used to track and assess the status or performance of a specific process, product, or initiative. Includes technical metrics and business KPIs. See also EAMM Performance Metrics.
- **Micro-batching:** Processing streaming data in small, discrete batches at very short intervals (e.g., every few seconds), offering a balance between true real-time processing and batch processing simplicity.
- **Microservices Architecture:** An architectural style structuring an application as a collection of small, independent, loosely coupled services that communicate over a network, often via APIs. Can enhance modularity and scalability for AI systems.
- **Microsoft:** A multinational technology corporation, developer of Azure cloud platform and AI tools like Copilot, cited as a source or example.
- **Microsoft Azure:** See Azure.
- **Microsoft Azure ML:** See Azure Machine Learning.
- **Mid-Sized Firm:** A company generally considered between small businesses and large enterprises based on criteria like revenue or number of employees. OmnioTech is positioned as such.
- **Mitigation (Risk):** Implementing controls or actions to reduce the likelihood or impact of an identified risk. A common risk treatment strategy.
- **ML (Machine Learning):** A subfield of artificial intelligence focused on developing algorithms that allow computer systems to learn patterns and make predictions or decisions from data without being explicitly programmed for the task.
- **ML Engineer (Machine Learning Engineer):** A specialized engineering role focused on designing, building, deploying, and maintaining robust, scalable machine learning systems and infrastructure in production environments. Bridges data science and software engineering. Key role for MLOps.
- **ML Platform:** See AI/ML Development Platforms.
- **MLSecOps:** See SecMLOps.
- **MLflow:** An open-source platform for managing the end-to-end machine learning lifecycle, including com-

ponents for experiment tracking (MLflow Tracking), model packaging, model registration (MLflow Model Registry), and deployment. Widely used in MLOps.

- **MLOps (Machine Learning Operations) (EAMM Dimension):** Assesses the maturity and automation level of practices for managing the ML lifecycle (versioning, testing, CI/CD/CT, monitoring). Core part of Compass South. See also Chapter 20.
- **MLOps Engineer:** See ML Engineer.
- **Model (AI/ML):** The artifact produced by the machine learning training process, consisting of learned parameters or structures that represent patterns in the data and can be used to make predictions or decisions on new inputs.
- **Model Card:** A documentation framework proposed by Google (and widely adopted) providing short summaries of an ML model's characteristics, intended use, performance metrics, training data, fairness evaluations, ethical considerations, and limitations, promoting transparency and responsible deployment.
- **Model Compression:** Techniques used to reduce the size (memory footprint) and computational requirements of trained ML models, making them more efficient for deployment, especially on edge devices. Examples: quantization, pruning, knowledge distillation.
- **Model Decay / Staleness:** The tendency for a deployed ML model's predictive performance to degrade over time as the real-world data it encounters drifts away from the data it was trained on (due to data drift or concept drift). Requires monitoring and retraining via MLOps.
- **Model Drift:** See Model Decay / Staleness.
- **Model Governance:** The subset of AI Governance focused specifically on establishing policies, processes, and controls for the development, validation, deployment, monitoring, and management of machine learning models.
- **Model Inversion Attack:** An AI security/privacy attack where an adversary attempts to reconstruct sensitive features or parts of the training data by analyzing a model's outputs or accessing its parameters.
- **Model Monitoring:** See Monitoring (Production).
- **Model Registry:** A centralized system within an MLOps platform used to store, version, manage metadata for, and govern the lifecycle of trained machine learning models before deployment.
- **Model Serving:** The process of deploying a trained ML model and making it available to receive inference requests and return predictions, often via an API endpoint.
- **Model Stealing:** See Extraction Attack.
- **Model Validation:** The process of evaluating a trained ML model's performance, fairness, robustness, and adherence to requirements using appropriate metrics and test datasets before deployment.
- **Modularity:** Designing systems from distinct components (modules) that can be developed, tested, deployed, and replaced independently, improving maintainability and flexibility. Key architectural principle.
- **Monitoring (Production):** The ongoing process of observing and checking the performance, health, reliability, security, and potentially fairness of deployed AI systems and their supporting infrastructure in the production environment. Critical MLOps practice. See also Continuous Monitoring (CM).
- **Monte Carlo:** A commercial Data Observability platform mentioned as an example tool.
- **MPP (Massively Parallel Processing):** An architecture used in some data warehouses and database systems where data and processing are distributed across many nodes working in parallel to handle large-scale analytical queries efficiently.
- **MTTR (Mean Time to Recovery):** See Mean Time to Recovery.
- **Multimodal AI:** AI systems capable of processing, understanding, and potentially generating information from multiple types of data modalities simultaneously (e.g., text, images, audio, video).
- **Multinational Corporation:** A company that operates in multiple countries.
- **MVP (Minimum Viable Product):** A version of a new product which allows a team to collect the max-

imum amount of validated learning about customers with the least effort. Concept sometimes applied to initial AI deployments.

- **Natural Language Processing:** See NLP.
- **Neptune.ai:** A commercial platform mentioned as an example tool for MLOps experiment tracking and model registry.
- **Net Promoter Score:** See NPS.
- **Network Segmentation:** Dividing a computer network into smaller subnetworks or segments, often used as a security measure to limit the impact of a breach in one segment and control traffic flow. Relevant for isolating AI sandbox environments.
- **Neural Information Processing Systems:** See NeurIPS.
- **Neural Network (Artificial Neural Network - ANN):** A type of machine learning model inspired by the structure and function of biological neural networks, consisting of interconnected nodes or 'neurons' organized in layers. Foundation of deep learning.
- **Neuro-Symbolic AI:** An emerging area of AI research aiming to combine the strengths of deep learning (pattern recognition from data) with symbolic reasoning (logic, knowledge representation) to create more robust and interpretable AI systems. Mentioned as future trend.
- **NeurIPS (Neural Information Processing Systems):** A major international academic conference focused on machine learning and computational neuroscience, mentioned as a source for monitoring AI research.
- **Newsletter:** A regularly distributed publication generally about one main topic that is of interest to its subscribers. Mentioned as an internal communication channel.
- **Niche Expertise:** Specialized skills or knowledge in a very specific, often narrow, area. May require "Buy" or "Borrow" talent strategies.
- **NIST (National Institute of Standards and Technology):** A non-regulatory agency of the U.S. Department of Commerce that develops technology, measurement, and standards. Developed the voluntary AI Risk Management Framework (AI RMF).
- **NIST AI RMF:** See AI RMF (NIST).
- **NLP (Natural Language Processing):** A subfield of AI focused on enabling computers to process, understand, interpret, and generate human language. Powers applications like chatbots, translation, sentiment analysis.
- **NLTK (Natural Language Toolkit):** A popular open-source Python library for working with human language data (NLP tasks).
- **No-Code Platforms:** See Low-Code / No-Code Platforms.
- **Normalization (Data):** Scaling numerical data features to a standard range (e.g., 0 to 1 or -1 to 1) or standardizing them to have zero mean and unit variance. Common pre-processing step for many ML algorithms.
- **North (Compass Domain):** The domain of the AI Office Compass™ Framework focused on Strategic Alignment, ensuring AI initiatives align with business strategy, including roadmap development and prioritization.
- **Notebook (Jupyter/Managed):** See Jupyter Notebook.
- **Not Invented Here Syndrome:** A cultural attitude of aversion to using or adopting ideas, products, standards, or knowledge developed outside one's own group or organization. Can be a barrier to adopting central AI platforms/standards.
- **NPS (Net Promoter Score):** A widely used market research metric measuring customer loyalty based on the likelihood of recommending a company, product, or service. Calculated based on responses to a single question on a 0-10 scale.
- **Null Value:** Represents a missing or unknown value in a database field or dataset feature. Handling nulls appropriately is a key data quality challenge.

- **Object Storage:** A data storage architecture that manages data as objects, as opposed to other architectures like file systems or block storage. Widely used for data lakes in the cloud (e.g., AWS S3, Azure ADLS, Google GCS).
- **Objective Function:** See Loss Function.
- **Objectives (Mandate):** Specific, measurable goals or outcomes defined as part of the AI Office's mandate (e.g., achieve EAMI Level 3 within 18 months).
- **Observability:** The ability to measure the internal states of a system by examining its outputs (logs, metrics, traces). Applied to data systems (Data Observability) and ML systems (ML Observability) to understand behavior and diagnose issues proactively.
- **OmnioTech:** The fictional mid-sized technology company used as a running case study throughout the book to illustrate concepts in a practical context.
- **ONNX (Open Neural Network Exchange):** An open format built to represent machine learning models, enabling interoperability between different ML frameworks and tools.
- **On-Premise Infrastructure:** IT infrastructure (servers, storage, networking) owned and operated by an organization within its own physical data centers, rather than using public cloud services.
- **OPA (Open Policy Agent):** An open-source, general-purpose policy engine that enables unified, context-aware policy enforcement. Mentioned for use in CI/CD pipelines ("Governance-as-Code") and Infrastructure as Code security checks.
- **Open Neural Network Exchange:** See ONNX.
- **Open Policy Agent:** See OPA.
- **Open Source:** Software for which the original source code is made freely available and may be redistributed and modified. Many key AI libraries (TensorFlow, PyTorch, Scikit-learn) and MLOps tools are open source.
- **OpenAI:** An AI research and deployment company, developer of models like GPT-4, mentioned as part of the AI landscape.
- **OpenPages (IBM):** A commercial enterprise Governance, Risk Management, and Compliance (GRC) platform mentioned as an example tool.
- **Operating Model (AI Office):** Defines how the AI Office functions and interacts with the rest of the organization. Common models discussed are Centralized, Decentralized (Embedded), and Hybrid/Federated (Hub-and-Spoke).
- **Operational Efficiency:** A dimension of AI value focused on improving the speed, reducing the cost, or optimizing the resource consumption of business processes. See Appendix B KPIs.
- **Operational Metrics:** Measurements tracking the technical performance and health of deployed systems (e.g., latency, throughput, error rates, uptime). Contrasted with business value KPIs.
- **Operational Risk:** The risk of loss resulting from inadequate or failed internal processes, people, systems, or from external events. Relevant to AI deployment.
- **Operationalization:** The process of putting a framework, plan, or strategy into practical action and integrating it into routine operations. Key focus of Chapter 9 for governance.
- **Operations (IT Operations / Ops):** The team or function responsible for managing and maintaining IT infrastructure and ensuring the reliability and availability of production systems. Key collaborator in MLOps.
- **Opportunity Cost:** The potential benefits an individual, investor, or business misses out on when choosing one alternative over another. Relevant when considering misaligned AI investments.
- **Opportunity Funnel:** A conceptual model representing the process of capturing, assessing, and prioritizing potential AI use cases or initiatives, starting with many ideas and narrowing down to those included in the roadmap.
- **Optimization:** The process of making something (like a process, decision, or model performance) as effective, perfect, or functional as possible. A common application area for AI/ML.

- **Optimized (EAMI Level 4):** The fourth level of AI maturity in the EAMM framework, characterized by integrated strategy, robust embedded governance, mature MLOps/data practices, systematic value tracking, and a developing AI-aware culture.
- **Oracle:** A multinational computer technology corporation known for its database software and enterprise solutions. Mentioned as a potential data source.
- **Oracle Exadata:** An engineered system from Oracle designed for running Oracle Database workloads, mentioned as a traditional data warehouse example.
- **Orchestration (Pipeline/Workflow):** The automated arrangement, coordination, and management of complex computer systems and services, such as data pipelines (e.g., using Airflow, ADF) or MLOps pipelines (e.g., using Azure Pipelines, Kubeflow Pipelines).
- **Orchestrator (AI Office Role):** Describes the AI Office's central function in coordinating and integrating diverse AI-related activities across the enterprise, guided by the Compass framework.
- **Organizational Change Management:** See Change Management.
- **Organizational Learning:** The process through which an organization acquires, shares, and utilizes knowledge to adapt and improve over time. Facilitated by knowledge sharing mechanisms like CoPs.
- **Organizational Readiness:** The state of preparedness of an organization (in terms of culture, skills, processes, technology) to successfully undertake a significant change or initiative, such as large-scale AI adoption.
- **Outlier:** A data point that differs significantly from other observations. Identifying and handling outliers is part of data quality management and model robustness assessment.
- **Outsourcing:** Obtaining goods or services from an outside or foreign supplier, especially in place of an internal source. Relevant to the "Borrow" talent strategy.
- **Overfitting:** A modeling error occurring when a machine learning model learns the detail and noise in the training data to the extent that it negatively impacts the performance of the model on new, unseen data (poor generalization). Regularization techniques and validation help mitigate this.
- **Oversampling:** A data pre-processing technique used to address class imbalance by creating copies of instances from the under-represented minority class.
- **Oversight:** The action of overseeing something; supervision. Key function of AI governance bodies.
- **OWASP (Open Web Application Security Project):** An online community producing freely-available articles, methodologies, documentation, tools, and technologies in the field of web application security. OWASP Top 10 risks often adapted for secure coding standards.
- **Ownership (Data/Model/Process):** Clearly assigned responsibility and accountability for a specific asset or activity within the organization. Critical for effective governance.
- **PaaS (Platform-as-a-Service):** A cloud computing model where a third-party provider delivers hardware and software tools – usually those needed for application development – to users over the internet. Cloud AI/ML platforms (SageMaker, Azure ML, Vertex AI) are examples.
- **Pain Point:** A specific problem or frustration experienced by customers or employees within a process or system. Identifying pain points often reveals opportunities for AI intervention.
- **Pandas:** A popular open-source Python library providing high-performance, easy-to-use data structures and data analysis tools. Widely used by data scientists.
- **Parameters (Model):** The internal variables learned by a machine learning model from data during the training process (e.g., weights in a neural network). Contrasted with hyperparameters.
- **Partnership:** A collaborative arrangement between two or more parties who agree to cooperate to advance their mutual interests. Relevant for AI Office collaboration (internal/external).
- **Patching:** Applying updates (patches) to software or operating systems to fix bugs, address security vulnerabilities, or improve functionality. Key cybersecurity practice.
- **Patent:** A government authority or license conferring a right or title for a set period, especially the sole right

to exclude others from making, using, or selling an invention. Relevant for protecting novel AI algorithms or applications.

- **Peer Review:** Evaluation of scientific, academic, or professional work by others working in the same field. Relevant for ensuring quality in technical CoPs or governance reviews.

- **Penetration Testing (Pen Testing):** An authorized simulated cyberattack on a computer system, performed to evaluate the security of the system by actively exploiting vulnerabilities. Used to validate AI endpoint security.

- **Performance (Model):** How well a machine learning model achieves its intended predictive or analytical task, typically measured by technical metrics (accuracy, F1-score, MAE, etc.) on validation data.

- **Performance Management:** The ongoing process of monitoring, measuring, analyzing, reporting on, and managing the performance and value delivery of systems or initiatives over time (Chapter 14). Key activity of Compass West.

- **Performance Metrics (EAMM Dimension):** Assesses the maturity of defining, tracking, and reporting relevant technical and business KPIs for AI initiatives. Key aspect of Compass West.

- **Personal Data / Personally Identifiable Information (PII):** Any information relating to an identified or identifiable natural person. Subject to strict data protection regulations like GDPR and CCPA. Requires careful handling in AI systems.

- **Personalization:** Tailoring services, products, communications, or experiences to individual users based on their characteristics, preferences, or past behavior. A major application area for AI.

- **PETs (Privacy-Enhancing Technologies):** Technologies and techniques designed to minimize the use of personal data, maximize data security, and empower individuals' privacy, without losing data utility. Examples: differential privacy, federated learning, homomorphic encryption.

- **Phased Approach / Rollout:** Implementing a project or initiative in distinct stages or phases (e.g., pilot -> scale -> optimize; rollout by BU/region), rather than all at once ("big bang"). Manages risk and allows for learning.

- **Pilot Project:** A small-scale, preliminary study or trial conducted to evaluate feasibility, duration, cost, adverse events, and improve upon the study design prior to performance of a full-scale project or deployment. Critical stage in AI lifecycle.

- **Pilot Purgatory:** A common failure mode where promising AI pilot projects never successfully transition to full-scale production deployment due to technical, organizational, financial, or governance challenges encountered during scaling.

- **Pipeline (Data/ML/CI/CD):** A set of automated data processing or workflow steps connected in sequence. Data pipelines handle ETL/ELT; MLOps pipelines handle model training, validation, deployment (CI/CD/CT).

- **PitchBook:** A financial data and software company providing information on private capital markets, mentioned for horizon scanning.

- **PII (Personally Identifiable Information):** See Personal Data.

- **Platform-as-a-Service:** See PaaS.

- **Platform Engineering:** A discipline focused on building and managing internal developer platforms (IDPs) that provide self-service capabilities, automated infrastructure operations, and standardized tooling to accelerate application delivery, increasingly applied to AI/ML platform management.

- **PM (Project Manager / Program Manager):** Role responsible for planning, executing, and closing projects or programs, managing resources, risks, timelines, and communication.

- **PMO (Project Management Office / Program Management Office):** A group or department within an organization that defines and maintains standards for project management, often providing project managers and tracking portfolio status.

- **PoC (Proof of Concept):** An early-stage realization of a certain method or idea in order to demonstrate its feasibility, or a demonstration in principle with the aim of verifying that some concept or theory has practical potential. Often precedes a pilot project.
- **Policy:** A course or principle of action adopted or proposed by a government, party, business, or individual. AI governance relies on clear, enforceable policies.
- **Policy-as-Code:** Defining policies (security, compliance, configuration) as code that can be version controlled, automatically tested, and applied consistently within automated pipelines (e.g., using OPA).
- **Portfolio Balancing:** Ensuring a strategic mix of initiatives within a program or roadmap across dimensions like risk, time horizon, business impact, and technology type.
- **Portfolio Management:** The centralized management of one or more portfolios to achieve strategic objectives, including identifying, prioritizing, authorizing, managing, and controlling projects, programs, and other related work. Relevant for AI Office managing multiple initiatives.
- **Post-hoc Explainability:** XAI techniques applied after a model has been trained (model-agnostic methods like LIME, SHAP) to understand its behavior, as opposed to using inherently interpretable models.
- **Post-Mortem:** A process, usually performed after a project or event has completed (or failed), designed to analyze what happened, identify successes and failures, and extract lessons learned to improve future performance. Encouraged as part of a learning culture.
- **Post-Processing (Bias Mitigation):** Bias mitigation techniques applied to adjust a model's predictions after it has been trained, aiming to improve fairness outcomes without retraining the model itself.
- **Power BI (Microsoft):** A business analytics service by Microsoft providing interactive visualizations and business intelligence capabilities with an interface simple enough for end users to create their own reports and dashboards. Mentioned frequently as a dashboarding tool.
- **Precision (Model Metric):** A classification metric representing the proportion of positive predictions made by the model that were actually correct (True Positives / (True Positives + False Positives)). Measures accuracy among predicted positives.
- **Predictive Maintenance:** Using sensor data and machine learning models to predict when equipment or machinery is likely to fail, allowing maintenance to be scheduled proactively before failure occurs, reducing downtime and costs. Key OmnioTech pilot use case.
- **Predictive Model:** An AI/ML model designed to forecast future outcomes or probabilities based on input data.
- **Prefect:** An open-source workflow orchestration tool, often used for building and managing complex data pipelines. Mentioned as alternative to Airflow.
- **Prejudice:** Preconceived opinion that is not based on reason or actual experience; bias. Can manifest in data or human interpretation.
- **Pre-processing (Bias Mitigation):** Bias mitigation techniques applied to modify the training data before model training begins, aiming to reduce inherent biases in the data itself.
- **Prescriptive Analytics:** A type of analytics that uses optimization and simulation algorithms to advise on possible outcomes and recommend actions that will lead to the optimal result for a specific business objective. Often builds on predictive models.
- **Primary KPI (EAMI):** The main Key Performance Indicator selected to represent performance against a specific EAMM dimension, used for calculating the initial dimension score in the EAMI methodology. Supported by Sub-KPIs and qualitative validation.
- **Principles (AI):** High-level guiding statements outlining an organization's core values and commitments regarding the responsible and ethical development and use of AI (e.g., fairness, transparency, accountability).
- **Prioritization:** The action or process of deciding the relative importance or urgency of things. Critical for managing the AI Opportunity Funnel and developing the roadmap.

- **Privacy (Pillar of Trustworthy AI):** Respecting individual rights regarding their personal data and complying with applicable data protection laws and regulations. Key aspect of Governance (East).
- **Privacy Officer:** A role responsible for overseeing data protection strategy and implementation within an organization, ensuring compliance with privacy laws like GDPR. Key collaborator for AI governance.
- **Privacy-Enhancing Technologies:** See PETs.
- **Proactive:** Creating or controlling a situation by causing something to happen rather than responding to it after it has happened. Key principle for governance, risk management, monitoring.
- **Process Mining:** A technique used to discover, monitor, and improve real business processes by extracting knowledge from event logs readily available in today's information systems. Can identify opportunities for AI automation/optimization.
- **Procurement:** The action of obtaining or securing something, typically referring to the corporate function responsible for purchasing goods and services, including technology vendors. Needs involvement in AI vendor risk management.
- **Product Development:** The process of creating a new product or improving an existing one, from conception through design, testing, and launch. AI is increasingly used within this process.
- **Product Management:** The business process of planning, developing, launching, and managing a product or service. Includes understanding market needs, defining vision, setting strategy, and managing the product lifecycle. Increasingly specialized for AI products.
- **Product Manager:** See AI Product Manager.
- **Production Environment:** The live IT environment where software applications, services, and AI models run and are accessed by end-users or other systems. Contrasted with development and testing environments.
- **Profiling (Data):** The process of examining data available in an existing data source and collecting statistics and information about that data (e.g., value distributions, frequencies, data types, patterns, quality issues). First step in understanding data quality.
- **Program Management:** The process of managing several related projects, often with the intention of improving an organization's performance. AI adoption often managed as a program.
- **Programmatic Labeling:** Using heuristics, rules, weak supervision models, or other computational techniques to automatically generate labels for training data, potentially augmenting or replacing manual human labeling, especially for large datasets.
- **Project Management (EAMM Dimension):** Assesses the maturity and effectiveness of methodologies and practices used to plan, execute, monitor, and control AI projects. Relevant to Compass North/Center.
- **Project Management Office / Program Management Office:** See PMO.
- **Prometheus:** An open-source systems monitoring and alerting toolkit originally built at SoundCloud, widely used for collecting time-series metrics, often paired with Grafana for visualization. Mentioned for MLOps monitoring.
- **Prompt Engineering:** The process of structuring text prompts provided to generative AI models (especially LLMs) to elicit desired outputs reliably and effectively. An emerging skill.
- **Proof of Concept:** See PoC.
- **Protected Attributes / Characteristics:** Personal characteristics defined by law as protected against discrimination (e.g., race, color, religion, sex, national origin, age, disability, veteran status). Assessing fairness often involves checking model outcomes across groups defined by these attributes.
- **Prototype:** An early sample, model, or release of a product built to test a concept or process. Used extensively in AI development and R&D.
- **Prototyping:** The process of creating prototypes.
- **Proxy Bias:** Bias occurring when a model relies on features that are highly correlated with protected attributes (but not the protected attributes themselves) to make predictions, inadvertently leading to discrimina-

tory outcomes.

- **Pseudonymization:** A data management and de-identification procedure by which personally identifiable information fields within a data record are replaced by one or more artificial identifiers, or pseudonyms. Reduces privacy risk but may allow re-identification with additional information.
- **Psychological Safety:** A shared belief held by members of a team that the team is safe for interpersonal risk-taking, allowing individuals to speak up with ideas, questions, concerns, or mistakes without fear of negative consequences. Crucial for innovation and ethical reporting in AI teams.
- **Public Cloud:** Cloud computing services offered by third-party providers (like AWS, Azure, GCP) over the public Internet, making them available to anyone who wants to use or purchase them.
- **Public Sector:** The part of an economy concerned with providing basic government services. AI applications in the public sector often have high impact and specific governance needs.
- **Pulumi:** An open-source Infrastructure as Code tool that allows users to define cloud infrastructure using familiar programming languages.
- **Purview (Azure Purview):** See Azure Purview.
- **PwC (PricewaterhouseCoopers):** A multinational professional services network, cited as a source for industry reports and statistics related to AI adoption and management.
- **PyTorch:** A popular open-source machine learning framework based on the Torch library, widely used for applications such as computer vision and natural language processing, particularly favored in research.
- **Python:** A high-level, interpreted, general-purpose programming language widely used in data science, machine learning, and AI development due to its extensive libraries (e.g., Pandas, NumPy, Scikit-learn, TensorFlow, PyTorch) and large community.
- **Pytest:** A popular framework for writing and running tests for Python code. Used for unit/integration testing in MLOps pipelines.
- **Qualitative:** Relating to, measuring, or measured by the quality of something rather than its quantity. EAMM provides qualitative descriptions.
- **Quality (Data):** See Data Quality.
- **Quantitative:** Relating to, measuring, or measured by the quantity of something rather than its quality. EAMI provides quantitative scores.
- **Quantitative Analyst (Quant):** A professional who specializes in the application of mathematical and statistical methods – quantitative techniques – to financial and risk management problems. Often overlaps with data science roles in finance.
- **Quantum Computing:** A type of computation that harnesses the collective properties of quantum states, such as superposition, interference, and entanglement, to perform calculations. Potentially offers speedups for certain problems, including some ML tasks (QML).
- **Quantum Machine Learning (QML):** An emerging interdisciplinary field exploring the interplay of quantum computing and machine learning, investigating how quantum algorithms might enhance ML or how ML might help analyze quantum systems. Mentioned as a long-term trend.
- **Query:** A request for data or information from a database table or combination of tables. SQL is a common query language.
- **Question Answering (QA):** An NLP task focused on building systems that can automatically answer questions posed by humans in a natural language. Key application of LLMs.
- **Quick Wins:** Initiatives identified during prioritization (often via Value vs. Effort matrix) that offer relatively high business value for relatively low implementation effort. Important for building early momentum.
- **Qlik Replicate:** A commercial data integration and replication tool mentioned as potentially supporting CDC.
- **Qlik Sense:** A commercial business intelligence and data analytics platform mentioned as a tool for creating

performance dashboards.

- **Quota (Resource):** A limit placed on the amount of a particular resource (e.g., cloud compute hours, storage space) that can be used, often applied in sandbox environments or for cost control.
- **R (Programming Language):** A free software environment for statistical computing and graphics. Popular among statisticians and data miners for developing statistical software and data analysis, an alternative to Python for some ML tasks.
- **RACI Chart:** A responsibility assignment matrix used in project management and organizational design to clarify roles and responsibilities for tasks or deliverables. Defines who is Responsible, Accountable, Consulted, and Informed.
- **Radar Chart / Spider Chart:** A graphical method of displaying multivariate data in the form of a two-dimensional chart of three or more quantitative variables represented on axes starting from the same point. Used to visualize EAMI dimension scores.
- **Random Forest:** An ensemble learning method for classification, regression and other tasks that operates by constructing a multitude of decision trees at training time and outputting the class that is the mode of the classes (classification) or mean prediction (regression) of the individual trees. A common and powerful ML algorithm.
- **Rate Limiting:** A control mechanism used to limit the number of requests a user or system can make to an API or service within a specific time period. Used as a defense against abuse and model extraction attacks.
- **RBAC (Role-Based Access Control):** A method of restricting network or system access based on the roles of individual users within an enterprise. Permissions are assigned to roles, and users are assigned roles. Fundamental security practice.
- **R&D (Research and Development):** Activities companies undertake to innovate and introduce new products and services or improve existing ones. AI R&D involves exploring emerging techniques and potential applications.
- **Reactive:** Acting in response to a situation rather than creating or controlling it. Often contrasted with Proactive, especially regarding governance and risk management.
- **Real-Time Processing:** Processing data immediately as it is received or generated, typically with very low latency (milliseconds or seconds). Contrasted with Batch Processing.
- **Recall (Model Metric):** Also known as Sensitivity or True Positive Rate. A classification metric representing the proportion of actual positive cases that the model correctly identified (True Positives / (True Positives + False Negatives)). Measures ability to find all relevant instances.
- **Recommendation System:** An information filtering system that seeks to predict the "rating" or "preference" a user would give to an item (e.g., product, movie, article). Major application area for AI/ML.
- **Recognition (Employee):** Acknowledging and showing appreciation for employee contributions, efforts, or achievements. Important for retaining AI talent and reinforcing desired cultural behaviors.
- **Record Linkage:** Techniques used to find records in a dataset that refer to the same entity across different data sources or within the same source where duplicates exist. Used in deduplication.
- **Red Team / Red Teaming:** An independent group that challenges an organization to improve its effectiveness by assuming an adversarial role or perspective. In AI security, involves simulating attacks to test defenses.
- **Redress Mechanisms:** Processes or channels through which individuals affected by an AI system's decision or outcome can seek explanation, challenge the decision, request correction, or obtain remedy. Important for accountability and fairness.
- **Redis:** An open-source, in-memory data structure store, used as a database, cache, message broker, and queue. Often used as the low-latency "online" store component in Feature Stores.
- **Red Hat OpenShift:** A family of containerization software products developed by Red Hat, built around Kubernetes. Mentioned as alternative platform.

- **Redshift (Amazon Redshift):** Amazon Web Services' fully managed, petabyte-scale cloud data warehouse service.
- **Refactoring:** The process of restructuring existing computer code—changing the factoring—without changing its external behavior, often done to improve readability, reduce complexity, or enhance maintainability. Important for managing technical debt.
- **Reference Architecture:** A predefined architectural pattern or template providing recommended structures and components for building a specific type of system (e.g., an MLOps platform), promoting consistency and best practices.
- **Reference Data:** Data used to classify or categorize other data, often relatively static and authoritative (e.g., country codes, product categories, approved customer segments). Used in data cleaning and validation.
- **Referential Integrity:** A property of data stating that all references within a database or dataset are valid. Ensures relationships between tables remain consistent; lack of it causes data quality issues.
- **Regression (ML Task):** A type of supervised learning task where the goal is to predict a continuous numerical value (e.g., predicting house prices, forecasting sales).
- **Regression Metrics:** Metrics used to evaluate the performance of regression models (e.g., MAE, RMSE, R-squared).
- **Regularization (ML):** Techniques used during model training to prevent overfitting by adding a penalty term to the loss function based on the complexity or magnitude of the model parameters (e.g., L1/Lasso, L2/Ridge regularization).
- **Regulatory Compliance:** Adhering to specific laws, regulations, and standards mandated by governmental or industry bodies. Key aspect of AI Governance (East). See also EAMM Compliance & Ethics.
- **Regulatory Intelligence:** The process of monitoring, gathering, analyzing, and disseminating information about evolving regulations, guidance documents, and enforcement trends relevant to an organization's operations. Critical for AI compliance.
- **Reinforcement Learning (RL):** An area of machine learning concerned with how intelligent agents ought to take actions in an environment in order to maximize the notion of cumulative reward. Learns through trial and error via feedback (rewards/penalties). Used in robotics, games, optimization.
- **Reliability (Pillar of Trustworthy AI):** Ensuring AI systems perform accurately, consistently, robustly, and predictably according to their intended function under expected operational conditions. Supported by strong MLOps and data quality. See also EAMM Reliability aspects.
- **Remediation (Risk/Issue):** The action of correcting or resolving a problem, deficiency, or identified risk.
- **Reporting:** The process of organizing and presenting information, often performance data or analysis results, in a structured format (reports, dashboards) for specific audiences. Key part of Performance Management (West).
- **Representativeness (Data Quality Dimension):** The degree to which a dataset accurately reflects the characteristics and diversity of the real-world population or phenomenon it is intended to represent. Lack of representativeness leads to sampling bias and models that don't generalize well. Critical for fairness.
- **Reproducibility:** The ability to duplicate the results of a process or experiment exactly, given the same inputs, code, environment, and configuration. A core MLOps principle, essential for debugging, validation, and trust.
- **Reputational Damage:** Harm to the perception and standing of an organization among its stakeholders (customers, employees, investors, public) resulting from negative events like ethical failures, security breaches, or compliance violations.
- **Requirements Gathering:** The process of identifying, documenting, and validating the needs and constraints of stakeholders for a system or project. Crucial translation step between business and AI teams.
- **Resilience:** The ability of a system or organization to withstand or recover quickly from difficulties or dis-

ruptions. Relevant for both technical AI infrastructure and organizational adaptability.

- **Resource Management (South - Compass Domain):** The domain of the AI Office Compass™ Framework focused on strategically managing the essential inputs required for AI: Talent, Technology, Data, MLOps, and Finance.

- **Resources:** The assets (people, money, materials, technology) available to an organization to undertake an endeavor.

- **Responsible AI:** A governance approach focused on developing and deploying artificial intelligence systems in a manner that is ethical, transparent, fair, secure, accountable, and beneficial to humanity, proactively mitigating potential harms. Encompasses the pillars in Chapter 7.

- **Responsible AI Lead:** A role often within the AI Office or governance structure dedicated to championing and implementing responsible AI practices across the organization. See also AI Ethicist.

- **Retail:** The sector involving the sale of goods to the public in relatively small quantities for use or consumption rather than for resale. Frequently cited industry example for AI applications (personalization, inventory, supply chain).

- **Retention (Customer):** Keeping customers engaged and continuing to purchase from a company over time. Often measured by churn rate.

- **Retention (Employee):** Keeping valuable employees engaged, motivated, and committed to staying with the organization long-term. Critical for scarce AI talent.

- **Retraining (Model):** The process of retraining a deployed machine learning model, typically using updated data or revised configurations, often triggered by monitoring detecting performance degradation (drift). A key MLOps practice (CT).

- **Return on Investment:** See ROI.

- **Reuse (Code/Data/Features):** Designing and managing technical assets (code libraries, data pipelines, ML features) in a way that allows them to be easily discovered and leveraged across multiple projects, improving efficiency and consistency. Promoted by Feature Stores and CoPs.

- **Revenue Enhancement:** A dimension of AI value focused on directly increasing top-line income through improved sales, marketing, pricing, or new offerings. See Appendix B KPIs.

- **Review Cadence:** A defined frequency (e.g., weekly, monthly, quarterly) for conducting specific review meetings related to AI performance, value, risk, or governance.

- **RICE Score:** A prioritization framework scoring opportunities based on Reach, Impact, Confidence, and Effort. Extends ICE score by considering scale.

- **Right-sizing:** An optimization technique (part of FinOps) involving selecting the most appropriate type and size of cloud compute instances or other resources to match workload performance requirements without overprovisioning (and incurring excess cost).

- **Risk Appetite:** The amount and type of risk that an organization is willing to pursue or retain in order to achieve its strategic objectives. Influences governance rigor.

- **Risk Assessment:** The process of identifying potential hazards or risks, analyzing or evaluating the associated risk (likelihood and impact), and determining appropriate ways to eliminate or control the hazard/risk. Key AI governance process.

- **Risk Management (EAMM Dimension):** Assesses the maturity of processes for identifying, assessing, mitigating, and monitoring risks specifically associated with AI systems. Key part of Compass East. See also Chapter 9.

- **Risk Tier / Risk Tiering:** Categorizing AI applications into different levels (e.g., Low, Medium, High) based on their assessed potential risk and impact, allowing governance requirements to be tailored appropriately (stricter controls for higher tiers).

- **Roadmap (AI):** A strategic plan outlining the vision, prioritized initiatives, key milestones, dependencies, resources, and timelines for an organization's AI program over a defined period (typically 1-3 years). Core artifact of Compass North (Chapter 6).

- **Robotic Process Automation:** See RPA.
- **Robotics:** The branch of technology dealing with the design, construction, operation, and application of robots. Increasingly integrates advanced AI capabilities (Embodied AI).
- **Robustness (AI Model):** The ability of an AI model to maintain its performance level even when faced with noisy, perturbed, or unexpected inputs, or changes in the operating environment. Key aspect of reliability and security.
- **ROI (Return on Investment):** A performance measure used to evaluate the efficiency or profitability of an investment. Calculated as (Gain from Investment - Cost of Investment) / Cost of Investment. Key metric for demonstrating AI value (Domain West).
- **Role-Based Access Control:** See RBAC.
- **Root Cause Analysis (RCA):** A systematic problem-solving method used to identify the underlying causes of faults or problems. Used in incident response and continuous improvement.
- **RPA (Robotic Process Automation):** Technology that allows configuring computer software, or a "robot," to emulate and integrate the actions of a human interacting within digital systems to execute a business process. Often simpler than AI, focused on automating rule-based tasks. Sometimes distinguished from AI scope.
- **R-squared (R^2):** A statistical measure representing the proportion of the variance for a dependent variable that's explained by independent variables in a regression model. Indicates goodness of fit.
- **RMSE (Root Mean Squared Error):** A common regression metric measuring the square root of the average of the squared differences between predicted values and actual values. Represents standard deviation of prediction errors.
- **S3 (Simple Storage Service - AWS):** Amazon Web Services' highly scalable, durable, and cost-effective object storage service, commonly used for data lakes.
- **SAFe® (Scaled Agile Framework):** An enterprise framework for implementing Agile, Lean, and DevOps practices at scale. Mentioned as relevant to project management maturity. (Note: This acronym itself doesn't appear widely used in the draft, but Agile is).
- **Safety (AI):** Ensuring AI systems operate without causing unintended harm to people, property, or the environment. Key aspect of Compass East and EAMM Security.
- **SageMaker (AWS):** Amazon Web Services' fully managed platform for building, training, and deploying machine learning models at scale. Includes components like SageMaker Studio (IDE), Experiments, Feature Store, Model Registry, Endpoints.
- **Salesforce:** A multinational cloud-based software company known primarily for its customer relationship management (CRM) product. Also provides AI capabilities (Einstein). Cited as source/example.
- **Salesforce Einstein:** Salesforce's integrated AI platform providing capabilities like predictive lead scoring, personalized recommendations, and automated service responses within the Salesforce ecosystem.
- **Sampling Bias:** Bias introduced when the data collected for training is not representative of the real-world population the model will encounter, often due to non-random sampling methods or under-representation of certain groups.
- **Sandbox (AI):** See AI Sandbox.
- **SAP:** A multinational software corporation that makes enterprise software to manage business operations and customer relations. Mentioned as data source/integration point.
- **SAP Signavio:** A business process management and process mining software suite, mentioned as an example tool.
- **SAST (Static Application Security Testing):** Security testing methodology that analyzes application source

code, bytecode, or binary code for security vulnerabilities without executing the program. Integrated into secure CI/CD pipelines.

- **Scalability:** The ability of a system, network, or process to handle a growing amount of work, or its potential to be enlarged to accommodate that growth. Critical requirement for AI infrastructure, platforms, MLOps, and governance.
- **Scale (AI Deployment):** Moving an AI model from a limited pilot or testing phase to full production deployment, handling real-world data volumes, traffic, and operational requirements reliably.
- **Scale AI:** A commercial company providing data annotation and labeling services and platforms, mentioned as an example vendor.
- **Scale-to-Zero:** A cloud infrastructure capability, often used with serverless or containerized applications, where compute resources automatically scale down to zero instances when there is no traffic, eliminating costs for idle time. Mentioned for inference endpoint optimization.
- **Scaling ROI Gates:** Formal checkpoints proposed in Chapter 25 where continued investment in scaling an AI initiative requires demonstrating validated pilot value, a realistic scaling plan/budget, and operational readiness.
- **Scanning (Horizon):** See Horizon Scanning.
- **Scenario Planning:** A strategic planning method organizations use to make flexible long-term plans based on exploring plausible alternative future scenarios and their potential implications. Mentioned for future-proofing AI strategy.
- **Schema (Data):** The structure or organization of data, defining tables, fields, data types, relationships, and constraints. Relevant for databases, warehouses, and enforced schemas in lakehouses (schema-on-write vs. schema-on-read).
- **Schema Enforcement / Evolution (Lakehouse):** Features of lakehouse table formats (Delta Lake, Iceberg, Hudi) allowing definition and enforcement of data schemas upon writing, while also supporting controlled schema changes over time without breaking pipelines.
- **Schema-on-Read:** A data handling approach typically used in data lakes where raw data is loaded without a predefined structure, and the schema is applied only when the data is read or queried for analysis. Offers flexibility but requires careful downstream processing.
- **Schema-on-Write:** A traditional data handling approach used in data warehouses where data must conform to a predefined schema before it is written or loaded into the system. Ensures structure but lacks flexibility for diverse data.
- **Science Project (AI Context):** Pejorative term for an AI initiative driven by technical curiosity ("technology push") without clear strategic alignment or demonstrable business value.
- **Scikit-learn:** A popular, comprehensive open-source Python library providing efficient tools for data analysis and machine learning, including algorithms for classification, regression, clustering, dimensionality reduction, model selection, and preprocessing.
- **Scope (AI Office Mandate):** The defined range of activities, responsibilities, and authority covered by the AI Office's mandate.
- **Scoring (EAMI):** The process of calculating dimension scores and the overall EAMI score based on assessing performance against defined KPIs and qualitative evidence, as detailed in Chapter 24 and source files.
- **Screening (Talent):** The process of reviewing job applications or candidate profiles to assess qualifications and suitability for a role. Part of AI talent acquisition.
- **Scrum:** An agile framework for managing complex projects, often used in software development, characterized by iterative sprints, defined roles (Product Owner, Scrum Master, Dev Team), and specific ceremonies (planning, daily scrum, review, retrospective). Can be adapted for AI projects.
- **SDR (Sales Development Representative):** A role often involved in lead generation and qualification in

sales processes.

- **Secrets Management:** Securely storing, managing access to, and rotating sensitive credentials like API keys, database passwords, and certificates used by applications and pipelines. Critical security practice.
- **Security (AI):** Protecting AI systems (models, data, infrastructure) from threats and vulnerabilities, ensuring confidentiality, integrity, availability, and robustness. Key pillar of Trustworthy AI (Chapter 7) and Governance (East, Chapter 11). See also EAMM Security.
- **Security Champion:** Individuals embedded within development teams who have extra security training and act as local advocates and liaisons for security best practices. Mentioned as potential model for AI governance champions.
- **Security Information and Event Management:** See SIEM.
- **SecMLOps / MLSecOps:** The practice of integrating security principles, practices, and automated testing throughout the entire MLOps lifecycle ("shifting security left") to build secure AI systems by default.
- **Seldon Core:** An open-source platform for deploying machine learning models on Kubernetes at scale, mentioned as a model serving option.
- **Self-Serve Infrastructure:** Platforms and tools designed to allow users (like data scientists or developers) to provision and manage required resources (compute, environments, pipelines) themselves through automated interfaces, reducing reliance on central IT/Ops teams. Key principle of Data Mesh.
- **Semantic Consistency:** Ensuring that data values representing the same real-world concept have a consistent meaning and interpretation across different systems or datasets. Addressed by MDM and data governance.
- **Semi-structured Data:** Data that does not conform to the strict structure of a relational database model but nonetheless contains tags or other markers to separate semantic elements and enforce hierarchies of records and fields within the data (e.g., JSON, XML, Avro). Handled well by lakes/lakehouses.
- **Sensitivity (Model Metric):** See Recall.
- **Sentiment Analysis:** Using natural language processing (NLP) techniques to identify, extract, quantify, and analyze subjective information, opinions, and attitudes expressed in text data (e.g., customer reviews, social media).
- **Serverless Computing:** A cloud computing execution model where the cloud provider dynamically manages the allocation and provisioning of servers. Code typically runs in stateless compute containers triggered by events. Can offer cost efficiency and scalability.
- **Service Level Agreement (SLA):** A commitment between a service provider and a client defining specific aspects of the service – quality, availability, responsibilities – and penalties for violation. Mentioned for interactions between AI Office and IT Ops.
- **Service Level Objective (SLO):** A specific, measurable target for a service level indicator (SLI), defining the desired level of reliability for a service. E.g., 99.9% API uptime.
- **ServiceNow:** A commercial cloud computing platform helping companies manage digital workflows for enterprise operations, including IT Service Management (ITSM). Mentioned as tool in OmnioTech case study.
- **SHAP (SHapley Additive exPlanations):** A popular game theory-based XAI approach used to explain the output of any machine learning model by computing the contribution of each feature to a specific prediction based on Shapley values. Provides both local and global explanations.
- **Shiny Object Syndrome:** The tendency to pursue new, trendy, or exciting things (like the latest AI technology) simply because they are new, often distracting from more important strategic goals or fundamental work.
- **Showback (Cost Allocation):** A FinOps practice involving tracking cloud resource consumption and reporting allocated costs back to the teams or business units that consumed them, increasing cost awareness without necessarily charging them directly. Contrasted with Chargeback.
- **SIEM (Security Information and Event Management):** Technology providing real-time analysis of se-

curity alerts generated by applications and network hardware. AI monitoring alerts should ideally feed into enterprise SIEM systems.

- **Significance of KPI (EAMI KPI Table):** Column in the detailed KPI table explaining why a specific KPI is important and what aspect of performance or value it reflects.
- **Silo (Data/Organizational):** An isolated system, process, department, or mindset that does not share information or collaborate effectively with others. Major barrier to enterprise AI success.
- **Simulation:** Imitating the operation of a real-world process or system over time, often using computer models. AI/ML can be used within simulations or to analyze simulation outputs.
- **Skill Development (EAMM Dimension):** Assesses the maturity of organizational programs and processes for upskilling/reskilling internal talent, fostering broad AI literacy, and supporting continuous learning in AI-related areas. Managed within Compass South/Center.
- **Skill Gaps:** A situation where the available workforce lacks the necessary skills required for current or future job roles. Significant challenge for AI talent.
- **Slack:** A popular cloud-based team collaboration tool and messaging platform. Mentioned as potential channel for CoPs or ideation.
- **SLA:** See Service Level Agreement.
- **SLO:** See Service Level Objective.
- **SMART Objectives:** A goal-setting acronym ensuring objectives are Specific, Measurable, Achievable, Relevant, and Time-bound. Recommended for AI Office mandates and uplift initiatives.
- **SME (Small and Medium-sized Enterprise / Subject Matter Expert):** Can refer to either: 1) Businesses below certain size thresholds (often used for context applicability). 2) Subject Matter Expert: Individuals with deep knowledge in a specific domain, crucial collaborators in AI projects.
- **Snorkel AI:** A platform/approach focused on programmatic data labeling using weak supervision techniques to create training data more efficiently than manual labeling.
- **Snowflake:** A popular cloud-based data warehousing company offering a platform that often supports Lakehouse architectural patterns.
- **SBOM (Software Bill of Materials):** A formal record containing the details and supply chain relationships of various components used in building software. Important for managing security risks from dependencies.
- **SOC 2 (System and Organization Controls 2):** A compliance standard developed by the American Institute of CPAs (AICPA) specifying how organizations should manage customer data based on five "trust service principles": security, availability, processing integrity, confidentiality, and privacy. Relevant for vendors handling data.
- **Social Responsibility:** An ethical framework suggesting that an entity, be it an organization or individual, has an obligation to act for the benefit of society at large. Relevant to AI ethics discussions.
- **Software Bill of Materials:** See SBOM.
- **Software Composition Analysis (SCA):** Automated tools that scan application code (source, binary, or dependencies) to identify open-source components used and detect known security vulnerabilities or license compliance issues within them. Key part of secure CI/CD.
- **Solution:** A means of solving a problem or dealing with a difficult situation. Often used to refer to AI systems designed to address specific business needs.
- **Sourcing (Talent):** The process of finding, identifying, and attracting potential candidates for job roles. Key part of AI talent acquisition.
- **South (Compass Domain):** The domain of the AI Office Compass™ Framework focused on Resource Management, encompassing Talent, Technology, Data, MLOps, and Finance.
- **Spam Detection:** Identifying unwanted or unsolicited electronic messages (email, comments). Classic application of ML classification.

- **Spark (Apache Spark):** A popular open-source unified analytics engine for large-scale data processing, with APIs for Java, Scala, Python, R, and SQL. Widely used for ETL/ELT, data preparation, and ML on big data.
- **spaCy:** An open-source software library for advanced Natural Language Processing (NLP) in Python.
- **Specialized Hardware (AI):** Compute hardware optimized for AI workloads, primarily GPUs (Nvidia) and TPUs (Google), offering significant speedups for deep learning training and inference compared to CPUs.
- **Spider Chart:** See Radar Chart.
- **Spoke (Hybrid Model):** In a Hybrid/Federated AI operating model, refers to the distributed AI specialists or teams embedded within business units or functions, focusing on domain-specific solutions while collaborating with the central Hub.
- **Sponsorship (Executive):** See Executive Sponsor.
- **Spot Instances:** A pricing model offered by cloud providers (AWS, Azure, GCP) allowing users to bid on unused compute capacity at significantly lower prices than on-demand rates, but with the caveat that instances can be terminated with little notice if capacity is needed elsewhere. Useful for cost-optimizing fault-tolerant workloads like some ML training jobs.
- **SQL (Structured Query Language):** A standard domain-specific language used in programming and designed for managing data held in relational database management systems (RDBMS), or for stream processing in relational data stream management systems (RDSMS). Widely used for querying data warehouses and lakehouses.
- **SRE (Site Reliability Engineering):** A discipline incorporating aspects of software engineering and applying them to infrastructure and operations problems, aiming to create scalable and highly reliable software systems. Often collaborates with MLOps teams.
- **Stabilization:** The process of making something stable or preventing fluctuations. Relevant to model performance and operational reliability.
- **Staging Environment:** An environment used for testing that exactly resembles the production environment. Used for final validation before deployment.
- **Stakeholder:** An individual, group, or organization who may affect, be affected by, or perceive themselves to be affected by a decision, activity, or outcome of a project, program, or portfolio.
- **Stakeholder Communication Plan:** See Communication Plan (Stakeholder).
- **Stakeholder Engagement:** The process by which an organization involves people who may be affected by the decisions it makes or can influence the implementation of its decisions. Critical for AI success.
- **Standard Deviation:** A measure of the amount of variation or dispersion of a set of values. Used in statistics and relevant to metrics like RMSE.
- **Standardization:** The process of implementing and developing technical standards based on the consensus of different parties. Important for AI platforms, tools, processes (MLOps, governance) to ensure consistency and efficiency.
- **Standards (AI Governance):** Detailed, actionable requirements and best practices derived from high-level AI policies, providing specific guidance for development and operations (e.g., data handling standard, model validation standard).
- **Startup:** A young company founded to develop a unique product or service, bring it to market and make it irresistible and irreplaceable for customers. The AI startup ecosystem is a key area for horizon scanning.
- **Static Application Security Testing:** See SAST.
- **Statistical Significance:** A determination by an analyst that the results in the data are not explainable by chance alone.
- **Steering Committee:** See AI Governance Council / Steering Committee.
- **Stitch Fix:** An online personal styling service company cited as an example of using AI for personalization.
- **Storage (Data):** See Data Storage.

- **Storytelling (Data):** The practice of weaving data and analysis into a compelling narrative to communicate insights, explain phenomena, and influence decisions effectively, especially for non-technical audiences.
- **Strategic Alignment (North - Compass Domain):** The domain of the AI Office Compass™ Framework focused on ensuring AI initiatives align with and drive business strategy, including opportunity identification/prioritization and roadmap development.
- **Strategic Alignment (EAMM Dimension):** Assesses the maturity of linking AI strategy and initiatives directly to core business objectives and ensuring C-level sponsorship and communication.
- **Strategic Bets:** High-value, high-effort AI initiatives identified during prioritization, requiring significant investment and careful planning due to their strategic importance and complexity.
- **Strategic Foresight:** The ability to anticipate future changes, challenges, and opportunities relevant to an organization's strategy. Key role for a mature AI Office Innovation Hub.
- **Strategic Goal / Objective:** A high-level, long-term aim that an organization plans to achieve to fulfill its mission and vision. AI strategy should derive from these.
- **Strategic Imperative:** Something that is extremely important or urgent for achieving strategic success. Used to describe the need for an AI Office and strategic alignment.
- **Strategic Planning:** An organizational management activity used to set priorities, focus energy and resources, strengthen operations, ensure employees/stakeholders work toward common goals, establish agreement around intended outcomes/results, and assess/adjust direction in response to changing environment. AI Office participates/informs this.
- **Strategy:** A plan of action designed to achieve a major or overall aim.
- **Stream Processing:** Processing data continuously as it arrives ("in motion"), typically with low latency. Used for real-time AI applications. See also Real-Time Processing.
- **Structured Data:** Data that adheres to a predefined data model and is therefore straightforward to analyze. Conforms to a tabular format with relationships between rows and columns (e.g., in relational databases, spreadsheets).
- **Subject Matter Expert:** See SME.
- **Sub-KPI (EAMI):** Secondary Key Performance Indicators used within the EAMI framework to provide additional context or validation for the Primary KPI score of a specific EAMM dimension.
- **Success Criteria:** Predefined standards or metrics used to determine whether a project or initiative has achieved its objectives. Should be clearly defined for AI projects, linking to business value.
- **Summarization (Text):** An NLP task focused on automatically creating a short, accurate, and coherent summary of a longer text document. Key application of LLMs.
- **Sun Tzu:** An ancient Chinese military general, strategist, and philosopher, author of The Art of War, quoted in Chapter 4 regarding strategy and tactics.
- **Supervised Learning:** A type of machine learning where algorithms learn from labeled training data, meaning each input example is paired with a known correct output label. Used for classification and regression tasks.
- **Supply Chain Management:** The management of the flow of goods and services, involving the movement and storage of raw materials, work-in-process inventory, and finished goods from point of origin to point of consumption. AI is often used for optimization (demand forecasting, logistics).
- **Support Vector Machine (SVM):** A type of supervised learning algorithm used for classification and regression tasks, effective in high-dimensional spaces.
- **Survey:** A research method used for collecting data from a predefined group of respondents to gain information and insights on various topics of interest. Used for EAMI data collection (culture, awareness).
- **Sustainability:** Meeting the needs of the present without compromising the ability of future generations to meet their own needs. Encompasses environmental, social, and governance (ESG) aspects. AI can both

impact (Green AI) and contribute to (AI for Green) sustainability goals. See Appendix B KPIs.

- **SVM:** See Support Vector Machine.
- **Symptom:** An indication or sign of the existence of something, especially of an undesirable situation. Used to describe the observable problems resulting from uncoordinated AI (e.g., duplication, unclear ROI).
- **Synergy:** The interaction or cooperation of two or more organizations, substances, or other agents to produce a combined effect greater than the sum of their separate effects. Emphasized for the interconnected Compass domains.
- **Synonym:** A word or phrase that means exactly or nearly the same as another word or phrase in the same language.
- **Synthetic Data:** Artificially generated data created algorithmically rather than being collected from real-world events. Can be used to augment training datasets, address privacy concerns, or simulate scenarios.
- **System Transparency:** Providing clear documentation about an AI system's overall design, purpose, data sources, performance characteristics, and known limitations.
- **Systematic:** Done or acting according to a fixed plan or system; methodical. Emphasized for approaches like assessment, uplift, governance.
- **Tableau:** A popular commercial interactive data visualization software company (owned by Salesforce), mentioned as a tool for creating performance dashboards.
- **Tabletop Exercise:** A discussion-based session where team members meet in an informal, classroom setting to discuss their roles during an emergency and their responses to a particular situation (e.g., an AI security incident). Used to validate plans and identify gaps.
- **Tactical:** Relating to or constituting actions carefully planned to gain a specific end, often short-term or localized, as opposed to strategic.
- **Tagging (Cloud Resources):** Applying labels (key-value pairs) to cloud resources (VMs, storage, databases) to organize them, track costs, manage access, and automate operations. Essential for FinOps cost allocation.
- **Talend:** A commercial software vendor specializing in data integration and data management tools, including ETL capabilities.
- **Talent (AI):** Individuals possessing the specialized skills and expertise required for developing, deploying, and managing AI systems (e.g., data scientists, ML engineers, data engineers, AI ethicists). Key resource managed within Compass South.
- **Talent Acquisition (EAMM Dimension):** Assesses the maturity and effectiveness of organizational processes for attracting, recruiting, and hiring individuals with necessary AI skills. Key part of Compass South.
- **Talent Development:** See Skill Development (EAMM Dimension).
- **Target Variable / Label:** In supervised machine learning, the known correct output or category associated with each input example in the labeled training dataset, which the model learns to predict.
- **TCO (Total Cost of Ownership):** A financial estimate intended to help buyers and owners determine the direct and indirect costs of a product or system over its full lifecycle (acquisition, deployment, operation, maintenance, decommissioning). Crucial for realistic AI budgeting and ROI calculation (Chapter 19).
- **Team Competency (EAMM Dimension in Source Tables):** Corresponds to Talent Acquisition and Skill Development dimensions, assessing overall team skills and sourcing/development maturity.
- **Technical Debt:** The implied cost of rework caused by choosing an easy (limited) solution now instead of using a better approach that would take longer. Accumulates over time, increasing maintenance costs and reducing agility. Relevant to ML systems ("ML Technical Debt").
- **Technical Literacy:** The ability to understand, use, and interact effectively with technology. AI literacy is a specific form.
- **Technical Metrics:** Quantitative measures focused on the performance, reliability, or operational characteristics of a system or model itself (e.g., accuracy, latency, error rate), as opposed to business value metrics.

- **Technology Ecosystem (AI):** The integrated set of infrastructure, platforms, tools, frameworks, and libraries used by an organization to support the end-to-end AI lifecycle (Chapter 17). Managed within Compass South.
- **Technology Selection (EAMM Dimension):** Assesses the maturity of processes for choosing, implementing, managing, and integrating appropriate AI/ML platforms, libraries, tools, and infrastructure. Key part of Compass South.
- **Technology Push:** An approach where new technology capabilities drive the search for potential applications or problems to solve, rather than starting from defined business needs. Contrasted with Business Pull.
- **Tecton:** A commercial enterprise Feature Store platform mentioned as an example tool.
- **Template:** A preset format or pattern used as a guide for making something (e.g., document templates like Model Cards, TCO templates, pipeline templates). Promotes consistency.
- **TensorFlow:** A popular open-source software library developed by Google for machine learning and artificial intelligence, particularly well-suited for deep learning model development and deployment.
- **TensorFlow Serving (TF Serving):** A flexible, high-performance serving system for machine learning models, designed for production environments, optimized for TensorFlow models.
- **Teradata:** A company providing database and analytics-related software, products, and services, known for its enterprise data warehouse solutions. Mentioned as traditional warehouse example.
- **Terraform:** An open-source Infrastructure as Code (IaC) software tool created by HashiCorp, enabling users to define and provision data center infrastructure using a declarative configuration language.
- **Testing (AI Model):** Evaluating an AI model against various criteria (performance, fairness, robustness, security) using predefined datasets and methodologies to ensure it meets requirements before deployment. Critical MLOps practice.
- **TF Serving:** See TensorFlow Serving.
- **Threat Modeling:** A structured process for identifying potential threats and vulnerabilities in a system, prioritizing them, and defining countermeasures to mitigate risk. Applied to AI systems for security.
- **Threshold (Decision):** A cutoff value used in classification models to convert predicted probabilities into discrete class assignments (e.g., predict "fraud" if probability > 0.8). Adjusting thresholds can impact precision/recall and fairness metrics.
- **Throughput:** The rate at which something can be processed (e.g., number of inference requests per second an AI endpoint can handle). Key operational metric.
- **Time Horizon:** The length of time into the future that is considered in planning or forecasting. Roadmaps typically have 1-3 year horizons; innovation might look further.
- **Time Series Data:** A sequence of data points indexed in time order. Common in forecasting, anomaly detection, IoT applications.
- **Time-to-Market:** The length of time it takes from a product being conceived until it is available for sale. MLOps aims to reduce time-to-market for AI features.
- **Timeliness (Data Quality Dimension):** The degree to which data is sufficiently up-to-date for its intended use and available within the required timeframe.
- **Tooling / Toolchain:** The set of software tools used to perform tasks in a particular process, such as AI development or MLOps (e.g., IDEs, libraries, platforms, orchestrators, monitoring tools).
- **Top-Down Approach:** Strategy or initiative driven from senior leadership levels down through the organization. Often combined with Bottom-Up.
- **Total Cost of Ownership:** See TCO.
- **Town Hall Meeting:** An informal public meeting, function, or event derived from the traditional town meetings of New England. Used in corporate settings for large-group communication from leadership.
- **TPU (Tensor Processing Unit):** Application-specific integrated circuits (ASICs) developed by Google spe-

cifically for accelerating neural network machine learning workloads. Alternative to GPUs.

- **Tracking (Experiment/Performance):** Systematically recording metrics, parameters, artifacts, or outcomes related to ML experiments or deployed model performance over time.
- **Training (AI Model):** The process of using a learning algorithm to adjust the internal parameters of an AI model based on patterns found in a training dataset, enabling it to perform a specific task (e.g., prediction, classification).
- **Training (Employee):** See AI Literacy & Training Programs.
- **Training Data:** The dataset used to train a machine learning model. Its quality and representativeness are critical.
- **Training-Serving Skew:** A situation where the performance of an ML model in production differs significantly from its performance during training/validation, often due to discrepancies between the data or feature calculations used in the training environment versus the live serving environment. Feature Stores help prevent this.
- **Transaction (Database):** A sequence of operations performed as a single logical unit of work in a database. ACID properties ensure transactional reliability. Relevant to Lakehouse architecture.
- **Transfer Learning:** An ML technique where a model developed for a task is reused as the starting point for a model on a second, related task, often speeding up training and improving performance when labeled data for the second task is limited. Commonly used with large pre-trained foundation models.
- **Transform (ETL/ELT):** The step in data integration processes where data is cleaned, validated, standardized, aggregated, enriched, or restructured to prepare it for analysis or model consumption. Occurs before Load in ETL, after Load in ELT.
- **Transformer Architecture:** A deep learning model architecture introduced in the paper "Attention Is All You Need," heavily reliant on self-attention mechanisms. It has become the dominant architecture for Large Language Models (LLMs).
- **Transformative (EAMI Level 5):** The fifth and highest level of AI maturity in the EAMM framework, where AI drives business innovation, governance is adaptive, processes are highly automated and optimized, and an AI-driven culture is pervasive.
- **Transparency (AI):** The principle and practice of ensuring that AI system operations, decision-making processes, data usage, and governance are understandable and accessible to relevant stakeholders to an appropriate degree. Pillar of Trustworthy AI (Chapter 7).
- **Trend Analysis:** The practice of collecting information and attempting to spot a pattern or trend in the information over time. Used in horizon scanning and performance monitoring.
- **Trust (Organizational):** The belief in the reliability, truth, ability, or strength of someone or something within the organization (e.g., trust in leadership, trust in data, trust in AI systems). Foundational for AI adoption and collaboration.
- **Trustworthy AI:** AI systems developed and deployed in a way that aligns with ethical principles and societal values, demonstrating characteristics like accountability, fairness, transparency, security, privacy, reliability, and human-centricity. The overarching goal of AI governance.
- **Tuning (Hyperparameter):** The process of selecting the optimal set of hyperparameters for a machine learning model to maximize its performance on a validation dataset.
- **UI (User Interface):** The means by which a user interacts with a computer system, application, or device, including graphical elements like screens, buttons, and menus, or command-line interfaces. Relevant for AI tools.
- **UiPath:** A commercial vendor providing a platform for Robotic Process Automation (RPA) and process mining. Mentioned as an example tool.
- **Unacceptable Risk (EU AI Act):** The highest risk category under the EU AI Act, pertaining to AI systems

deemed a clear threat to the safety, livelihoods, and rights of people (e.g., social scoring by governments, real-time remote biometric identification in public spaces with exceptions). Such systems are generally prohibited.

- **Uncertainty:** A state of limited knowledge where it is impossible to exactly describe the existing state, a future outcome, or more than one possible outcome. AI development often involves managing uncertainty.
- **Undersampling:** A data pre-processing technique used to address class imbalance by removing instances from the over-represented majority class. Use with caution as it discards potentially useful data.
- **Unified Customer Profile:** A consolidated, consistent view of customer data aggregated from multiple sources, often created through Master Data Management (MDM). Essential for effective personalization.
- **Unified Platform:** A technology architecture approach aiming to consolidate multiple functionalities (e.g., BI, data science, ML) onto a single, integrated platform, like a Data Lakehouse, to reduce complexity and improve data consistency.
- **Uniqueness (Data Quality Dimension):** The degree to which each record or entity within a dataset represents a distinct real-world entity without duplication.
- **Unit Testing:** A software testing method where individual units or components of source code are tested to determine if they are fit for use. Applied to ML model code and pipeline components.
- **Unstructured Data:** Information that either does not have a pre-defined data model or is not organized in a pre-defined manner (e.g., text documents, images, audio, video). Handling unstructured data is a key capability for modern AI and data platforms like data lakes/lakehouses.
- **Unsupervised Learning:** A type of machine learning where algorithms learn patterns from unlabeled data, without explicit input-output pairs. Used for tasks like clustering, dimensionality reduction, and anomaly detection.
- **Uplift (AI Maturity):** The process of improving an organization's AI capabilities and maturity level based on targeted strategies derived from EAMI assessments (Chapter 25). See also AI Maturity Uplift Strategy.
- **Uplift Strategy (EAMM Dimension):** Assesses the maturity of processes for defining, implementing, and tracking targeted initiatives designed to improve AI capabilities based on maturity assessment findings. Key activity of Compass Center.
- **Uptime:** A measure of the time during which a computer or system is operational and available for use. Key operational metric.
- **Urgency:** Importance requiring swift action. Creating a sense of urgency is the first step in Kotter's change model.
- **URL (Uniform Resource Locator):** The address of a resource (like a webpage) on the World Wide Web. Used in references.
- **Usability:** The ease with which users can employ a tool or system to achieve a particular objective. Important for AI tool adoption.
- **Use Case:** A specific situation or scenario in which a product, service, or system (like AI) could potentially be used to solve a problem or achieve a goal.
- **Use Case Registry:** A central inventory used to track AI use cases being explored or implemented across the organization, often including details like purpose, status, risk tier, and owners.
- **User Acceptance Testing (UAT):** A phase of software testing in which actual users test the software in a realistic scenario to ensure it can handle required tasks and meets business requirements.
- **User Experience (UX):** Encompasses all aspects of the end-user's interaction with the company, its services, and its products. Important consideration for designing AI-powered tools and interfaces.
- **User Interface:** See UI.
- **Validation (Data):** See Data Validation.
- **Validation (Model):** See Model Validation.

- **Validity (Data Quality Dimension):** The degree to which data values conform to defined business rules, constraints, acceptable value ranges, data types, or formats. Ensures data fits expected schemas and constraints.
- **Valley of Death (Innovation):** A common metaphor describing the challenging phase between successful research/prototyping and achieving commercial viability or scaled deployment, where many promising innovations fail due to lack of funding, operational challenges, or market fit issues. Relevant to transitioning AI R&D to value.
- **Value (AI):** See Business Value (in AI Context).
- **Value Chain:** The sequence of activities a company performs to create and deliver a product or service, from conception to end use. Analyzing the value chain can identify AI opportunities.
- **Value Delivery (Compass Domain - West):** See Performance & Value Delivery.
- **Value Proposition:** A statement summarizing why a consumer should buy a product or use a service, focusing on the value it provides. Essential for AI business cases.
- **Value Realization (EAMM Dimension):** Assesses the maturity of processes for quantifying, tracking, validating (including ROI/TCO), and communicating the tangible business value delivered by deployed AI initiatives. Key aspect of Compass West.
- **Value Range (EAMI KPI Table):** Column in the detailed KPI table indicating the typical measurement scale or range for a specific KPI (e.g., 0-100%).
- **Value Source (EAMI KPI Table):** Column in the detailed KPI table suggesting potential sources (systems, logs, records, surveys) from which data for measuring a specific KPI might be obtained.
- **Value Stream Mapping:** A lean management method for analyzing the current state and designing a future state for the series of events that take a product or service from its beginning through to the customer. Can help identify AI optimization points.
- **Value vs. Effort Matrix:** A prioritization technique plotting initiatives on a 2x2 grid based on their assessed business value and implementation effort, helping identify quick wins and strategic bets.
- **Variability:** Lack of consistency or fixed pattern; liability to vary or change. EAMI aims to reduce variability in AI capabilities across the organization.
- **VC (Venture Capital):** Financing that investors provide to startup companies and small businesses that are believed to have long-term growth potential. Monitoring VC trends is part of horizon scanning.
- **Vendor:** A company or individual that sells goods or services. Relevant for AI platform, tool, data, or consulting providers.
- **Vendor Lock-in:** A situation where a customer using a product or service cannot easily transition to a competitor, often due to proprietary interfaces, data formats, dependencies, or significant switching costs. A risk to consider with commercial AI platforms.
- **Vendor Management / Vendor Risk Management:** Processes for managing relationships with third-party vendors, including selection, contract negotiation, performance monitoring, and assessing/mitigating associated risks (security, compliance, financial, operational). Critical for AI when using external tools/platforms/data.
- **Version Control:** The practice of tracking and managing changes to software code, documents, datasets, or models over time, typically using systems like Git or DVC. Essential for reproducibility, collaboration, and rollback capabilities in MLOps/DataOps.
- **Vertex AI (Google Cloud):** Google Cloud's unified platform for machine learning development, offering tools for data preparation, model building, training, deployment (MLOps), and management.
- **Vertex AI Experiments:** Feature within Google Cloud's Vertex AI for tracking ML experiments.
- **Vertex AI Feature Store:** Google Cloud's managed service for storing, sharing, and serving ML features.
- **Vertex AI Model Registry:** Google Cloud's service for managing the lifecycle of ML models within Vertex

AI.

- **Violation (Compliance/Ethical):** An act or instance of failing to observe or comply with a rule, law, regulation, or ethical principle.
- **Virtual Machine (VM):** An emulation of a computer system, providing the functionality of a physical computer. Common unit of cloud compute resources.
- **Virtual Private Cloud (VPC):** An on-demand configurable pool of shared computing resources allocated within a public cloud environment, providing a certain level of isolation between the different organizations using the resources. Used for network segmentation/security.
- **Vision (AI):** A clear, aspirational description of the desired future state an organization aims to achieve through the strategic application of artificial intelligence.
- **Visual Aid:** An item of illustrative matter, such as a film, slide, or model, designed to supplement written or spoken information so that it can be understood more easily. Placeholders used in the revised book draft.
- **Visualization (Data):** The graphical representation of information and data using visual elements like charts, graphs, and maps. Used extensively for data exploration, performance dashboards, and communicating results.
- **VM:** See Virtual Machine.
- **Volume (Big Data Characteristic / Data Quality):** Refers to the large amount of data generated and stored. Also a dimension monitored by Data Observability tools (tracking unexpected changes in data volume).
- **VPC:** See Virtual Private Cloud.
- **VP (Vice President):** A senior management title in organizations.
- **Vulnerability:** A weakness in a system, process, or control that could be exploited by a threat actor to cause harm or gain unauthorized access.
- **Vulnerability Scanning:** The automated process of proactively identifying security vulnerabilities in computer systems, networks, or applications. Part of secure MLOps.
- **WAF (Web Application Firewall):** A firewall that monitors, filters, or blocks HTTP traffic to and from a web application, distinct from a regular network firewall. Mentioned as a potential security layer for AI inference endpoints.
- **Walmart:** A multinational retail corporation cited as a source for examples or statistics related to AI implementation (e.g., inventory management).
- **Warehouse (Data):** A central repository of integrated data from one or more disparate sources, primarily used for reporting and data analysis (BI). Typically stores structured, transformed data using a predefined schema (schema-on-write). See also Data Warehouse.
- **Watermarking (Model):** Techniques for embedding unique, detectable patterns or identifiers into machine learning models, potentially helping to trace model provenance or detect unauthorized copying (model stealing).
- **Watson (IBM):** IBM's portfolio of AI software, services, and applications, often associated with question answering, natural language processing, and healthcare applications. Cited as source/example.
- **Weights & Biases (W&B):** A commercial MLOps platform providing tools for experiment tracking, model visualization, and collaboration. Mentioned as an example tool.
- **Weighted Average / Score:** A calculation that takes into account the varying degrees of importance (weights) of the numbers in a data set. Used as an optional method for calculating the overall EAMI score.
- **West (Compass Domain):** The domain of the AI Office Compass™ Framework focused on Performance & Value Delivery, encompassing KPI definition, value tracking, ROI calculation, and communication of AI impact.
- **Whistleblower:** A person who informs on a person or organization engaged in an illicit, illegal, or unethical

activity. Protecting whistleblowers is mentioned in the context of fostering ethical dialogue.

- **WIIFM ("What's In It For Me?"):** An acronym representing the question individuals implicitly ask when faced with change, focusing on the personal benefits or impacts. Addressing WIIFM is key in change communication.
- **Word Embedding:** A technique in Natural Language Processing (NLP) where words or phrases from a vocabulary are mapped to vectors of real numbers, capturing semantic relationships. Examples: Word2Vec, GloVe.
- **Word2Vec:** A popular technique for learning word embeddings from text data.
- **Workforce Productivity:** A dimension of AI value focused on improving the efficiency, output, or effectiveness of employees. See Appendix B KPIs.
- **Workflow:** The sequence of steps involved in moving work from initiation to completion. AI often aims to automate or optimize workflows. MLOps/DataOps focus on automating AI-related workflows.
- **Workflow Orchestration:** See Orchestration.
- **Working Group:** A group of experts working together to achieve specified goals. The AI Review Board functions as a working group.
- **Workshop:** A meeting at which a group of people engage in intensive discussion and activity on a particular subject or project. Used for EAMI assessments, ideation, strategy development.
- **XAI (Explainable AI):** Methods and techniques in artificial intelligence that attempt to make the results and decision-making processes of AI systems understandable to humans. Key for transparency and trust. See also Explainability.
- **XGBoost:** An open-source software library providing a gradient boosting framework, known for its performance and widely used for supervised learning tasks (classification, regression).
- **XML (Extensible Markup Language):** A markup language and file format for storing, transmitting, and reconstructing arbitrary data using human-readable tags. Example of semi-structured data.
- **YAML (YAML Ain't Markup Language):** A human-readable data serialization language often used for configuration files (e.g., in Kubernetes, CI/CD pipelines).
- **YoY (Year-over-Year):** Comparing a statistic for one period to the same period the previous year, often used to track trends while accounting for seasonality.

Acknowledgements

Writing this book was a significant undertaking, and it would not have been possible without the support, insights, and encouragement of many individuals.

I owe a debt of gratitude to the pioneers and thought leaders in the fields of artificial intelligence, data management, governance, and organizational strategy whose work formed the foundation upon which these frameworks are built.

I am particularly grateful to Alvaro Garcia, Harris Apostolopoulos, Eddy Perez, Vikas Yadav, Salvatore Rubino, Mouhsine Kebbaj, Norbert Heinemann for their insightful discussions, constructive review feedback on early drafts, and shared experiences in navigating the complexities of enterprise AI, which significantly shaped the practical aspects of this guide.

I must also acknowledge the vibrant AI community – the researchers, engineers, ethicists, and practitioners whose constant innovation and open sharing of knowledge (through publications, conferences, and open-source contributions) push the boundaries of what's possible and inform best practices globally.

On a personal note, my deepest appreciation goes to my wife, Manisha, for their unwavering patience, understanding, and support throughout the demanding process of research and writing.

Finally, thank you to the reader. My hope is that this book provides tangible value as you lead your organization's journey into the AI-driven future.

About the Author

Jitesh Goswami is a seasoned technology executive, AI strategist and management consultant with over 26 years of experience helping organizations navigate complex technological transformations and leverage data and AI (9 years) for strategic advantage.

His expertise spans leadership in strategy, technology, and project management. He has advised or worked with leading companies across various sectors, including technology, professional services, and financial services, guiding them in establishing effective AI capabilities and realizing tangible business value from their AI investments.

Jitesh is passionate about demystifying AI for business leaders, promoting responsible AI practices, helping organizations avoid the negative impacts of AI, and guiding them to leverage AI effectively for business transformation, steeper growth, and enduring success. He holds an Engineering Bachelor's degree from Gujarat University, India, is pursuing an E-MBA from UNSW, Australia, holds advanced project management qualifications from PMI/Axelos, and frequently speaks at industry conferences on topics related to business/digital transformation, enterprise AI maturity, and governance.

He wrote AI Office - The AI Center of Excellence to provide leaders with the practical frameworks and actionable guidance often missing in navigating the complexities of enterprise-wide AI adoption.

You can connect with Jitesh on LinkedIn: https://www.linkedin.com/in/jgoswami/

Notes

www.ingramcontent.com/pod-product-compliance
Lightning Source LLC
Chambersburg PA
CBHW061759210326
41599CB00034B/6814